Environmental Pollution: Causes, Impacts and Management Techniques

Environmental Pollution: Causes, Impacts and Management Techniques

Edited by Michelle Randall

SYRAWOOD
PUBLISHING HOUSE

New York

Published by Syrawood Publishing House,
750 Third Avenue, 9th Floor,
New York, NY 10017, USA
www.syrawoodpublishinghouse.com

Environmental Pollution: Causes, Impacts and Management Techniques
Edited by Michelle Randall

International Standard Book Number: 978-1-64740-142-9 (Hardback)

Cataloging-in-Publication Data

Environmental pollution : causes, impacts and management techniques / edited by Michelle Randall.
    p. cm.
Includes bibliographical references and index.
ISBN 978-1-64740-142-9
1. Pollution. 2. Pollution prevention. 3. Environmental impact analysis. 4. Environmental quality. I. Randall, Michelle.
TD174 .E58 2022
363.73--dc23

TABLE OF CONTENTS

PREFACE

Environmental pollution refers to the contamination of the natural environment due to the introduction of chemical substances or energy in the form of heat, noise or light. Such chemical and energy species, or pollutants, can be artificially produced or occur in nature. Environmental pollution can be classified as air pollution, water pollution, soil contamination, thermal pollution, noise pollution, radioactive contamination, etc. Motor vehicle emissions, nuclear waste disposal activity, petrochemical plants, burning of natural vegetation, use of pesticides and herbicides, nuclear waste disposal activity, etc. are primary causes of environmental pollution. Pollution is harmful to the environment and to humans in several ways. Adverse air and water quality can kill organisms and cause diseases, while mercury ingestion is associated with neurologic symptoms and developmental deficits in children. Through adroit pollution control measures, particularly by adopting the practices of reusing, recycling, composting and minimizing wastes, pollution can be effectively managed. This book covers in detail some existent theories and innovative concepts revolving around environmental pollution. Some of the diverse topics covered in this book address the varied causes, impacts and management techniques of environmental pollution. This book is a complete source of knowledge on the present status of this important field.

Various studies have approached the subject by analyzing it with a single perspective, but the present book provides diverse methodologies and techniques to address this field. This book contains theories and applications needed for understanding the subject from different perspectives. The aim is to keep the readers informed about the progresses in the field; therefore, the contributions were carefully examined to compile novel researches by specialists from across the globe.

Indeed, the job of the editor is the most crucial and challenging in compiling all chapters into a single book. In the end, I would extend my sincere thanks to the chapter authors for their profound work. I am also thankful for the support provided by my family and colleagues during the compilation of this book.

Editor

PROGNOSTICATION OF NOISE EXPOSURE RISK ON WORKERS' SAFETY AND HEALTH IN LITHUANIA

Ričardas BUTKUS, Alvidas ŠARLAUSKAS, Gediminas VASILIAUSKAS

Institute of Agricultural Engineering and Safety, Aleksandras Stulginskis University, Studentų g. 15b, 53361 Akademija, Kaunas distr., Lithuania

Abstract. This study explores the association between the levels of noise exposure at various sectors of economic activity and the percentage distribution of workplaces where these levels can occur. The results of the research are based on statistical data which was collected at various workplaces at the sectors of construction, transport, agriculture and forestry, electricity, water and gas supply etc. These results include the mathematical analysis of noise levels at 748 workplaces. These workplaces were sectioned by economic activity sector and percentage distribution was calculated as a ratio between the actual and total number of places where the respective noise level was exceeded. Probability index was calculated as a descriptive parameter for the evaluation of workplaces at various levels of noise exposure level normalized to a nominal 8 hour working day ($L_{EX, 8h}$). Results show that highest number of workplaces where $L_{EX, 8h}$ was exceeded was at the companies of wood processing and furniture manufacturing. $L_{EX, 8h}$ of 80 dB(A) was exceeded at 77%, $L_{EX, 8h}$ = 85 dB(A) – 72% and $L_{EX, 8h}$ = 87 dB(A) – 68% of all the workplaces. This shows that hearing loss occurrence is likely and it can be assessed as "very risky" or "potentially risky" at the companies of wood, metal and textile sectors (probability index's values from 0,087 to 0,032 respectively).

Keywords: noise, noise exposure, risk assessment, sector of economic activity.

Introduction

Noise is one of the most common environmental issues in both working and living environments. Environmental noise is usually discussed from various perspectives, but commonly narrows to the analysis of single noise sources such as transport (Oškinis *et al.* 2004; Vasarevičius, Graudinytė 2004), noise in populated agglomerations (Baltrėnas *et al.* 2010) or to biological and psychological effects on humans (Selander *et al.* 2009). Majority of the scientific research are related to the noise exposure at various workplaces and reviews the guidelines for noise reduction (Miyakita *et al.* 2004; Granneman *et al.* 2004), improvement of noise assessment methodology (Cagno *et al.* 2005) or the use of hearing protection and assessment of its effectiveness (Arezes, Miguel 2005).

There are very few scientific studies, where noise risk on operator was assessed and most of these publications usually narrow to the analysis of single measurements or the examples of a good practice. This shows the necessity to perform deeper analysis of expected noise levels on workers' health and to analyse these cases statistically.

However, the registered cases of noise induced hearing loss (NIHL) at various countries show the lack of vibroacoustic safety at work.

According to the data obtained by State Labour Inspectorate of the Republic of Lithuania, the dynamics of occupational diseases during the period 1998–2011 show the domination of NIHL (approximately 25–40% of all occupational diseases). According to the data of European Agency for Safety and Health at Work these results are similar to the cases of NIHL registered in Finland (34.5% at 2009 and 26.5% at 2010 respectively) (Oksa *et al.* 2012; Yränheikki, Savolainen 2000).

Noise exposure levels and its effects on workers' health are restricted by legal legislations such as the EU directive 2003/10/EC "On the minimum health and safety requirements regarding the exposure of workers to the risks arising from physical agents (noise)". The requirements of this directive are transferred to the laws of EU member states and obligate the employers to assess the occupational risk and to foresee the possibilities to reduce noise risk on workers. However, exposure to noise can

---

Corresponding author: Gediminas Vasiliauskas
E-mail: gediminas.vasiliauskas@asu.lt

have a number of physiological and psychological effects, therefore in situ analysis is usually necessary. According to the measurement results in a particular workplace it is possible to plan preventative actions and to calculate the expenses considering the effect of most probable risk factors and its effects on workers.

Assessment of risk at workplaces can be calculated by adapting statistical methods that enable to express expected risk level in numerous values. There has been an investigation carried out by Merkevičius *et al.* (2011) where these guidelines were used for prognostication of vibration risk on operators of agricultural machinery.

According to Ising and Kruppa (2004), quality of the evidence associating noise exposure and health hazards can be classified to one of three categories:
 – Sufficient;
 – Limited;
 – Inadequate.

International standard ISO 1999 indicates that occupational NIHL is not expected to occur below $L_{EX, 8h} = 80$ dB(A) with reference to 40 working hours per week. Higher exposure levels will increase the risk of permanent hearing threshold shift. Higher noise levels can cause different stress reactions that may lead to derangement of normal neuro-vegetative and hormonal processes and cause adverse effects on the vital body functions. These include cardiovascular parameters such as blood pressure, cardiac function, serum cholesterol and others (Babisch 2000). For noise levels exceeding 60 dB(A), the myocardial infarction risk increases continuously, and is equal or greater than 1.2 for noise levels of 70 dB(A).

Exposure to noise according to Haines *et al.* (2001) is associated with higher number of accidents (low level noise increase the number by 7.4%, high level noise by 16.5%), frequent injuries (low level noise 9.1% and high level noise 26.2%) and cognitive failures (11.2% and 17.3% respectively).

The aim of this study was to prognosticate the noise risk on workers of various economic activity sectors in Lithuania.

1. Methodology

Analysis of noise exposure was done by grouping the statistical data of physically measured A-weighted noise levels by the sector of economic activity. These results were compared to the levels given by EU directive 2003/10/EC which defines $L_{EX, 8h}$, peak sound pressure level and C-weighted instantaneous noise pressure as risk predictors. Calculated values of $L_{EX, 8h}$ were taken as a base for assessment while peak values were considered insignificant. Main reason for elimination of peak values was that some collected data did not have all required acoustic parameters. Analysis of collected data also showed that only for ≤5% of all workplaces peak values were significant. This allowed using equivalent continuous A-weighted sound pressure level (SPL) as a single measure for the calculation of $L_{EX, 8h}$. These results enabled to define total noise parameter, which describes the overall number of workplaces (N, %) where $L_{EX, 8h}$ may exceed the levels of 80, 85 and 87 dB(A). These values were described as a lower exposure, upper exposure action and exposure limit value respectively.

As the base of this study all workplaces were partitioned into 5 subcategories by economic activity sector. Manufacturing sector was additionally divided into 5 minor groups. These groups were selected because of high number of workers under the sway of noise in manufacturing.
1. *Construction* – manufacturing and installation of concrete and its constructions;
2. *Manufacturing*:
    1) food (including drinks);
    2) wood processing and furniture trade;
    3) metal processing (without foundries);
    4) the garment industry;
    5) textile sector;
3. *Transport* – workplaces of truck drivers;
4. *Agriculture and Forestry* – workplaces of tractors and self-propelled agricultural machinery;
5. *Electricity, water, gas and heat supply services* – repair and maintenance.

Statistical analysis was performed by analysing the data of 748 workplaces, where the noise exposure was investigated under the requirements of ISO 9612:2009. Percentage distribution of workplaces is shown in Figure 1.

Software package "Statistica" was used for the data analysis. Equivalent SPL's as well as number of workplaces were grouped in a frequency tables for different economic

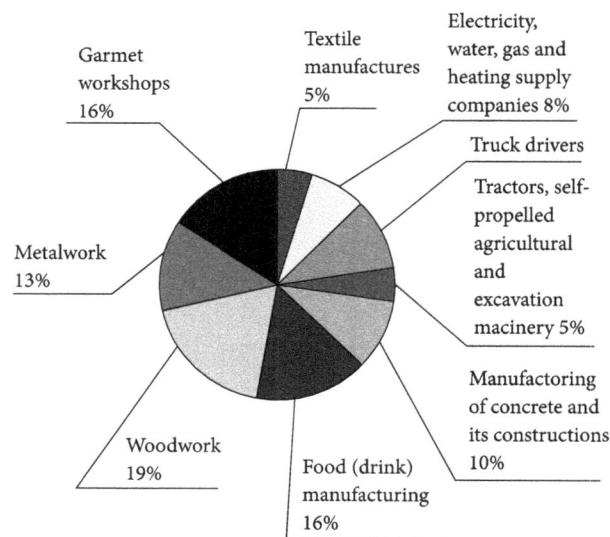

Fig. 1. Percentage distribution of workplace data used for statistical analysis (*n* = 748)

activities. These characteristics were later mathematically described as a third degree polynomial equation as follows:

$$N = a \cdot L_{A,eq}^3 + b \cdot L_{A,eq}^2 + c \cdot L_{A,eq} + d, \qquad (1)$$

where a, b, c ir d – regression coefficients.

For the conformance of the model to the data, determination coefficients R^2 were calculated (Table 1). Further investigation of equivalent SPL distribution at workplaces (N_L, %) was carried out by implementing the descending percentage distribution model. Intermediate results of this analysis are shown in Figure 2.

Generalization of research results was done by calculating percentage distribution of the workplaces according to calculated noise parameters N80, N85 and N87 i.e. where $L_{EX, 8h}$ of 80, 85 and 87 dB(A) was exceeded. These results were based on assumption that particular $L_{EX, 8h}$ is typical for 75% of all workplaces in that economic activity sector. Results of our study were compared to the findings of other authors. According to Haines et al. (2001), intensive noise causes the increase of accidents at work by 9% and number of errors by 6% respectively. The effects of noise were also reported in the research of Babisch (2000). It was concluded that even moderate noise levels as low as 50 dB(A) causes adverse human reaction of sensitive people while the noise level of 70 dB(A) increases the infarction risk by 45% if compared to the level of 50 dB(A). Lusk et al. (2004) found that a 2-mm Hg increase in

Table 1. Values of the regression coefficient for the determination of statistical distribution of noise exposure

Workplace categories in conformity to the economic activity sector	Regression coefficients				Determination coefficient R^2
	a	b	c	d	
Construction – Manufacturing and installation of concrete and its construction	0.000298	−0.02168	−7.21088	663.2143	99.9
Manufacturing:					
Food (including drinks)	−0.00087	0.138202	−7.3423	230.1032	99.7
Wood processing and furniture trade	0.006023	−1.75421	165.3267	−4980.93	98.3
Metal processing (without foundries)	0.010628	−2.97101	271.1111	−8015.94	100
The garment industry	0.019342	−4.5761	353.8051	−8879.85	99.8
Textile sector	0.003205	−0.89698	79.57921	−2167.21	97.3
Transport – Workplaces of truck drivers	0.008568	−2.11194	167.8367	−4237.88	96.8
Agriculture and Forestry – Workplaces of tractors and self-propelled agricultural machinery	0.018576	−4.94195	431.3198	−12290.8	99.7
Electricity, water, gas and heat supply services – Repair and maintenance	0.000198	−0.05748	3.701928	39.9483	99.9

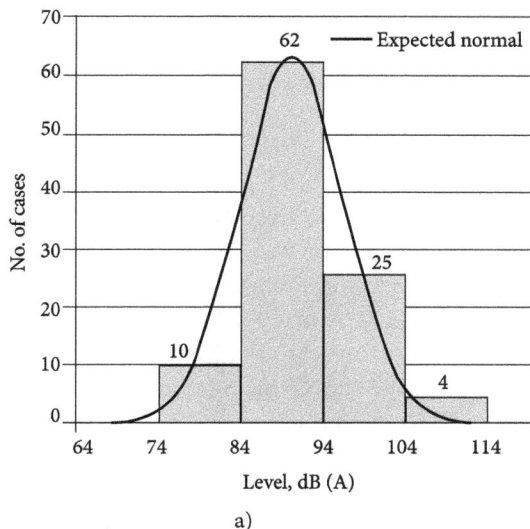

a)

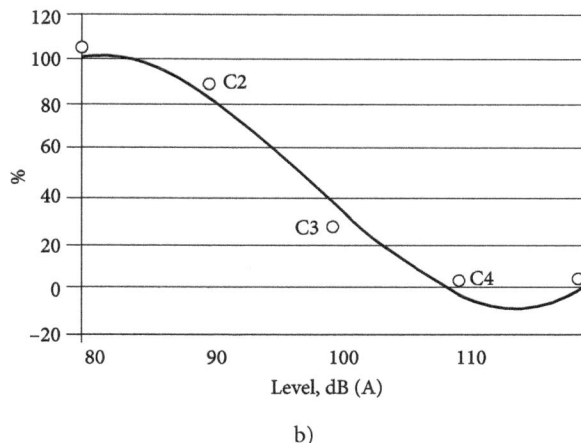

b)

Fig. 2. Histogram of the A-weighted SPL distribution in wood processing companies (a) and descending total percentage distribution and its polynomial expression (b)

systolic blood pressure (SBP) was associated with each 10 dB(A) increase in average noise, and a 2 mm Hg increase in diastolic blood pressure (DBP) was associated with each 13 dB(A) increase in average noise. This study showed that a long-term decrease of 5 to 6 mm Hg in usual DBP was associated with 35–40% less stroke and 20–25% less coronary disease. Banbury and Berry (2005) discussed how concentration of office workers was affected that sound level of the office changed from 55 dB(A) to 60 dB(A). These results show that the noise risk at workplaces should be considered even if the level does not exceed the maximum permissible. McReynolds (2005) stated in his paper that approximately 10% of all workers are under the influence of intense (>85 dB(A)) noise. These results are in agreement with the results of Vilnius Public Health Centre. They investigated 12457 workplaces in total and found that 3306 workers (27%) work in noisy conditions. 18.5% from this number work under the noise varying from 75 to 85 dB(A), 4% under the conditions of 86–90 dB(A) and 4% in the noise of ≥90 dB(A). Results of these studies suggest that noise effects on workers and noise induced outcomes should be assessed even when the noise does not exceed the lower exposure value. Noise of 75 dB(A) according to the recommendations of OSHA and NIOSH should be considered as risk factor to workers in at least 10% of all workplaces.

Total relative number $P(L_{p,x})$ of the possible accident, health or hearing damage in separate sector of economic activity was described as the hundredth of total number of cases at various noise levels $L_{80,85,87}$ for 10% of workers as follows:

$$P(L_{p,x}) = 0,001 \cdot N_L. \qquad (2)$$

The component 0.001 in Equation (2) is related to percentage indicator of all cases (0.01) at considered sound pressure level which will give at least ten percent of workers affected. The prognostication of possible number of noise induced health impairment, accident or work-related error (p_{HAE}) was described as a probability of independent events as follows:

$$p_{HAE} = P(L_{p,75}) \cdot P(L_{p,76}) ... \cdot P(L_{p,79}), \qquad (3)$$

where $P(L_{p,75})...P(L_{p,79})$ – total relative number of cases in each group when the noise levels are in the range from 75 dB(A) to 79 dB(A) respectively.

For higher exposure levels, when the preventative value of 80 dB(A) is reached, probability of noise induced hearing impairment P_{HI} is described as follows:

$$p_{HI} = P(L_{p,80}) \cdot P(L_{p,81}) ... \cdot P(L_{p,84}), \qquad (4)$$

where $P(L_{p,80})...P(L_{p,84})$ – total relative number of cases in each group when the noise levels are in the range from 80 dB(A) to 84 dB(A) respectively.

Considering the exposure limit value of 85 dB(A) as a level where serious hearing loss may occur, its probability P_{HL} can be proposed as follows:

$$p_{HL} = P(L_{p,85}) \cdot P(L_{p,86}) ... \cdot P(L_{p,90}), \qquad (5)$$

where $P(L_{p,85})...P(L_{p,90})$ – total relative number of cases in each group when the noise levels are in the range from 85 dB(A) to 90 dB(A) respectively.

On the basis of noise exposure data at various economic activity sectors and Equations (3), (4) and (5), the indexes of various probabilities were determined. These results as probability index p stands for the risk situations in analogy with Moriyama and Ohtani (2009) and Aven et al. (2007) estimation as described in Table 2.

Table 2. Probability index p and its assessment criteria

Range of p value	Assessment criteria
$...-10^{-6}$	Insignificant
$10^{-6}-10^{-5}$	Normal working conditions
$10^{-5}-10^{-4}$	Potentially risky situation
$10^{-4}-10^{-3}$	Very risky situation
$10^{-3}-10^{-2}$	Portentous situation
$10^{-2}-10^{-1}$	Unacceptable

Value of probability index should be related to calculated probabilities p_{HAE}, p_{HI} and p_{HL} individually by attributing this number to one of suggested arbitrary assessment criteria. These criteria should outline the acoustic situation and were selected according to the practice used in safety systems. Considering that the p value include two main parameters, i. e. noise level and number of workers (workplaces) affected it could be used for risk assessment in various objects where physical agent noise is prevailing.

2. Results

The results of accidents at work that are related with noise induced hearing impairment identify manufacturing as one of the most hazardous economic activity sectors. Noise exposure of 80 dB(A) (N_{80}) at wood processing and furniture manufacturing companies may be exceeded at 77% of all workspaces while N_{85} at 72% and N_{87} at 68% respectively (Fig. 3). In this regard, similar results and percentage distribution was found at individual wood processing company after the analysis of noise exposure levels where the level of 80 dB(A) was exceeded at all workplaces. In 30 workplaces noise exposure level was between 80–85 dB(A) while the noise of 85–90 dB(A) was found at 250 workplaces and the level of 90–95 dB(A) was exceeded at 3 workplaces.

According to the data provided by Vilnius Public Health Center the SPL's of wood processing machinery is as follows: universal machinery – 89–108 dB(A) (2 hour work shift at this level would give the $L_{EX, 8h}$ value of

83–102 dB(A)), planer – 96–99 dB(A) (2 hour $L_{EX, 8h}$ = 90–93 dBA), double acting machinery – 84–88 dB(A) (4 hour $L_{EX, 8h}$ = 81–85 dB(A)), quadri acting machinery – 94 dB(A) (2 hour $L_{EX, 8h}$ = 88 dB(A)), saw-frame – 89 dB(A), parquet production line – approx. 92 dB(A). According to Davies *et al.* (2009) in lumber mills the noise mean predicted value was 91.7 dB(A); minimum and maximum – 80.2 dB(A) and 103.6 dB(A) respectively.

Metal processing machinery and equipment also generates high levels of noise. The number of workplaces where N_{80} is exceeded was 75%, N_{85} – 67% and N_{87} – 61% (Fig. 3). In the research of Dutch researchers Granneman *et al.* (2004) the metal grinding $L_{EX, 8h}$ = 93 dB(A) over 1.5 hours, welding (over 3 hours) and pressing (over 6.5 hours) $L_{EX, 8h}$ = 88 dB(A), grinding with compressed air flow $L_{EX, 8h}$ = 85 dB(A) over 30 minutes.

According to the results obtained by measuring the noise levels at punch pressing company it was found that 100 (out of 200) workers were exposed to the noise level of 87–90 dB(A) SPL, 80 workers to the noise of 90–95 dB(A) SPL and 10 workers to the noise of more than 95 dB(A). For the remaining 10 workplaces the noise level of lower than 85 dB(A) was identified. Following values of noise measurements were found at the workplaces of metal forming, grinding and plastic shredding in the company of domestic appliances: 87–90 dB(A) – 8%, 90–95 dB(A) – 89%, >95 dB(A) – 3%. Calculated $L_{EX, 8h}$ (over 2 hours) exceeded the lower exposure value in all cases and the limit value when the noise level was >93 dB(A).

The analysis of noise exposure at the companies of concrete and its construction manufacturing was done analogically. Noise level of 80 dB(A) was exceeded at 75% of all places, 57% exceed the 85 dB(A) and 51% – the limit value (Fig. 3). Noise exposure level for all these cases would be as high as 81–90 dB(A) working only two hours per day. Similar results were found by Neitzel *et al.* (1999) at the construction sector in USA (workplaces of carpenters, metal construction assemblers, technicians and helpers). It was found that the upper exposure value of 85 dB(A) was exceeded at 40% of all workplaces (out of 338 investigated).

Similar situation and results were found at the textile company, where the risk assessment was accomplished. From total number of 600 workers 200 were under the influence of 80–85 dB(A) noise, 300 of 85–90 dB(A) and 100 of 90–95 dB(A) level noise. Generalized results at various workplaces are shown in Figs 3 and 4.

Agriculture is acknowledged as one of the most potentially harmful economic activity sectors worldwide. The analysis of noise exposure in the cabs of agricultural tractors and self-propelled agricultural machinery showed that 73% of these places exceed the value of N_{80}, 55% the N_{85} and 46% of N_{87} respectively. High number of noisy tractors can be justified by large number of tractors

made in Belarus, Russia and Ukraine. This old machinery is prevailing in Lithuania (approx. 50% of total number) and the noise levels at different models of tractors distributes as follows: MTZ-50/80/82 ($L_{EX, 8h}$ = 82 dB(A)), T-150K ($L_{EX, 8h}$ = 84 dB(A)), T-40AM and T-130 ($L_{EX, 8h}$ ≥ 90 dB(A).

The situation of noise exposure is slightly different in the sector of food and drink manufacturing. The lower exposure value of 80 dB(A) was exceeded at 61%, 85 dB(A) at 53% and 87 dB(A) at 48% of workplaces (Fig. 4). According to the statistical data provided by the British Health and Safety Executive the noise levels at the workplaces of bottle filling and packing varies from 85 to 95 dB(A) and 95–100 dB(A) at the workplaces of hammer mills. Noise measurement results at sewing workshops show that the level of 80 dB(A) is exceeded at 41% of all workplaces, while approximately 10 percent of the workers were exposed to the noise of 85 dB(A) and 1% to the noise level, higher than the exposure limit value.

The results from transportation sector and noise exposure measurements at truck driver workplaces show that the lower exposure value N_{80} can be exceeded at 45% of workplaces, (N_{85}) – 23% and (N_{87}) at 16% of all workplaces (Fig. 4). High percentage of workplaces affected is mainly caused by a large number of old CIS made machinery which usually lacks vibro-acoustic safety. Additional noise measurements were carried out in 9 workplaces of locomotive operators. Noise level varied within the level range from 76 dB(A) to 90 dB(A) and only approximate

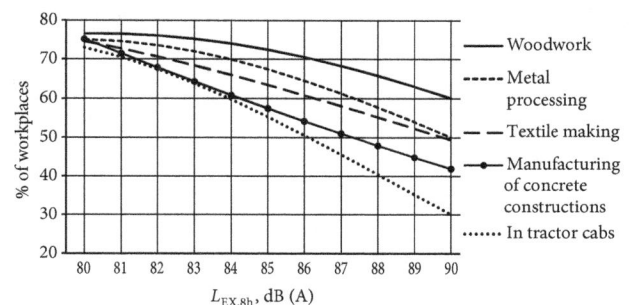

Fig. 3. Percentage of workplaces as a function of $L_{EX, 8h}$ at different sectors of economic activity

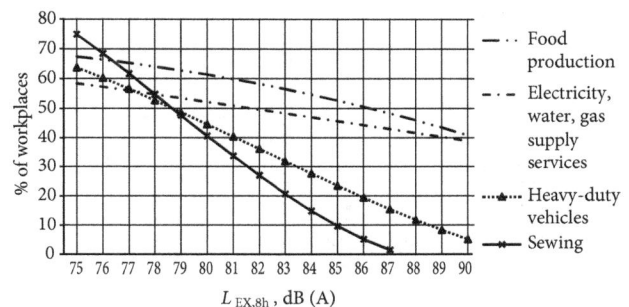

Fig. 4. Percentage distribution of workplaces as a function of noise exposure level normalized to a nominal 8 hour working day

prognostication was considered according to the values of N_L: N_{80} – 55%, N_{85} – 35% ir N_{87} – 27%.

In the companies of water, gas, electricity and heat supply where the operators use typical equipment and tools prognosticated value N_{80} is exceeded at 52%, N_{85} – 46% and N_{87} – 43% of all workplaces. According to the results provided by Vilnius Health Center compressor rooms and pump-houses are the noisiest workplaces where noise levels can be as high as 100–110 dB(A) ($L_{EX, 8h}$ = 94–104 dB(A) over 2 hours). Hacksawing and wringing are also frequent operations at these workplaces when the SPL is as high as 90–100 dB(A) for hacksawing and 82–89 dB(A) for wringing respectively.

The results reviewed above show that the exposure limit value of 87 dB(A) is exceeded at the workplaces of wood processing equipment (77%), metal processing workshops (forges, metal pressing and formation) approx. 61%, textile production (58%). The noise exposure values of slightly lower risk on workers were found at the workplaces of concrete and its construction (51%), tractor drivers (46%), food making and public utilities (approx. 43%).

Special attention should be addressed to these workplaces where parameter $L_{EX, 8h}$ is calculated for lower effective durations of the working day than the reference duration of 8 hours. This might give uncertainty at various durations therefore $L_{EX, 8h}$ was calculated for ordinary work operation durations at particular economic activities.

Noise effect on workers was assessed by performing probability analysis when the accident can occur or hazard to health or hearing is credible. As a result of these impairments probability index was determined for above mentioned economic activity sectors (Table 3). Percentage of workplaces was calculated for the levels of 80, 85 and 87 dB(A) while the calculations of probability index p was performed by using Equations (3), (4) and (5).

Considering the annotation in EU directive that exposure limit value can be lowered if the personal hearing protection is used, actual noise exposure can be decreased to lower exposure value of 80 dB(A) or even more. However a major problem rises fulfilling the agreement for the preventative actions when lower and upper exposure values are exceeded. The attenuation effect of hearing protection is then not considered and employee is obliged to apply any other preventative actions such as replacement of noisy machinery or equipment, shielding it, re-managing the workplace or to apply organizational restructuration (shorten work shift duration, managing breaks) and to improve the occupational understanding of work safety. However the priority should be oriented towards technical solutions.

Calculated values of hearing loss probability index in Table 3 indicate that risky conditions should be expected at wood processing (p = 0.0087), metal processing (0.0042) and textile (0.0032) sectors while less risky in construction (0.0014) and food processing (0.0011).

Table 3. Data of percentage distribution and probability index at various economic activity sectors for the $L_{ex, 8h}$ of 80 dB(A), 85 dB(A) and 87 dB(A) respectively

Sector of economic activity and particular workplaces	No. of workplaces, where the noise exposure levels (N_L) may be exceeded,%			Probability index (p)		
	N_{80}	N_{85}	N_{87}	Health impairment, accident or work-related error ($L_{75}...L_{79}$)	Hearing impairment ($L_{80}...L_{84}$)	Hearing loss ($L_{85}...L_{90}$)
Manufacturing:						
Wood processing and furniture manufacturing	77	72	68	–*	0.0248	0.00869
Metal processing (without foundries)	75	67	61	–*	0.0207	0.00416
Textile industry	75	63	58	0.0304	0.0173	0.00320
Food processing (including drinks)	61	53	48	0.0117	0.0066	0.00105
The garment industry	41	10	1	0.0083	0.0010	–
Construction – Manufacturing of concrete and installation of its construction	75	57	51	0.0484	0.0143	0.00143
Transport – Workplaces of truck drivers	45	23	16	0.0056	0.0006	–
Agriculture and Forestry – Drivers of tractors and self-propelled agricultural machinery	73	55	46	–*	0.0133	0.00055
Electricity, water, gas and heat supply services – repair and maintenance	52	46	43	0.0055	0.0030	0.00057

* Note: unspecified but supposedly >0.04.

Acoustic situation for hearing impairment in these sectors as well as in agricultural tractors ($p = 0.013$), construction (0.014) and textile industry (0.017) were identified as portentous. Lowest expected noise risk index values were at transport, electricity, water, gas and heat supply services, and garment industry.

The results of our study and in the researches of other authors create theoretical premises to prognosticate the approximate number of workplaces, where the noise-related risk is prevailing. According to the distribution of this data at different economic activity sectors it becomes reasonable to apply preventative actions for the noise reduction or workers' protection.

Conclusions

On the base of statistical data and using the methods of mathematical statistics a model for the prediction of noise exposure was created at workplaces of different economic activity sectors. This model allowed to predict the percentage distribution of workplaces where the noise exposure levels of 80, 85 and 87 dB(A) were exceeded.

Highest number of noise level at workplaces was found at the companies of wood processing and furniture manufacturing. The exposure value of 80 dB(A) was exceeded at 77%, 85 dB(A) – 72% and 87 dB(A) – 68% of all workplaces. Similar results were found at the workplaces of metal processing where the results were as follows: N_{80} – 75%, N_{85} – 67% and N_{87} – 61%.

Highest risk probability on workers to undergo the NIHL was found at the workplaces of wood, metal and textile sectors (probability index from 0.0087 to 0.0032 respectively). Construction sector and operation of agricultural tractors should be also attributed as having potential risk on workers.

Noise induced health impairment and accidents or work-related errors were identified as portentous in construction sector and agricultural tractors, wood and metal processing companies (probability index 0.04), textile sector (0.03) and food processing (0.012).

References

Arezes, M. P.; Miguel, A. S. 2005. Hearing protection use in industry: the role of risk perception, *Safety Science* 43(4): 253–267. http://dx.doi.org/10.1016/j.ssci.2005.07.002

Aven, T.; Vinnem, J. E.; Wiencke, H. S. 2007. A decision framework for risk management, with application to the offshore oil and gas industry, *Reliability Engineering and System Safety* 92(4): 433–448. http://dx.doi.org/10.1016/j.ress.2005.12.009

Babisch, W. 2000. Traffic noise and cardiovascular disease: epidemiological review and synthesis, *Noise Health* 2(8): 9–32.

Baltrėnas, P.; Petraitis, E.; Januševičius, T. 2010. Noise level study and assessment in the southern part of Panevėžys, *Journal of Environmental Engineering and Landscape Management* 18(4): 271–280. http://dx.doi.org/10.3846/jeelm.2010.31

Banbury, S. P.; Berry, D. C. 2005. Office noise and employee concentration: identifying causes of disruption and potential improvements, *Ergonomics* 48(1): 25–37. http://dx.doi.org/10.1080/00140130412331311390

Cagno, E.; Di Giulio, A.; Trucco, P. 2005. Statistical evaluation of occupational noise exposure, *Applied Acoustics* 66: 297–318. http://dx.doi.org/10.1016/j.apacoust.2004.08.001

Davies, H. W.; Teschke, K.; Kennedy, S. M.; Hodgson, M. R.; Demers, P. A. 2009. Occupational noise exposure and hearing protector use in Canadian lumber mills, *Journal of Occupational and Environmental Hygiene* 6(1): 32–41. http://dx.doi.org/10.1080/15459620802548940

Granneman, J. H.; Oostdijk, J. P. J.; Schermer, F. A. G. M. 2004. Extensive survey of occupational noise exposure in the metal working industry, in *The Proceedings of Inter Noise 2004: Progress in Noise Control for the 21st Century/The 33rd International Congress and Exposition on Noise Control Engineering*, 22–25 August, 2004, Prague, 1–8.

Haines, M. M.; Stansfeld, S. A.; Job, R. F. S.; Berglund, B.; Head, J. 2001. Chronic aircraft noise exposure, stress responses mental health and cognitive performance in school children, *Psychological Medicine* 31: 265–277. http://dx.doi.org/10.1017/S0033291701003282

Ising, H.; Kruppa, B. 2004. Health effects caused by noise: Evidence in the literature from the past 25 years, *Noise & Health* 6(22): 5–13.

ISO 1999:2013. Acoustics – Estimation of noise – induced hearing loss.

ISO 9612:2009. Acoustics – Determination of occupational noise exposure – Engineering method.

Lusk, S. L.; Gillespie, B.; Hagerty, B. M.; Ziemba, R. A. 2004. Acute effects of noise on blood pressure and heart rate, *Archives of Environmental Health: An International Journal* 59(8): 392–399. http://dx.doi.org/10.3200/AEOH.59.8.392-399

McReynolds, M. C. 2005. Noise-induced hearing loss, *Medical Journal* 24(2): 73–78. http://dx.doi.org/10.1016/j.amj.2004.12.005

Merkevičius, S.; Strelkauskis, R.; Butkus, R. 2011. Metodologiniai mobilių mašinų generuojamų vibracijų operatoriui keliamos rizikos nustatymo aspektai [Methodological aspects of evaluation of vibration risk on the operators of mobile machinery], in *The Proceedings of the International Scientific Conference "Human and Nature Safety 2011"*, 11–13 May, 2011, Akademija, Lithuania, 39–41.

Miyakita, T.; Ueda, A.; Futatsuka, M.; Inaoka, T.; Nagano, M.; Koyama, W. 2004. Noise exposure and hearing conservation for farmers of rural Japanese communities, *Journal of Sound and Vibration* 277(3): 633–641. http://dx.doi.org/10.1016/j.jsv.2004.03.026

Moriyama, T.; Ohtani, H. 2009. Risk assessment tools incorporating human error probabilities in the Japanese small-sized establishment, *Safety Science* 47: 1379–1397. http://dx.doi.org/10.1016/j.ssci.2009.01.005

Neitzel, R.; Seixas, N. S.; Camp, J.; Yost, M. 1999. An assessment of occupational noise exposures in four construction trades, *American Industrial Hygiene Association Journal* 60(6): 807–817. http://dx.doi.org/10.1016/j.jsv.2004.03.026

Oksa, P.; Palo, L.; Saalo, A.; Jolanki, R.; Mäkinen, I.; Kauppinen, T. 2012. *Ammattitaudit ja ammattitautiepäilyt 2010. Työperäisten sairauksien rekisteriin kirjatut uudet tapaukset* [Occupational Diseases and suspected Occupational Diseases 2010. New cases registered in the Register of Occupational Diseases] [online], [cited 5 June 2013]. Available from Internet: http://www.ttl.fi/fi/verkkokirjat/ammattitaudit/Documents/Ammattitaudit_2010.pdf

Oškinis, V.; Kindurytė, R.; Butkus, D. 2004. Automobilių triukšmo tyrimų magistralėje Vilnius–Kaunas–Klaipėda rezultatai [Evaluation of car noise on the highway Vilnius-Kaunas-Klaipėda], *Journal of Environmental Engineering and Landscape Management* 12(1): 10–18.

http://dx.doi.org/10.1016/S0022-4375(00)00039-6

Selander, J.; Nilsson, M. E.; Bluhm, G.; Rosenlund, M.; Lindqvist, M.; Nise, G.; Pershagen, G. 2009. Long-term exposure to road traffic noise and myocardial infarction, *Epidemiology* 20(2): 272–279.
http://dx.doi.org/10.1097/EDE.0b013e31819463bd

Vasarevičius, S.; Graudinytė, J. 2004. Transporto triukšmo lygio automobilių kelių ir geležinkelio sankirtose tyrimai ir įvertinimas [Investigation and evaluation of noise level at motorway and railway crossings], *Journal of Environmental Engineering and Landscape Management* 12(1): 19–24.

Yränheikki, E.; Savolainen, H. 2000. Occupational safety and health in Finland, *Journal of Safety Research* 31(4): 177–183.

Ričardas BUTKUS, Dr Assoc. Professor at the Institute of Agricultural Engineering and Safety, Aleksandras Stulginskis University (ASU). Publications: single-authored of co-written more than 50 scientific articles. Research interests: environment and mechanics engineering, occupational safety and ergonomics, sound and vibration.

Alvidas ŠARLAUSKAS, Lecturer at the Institute of Agricultural Engineering and Safety, Aleksandras Stulginskis University. Doctor of Sciences (Environmental Engineering and Landscape Management), ASU, 2000. Publications: author of >10 scientific research papers. Research interests: standardization, occupational safety, physical measurements.

Gediminas VASILIAUSKAS, Junior Researcher at the Institute of Agricultural Engineering and Safety, Aleksandras Stulginskis University. Doctor of Sciences (Environmental Engineering and Landscape Management), ASU, 2012. Publications: author of 10 scientific research papers. Research interests: environmental protection, occupational safety, sound quality, speech intelligibility.

2

MODELLING AND SIMULATION OF DISPERSIONS OF POWDER EMISSIONS FROM MULTIPLE SOURCES WITH THE MATHEMATICAL MODEL *POL 15SM*

Mihaela Budianu[a], Valeriu Nagacevschi[b], Matei Macoveanu[b]

[a] Environmental Protection Agency Vaslui, Str. Calugareni, no.63, 730149 Vaslui, Romania
[b] Department of Environmental Engineering and Management, Faculty of Chemical Engineering,
Technical University of Iasi, 71A D. Mangeron Bd., 700050 Iasi, Romania

Abstract. Over the last decades, air pollution has become one of the greatest challenges negatively affecting human health and the entire environment, including air, water, soil, vegetation, and urban areas. Lately, special attention has been given to mathematical modelling for diffusion of pollutants in the atmosphere as a particularly effective and efficient method that can be used to study, control and reduce air pollution. The diversity of models developed by different research groups imposed a rigorous understanding of model types in order to apply them correctly according to local or regional problems of air pollution phenomenon. Thus the authors have developed and improved two mathematical models for dispersion of air pollutants. This paper presents a case study of dispersion of powders in suspension originating from 14 point sources that correspond to 5 economic agents in the agroindustrial area of Vaslui city using a computer simulation based on the mathematical model *Pol 15sm*, for multiple point sources of pollution, designed by the authors.

Keywords: pollutant dispersion, dispersion models, air pollution, suspended powders, heavy metals, computer simulation.

Reference to this paper should be made as follows: Budianu, M.; Nagacevschi, V.; Macoveanu, M. 2014. Modelling and simulation of dispersions of powder emissions from multiple sources with the mathematical model *Pol 15sm*, *Journal of Environmental Engineering and Landscape Management* 22(02): 151–160.

Introduction

Lately, a special attention has been given to studying the dispersion of pollutants emitted into the atmosphere by multiple or isolated pollution sources, operating continuously or accidentally, as it is increasingly evident that human activities have already produced a "disturbance" of the environmental balance (Carpentieri *et al.* 2011; Hanna 1981; Huang *et al.* 2011; Mezdrea-Cojocareanu-Pata *et al.* 2010; Miclaus *et al.* 2006).

Atmospheric emissions of heavy metals are deposited in the form of suspended powders on the ground at varying distances from the emission source, distances that depend on particle sizes and the intensity of air currents (Radulescu *et al.* 2010). Problems related to the presence in the environment of metallic elements in the

environment are triggering a growing interest (Nisulescu *et al.* 2011; Stătescu, Cotiuşcă-Zauca 2006).

Areas with heavy metals contaminated soils around the industrial units differ in expansion due to intensity and duration of emissions, weather and geomorphologic aspects (Baldauf *et al.* 2009; Carpentieri *et al.* 2011; Gaba 2010; Miclaus *et al.* 2006).

In Romania, the calculation of pollutant emissions into the atmosphere is reflected in the methodology EEA/ EMEP/CORINAIR and U.S. EPA/AP-42 for all business groups, including industrial point sources.

One way of knowing and subsequently taking action against air pollution is the mathematical modelling of pollutant dispersion into the atmosphere (Hanna 1981; Vătăşescu *et al.* 2011; Ani *et al.* 2012). This consists of

Corresponding author: Matei Macoveanu
E-mail: matei_macoveanu@yahoo.com

estimating pollutants concentrations on soil and height accordingly with pollution sources characteristics, weather and geographic conditions, physical and chemical transformation processes that pollutants might have in the atmosphere and their interaction with the soil surface (Baldauf *et al.* 2009; Budianu, Macoveanu 2010; Carpentieri *et al.* 2011).

Insidious effects on medium and long term, regionally and even worldwide, can be caused by local polluting phenomena from a variety of "quasi-harmless" point sources. Considering them, especially in the last 30 years, scientific efforts have focused on detailed studying of transport and dispersion mechanisms of pollutants into the atmosphere and the consequences of these phenomena, without neglecting the measurement techniques of pollutants at source, adapted to various situations (Branquinho *et al.* 2008; Hanna 1981; Nisulescu *et al.* 2011; Stefan *et al.* 2013). The development and importance of mathematical models of pollutant dispersion are presented widely in specialized literature. Of the many models studied, four main types of models are distinguished: statistical/empirical, Gaussian (most used), Lagrangian and Eulerian (most complex) (Candelieri *et al.* 2012; Moreira *et al.* 2010; Pasquill 1961; Popescu *et al.* 2011; San Jose 1997; Smaranda, Gavrilescu 2008).

After analyzing these models, assessing the advantages, disadvantages, features, limitations and accessibility degree offered, computer simulations based on the mathematical models ECO95sp and *Pol 15sm* were used to calculate the pollutant dispersion in the atmosphere of Vaslui city (Nagacevschi, Macoveanu 1994, 1995, 1996, 2002; Nagacevschi *et al.* 1997).

The paper presents a case study on dispersion modelling of suspended powders emitted jointly by many point sources in the agroindustrial area of Vaslui, with calculating pollutant concentrations on immission for several set receptors, using *Pol 15sm* model.

1. Description of mathematical model *POL 15sm* and the accordingly simulation program

Using this Gaussian model for studying the dispersion of pollutants emitted by multiple sources is possible, continuing the study of pollutant dispersion originating from each individual point source, done using ECO95sp model. This is a logical follow-up of the fact that these 2 models have a complementary character; in addition, both use the same type of classification of atmospheric stability, the same division into 2 distinct types of terrain and, broadly, the same input requirements.

Like the mathematical model ECO95sp, *Pol 15sm* is a model for calculating dispersion of pollutants into the atmosphere, developed and tested as a computer simulation program in the Department of Chemical Engineering and Environmental Protection of Chemical Engineering and Environmental Protection Technical University "Gheorghe Asachi" of Iasi.

POL 15sm model is the base of a simulation program of concentration distribution for a pollutant emitted by several sources, well located, to a number of pollution receptors, also precisely located:
- maximum number of emitting point sources that the program can take into account – 15;
- maximum number of receptors that can be considered – 15.

Performances of the model can be best summarized by specifying the types of outputs; at each receiver point the following output are measured and played in the form of data:
- local concentration of pollutant due to emission of each source;
- total local concentration of pollutant, emitted by all sources considered;
- average local concentration, in the time interval of interest, due to each emission source;
- total average local concentration, in the time given.

Note that the first two categories of results are given periodically, with a period of maximum 15 hours, the modelling assumption formulated being that the average frequency weather changes is also periodically.

The stages of rolling are the classic ones, while the input required by *Pol 15sm* model needs:
- data describing the characteristics of each emission source in part (mass flow of pollutant emitted, total volumetric flow of exhaust gases, exhaust gas temperature, physical height of pollutant source, source coordinates to a fixed reference point);
- data describing the receptors considered (the number of receptors to pollutant immission concentration taken into account, the height of placement of every receptor considered, landmark coordinates to a fixed point, the nature of land);
- climatologically data (weather conditions, wind speed, air temperature).

To describe the model, it is said that both the input and the output are presented in tables in type ASCII text files: "POL.DAT" – for input data and "POL.REZ" – for output data. In term of input data required by the *Pol 15sm* model, these can be grouped in three categories:

a. "Sources" (valid for each emission source)

n_s = number of emission sources considered (n_s max. = 15);

Q = mass flow of pollutant emitted (mg/s);

Q_f = volumetric flow of "smoke" at emission (total gas flow emitted by each source), (m³/s);

T_f = temperature of "smoke" at emission (exhaust gas temperature), (°C);

h = physical height of pollutant source, (m).

Sources coordinates set to a fixed reference point, were chosen accordingly:

- cxs = West-East coordinate, (m);
- cys = South-North coordinate, (m).

b. *"Receptors"* (valid for each receptor)

n_r = receptors number of pollutant concentration considered at emission (n_r max. = 15);

z = height of every receptor considered, (m).

Receptors coordinates, to the same fixed reference point chosen:

- cxs = West-East coordinate, (m);
- cyr = South-North coordinate, (m).

"Rural / Urban" = option regarding the terrain conditions for each source

c. *"Hourly weather conditions"*

n_o = number of considered hours (n_o max. = 15);

u = wind speed, (m/s);

θ = wind direction, from North, measured clockwise, (degrees);

(1–6) = class selected to define the atmospheric conditions (using all 6 Pasquill stability classes, as in the case of "ECO95sp" model);

T = atmospheric temperature, (°C).

To facilitate the use of the simulation program by people who do not have sufficient knowledge regarding the characterizing criteria stability/atmospheric turbulence state, within the program recommendations are stated regarding choosing classes for weather conditions, with values from "1" to "6". Thus, we considered I–VI cases, characterized below by the following conditions:

- case I – day, sun shining brightly, solar radiation within an angle wider than 60°;
- case II – day, sun shining moderately, solar radiation within an angle between 35° and 60° (so-called the "slightly covered" sky);
- case III – day, sun shining weakly, solar radiation within an angle slighter than 35° ("partially covered" sky);
- case IV – day/night (night/day) interval, cloudy sky
- case V – night, cloudy sky (cloudiness over 50%);
- case VI – night, partially cloudy sky (cloudiness below 50%).

Then, the appropriate class for characterization of certain atmospheric conditions is set, accordingly with the specifics of every situation of dispersion modelling considered, using Table 1.

Note that in the case of this model also, the value "1" (or "A") corresponds to the category "Highly unstable", and the value "6" (or "F") corresponds to the category "Highly stable", according to the classification made by Pasquill (1961).

In the input data for this program the main geographical characteristics (natural or anthropic) are not found. These can influence the dispersion of pollutants emitted in the atmosphere from a given source and they are:

- significant bumps of the terrain, located in the main propagation direction of the pollutant plume (geological formations of hill, mountains, deep and narrow valley, etc.);
- the existence in the area of significant water courses, lakes of important volume and surface, rich in vegetation areas (e.g. forests), whose presence somehow influence the micro-climate of areas more or less extensively;
- the presence, near the source, of constructions of important dimensions, especially in terms of height, which can induce local changes of direction and even speed of air flow with a role in training and dispersal of pollutants.

Unlike the U.S. AERMOD model, which is a model that applies to industrial sources and of the last generation, that already contains principles of planetary boundary layer, and *Pol 15sm* is a much simpler, easy to use software that requires no complex and expensive sites.

There are two input data processors that are regulatory components of the AERMOD modelling system: AERMET, a meteorological data preprocessor that incorporates air dispersion based on planetary boundary layer turbulence structure and scaling concepts, and AERMAP, a terrain data preprocessor that incorporates complex terrain using USGS Digital Elevation Data. Other non-regulatory components of this system include: AERSCREEN, a screening version of AERMOD; AERSURFACE, a surface characteristics preprocessor, and BPIPPRIME, a multi-building dimensions program incorporating the GEP technical procedures for PRIME applications.

For *Pol 15sm* program selection in order to apply it with conclusive results for an industrial area (only for industrial point sources) a number of specific criteria were taken into account:

Criteria for the weather:

- the presence of atmospheric calm periods in over 24%;
- significant frequency of total and partial thermal inversions;
- the existence of periodic winds (NW and N wind);
- relatively high humidity conditions (favoring the occurrence of side effects for pollutants emitted).

Criteria for relief conditions:

- location in the Bârlad valley area – relatively flat surface;
- proximity to the relief of sub-Carpathians, with heights not exceeding 400 m and below 150 m altitude difference from the platform.

Criteria regarding emissions:

- the existence of emission sources with significantly different heights (between 12 m and 218 m);

– the existence of industrial buildings and constructions of moderate size, in the vicinity of emission sources of medium height, but at distances for which the existence of the building is not appreciated;

– different exhaust emissions, in favorable conditions, can lead to secondary reactions into the atmosphere, as the formation of another pollutant.

Other criteria:

– the importance of extending the assessment area within a radius of at least 40 km (to the border with Moldova);

– the opportunity to formulate conclusions based on modelling results, and consequences of historical pollution (e.g. damage quantification of buildings and civil engineering, possibly while assessing the impact on current health and the integrity of ecosystems potentially affected);

– the availability of models, both financially (since most of these models are only the demonstration, completing their purchase involves payment of substantial amounts) and in terms of logistics resources to be mobilized for their use.

The underlying mathematical model to the simulation program *Pol 15sm*, to calculate the dispersion of emitted pollutants into the atmosphere from multiple point sources, is described below:

1. Calculation of source – receptor distance in wind direction:

$$x_{rel} = cxr_j - cxs_i;\qquad(1)$$

$$y_{rel} = cyr_j - cys_i;\qquad(2)$$

$$x = x_{rel} \cdot \cos(\theta_k) + y_{rel} \cdot \sin(\theta_k);\qquad(3)$$

$$y = \sqrt{\left(x_{rel}^2 + y_{rel}^2 - x^2\right)},\qquad(4)$$

cxr = West-East Cartesian coordinate, for receptors, (m);
cyr = South-North Cartesian coordinate, for receptors, (m);
cxs = West-East Cartesian coordinate, for point sources, (m);
cys = South-North Cartesian coordinate, for point sources, (m);
i = current point sources number;

j = current receptors number of pollutant concentration at imission considered;
θ = wind direction, from North, measured clockwise, (degree);
$xrel, yrel$ = Cartesian projections of point source to receptor distance, (m);
x, y = projections of point source to receptor distance, on wind direction, (m).

2. Calculation of pollutant dispersion into the atmosphere:

$$\sigma_y = f_1(x, k);\qquad(5)$$

$$\sigma_z = f_2(x, k),\qquad(6)$$

σ_y, σ_z = pollutant dispersion on y, z directions, respectively.

3. Calculation of smoke plume height:

$$dh = f_3(x, k, i);\qquad(7)$$

$$H_t = h_0 + dh,\qquad(8)$$

H_t, h_0, dh = height of pollutant smoke plume (total, initial, increment), (m).

4. Calculation of concentration at receptor:

$$\psi_{i,j,k} = \frac{Q}{2\pi \cdot \bar{u} \cdot \sigma_y \cdot \sigma_z} e^{-0,5\left(\frac{y}{\sigma_y}\right)^2} \cdot \left[e^{-0,5\left(\frac{z-Ht}{\sigma_z}\right)^2} + e^{-0,5\left(\frac{z+Ht}{\sigma_z}\right)^2} \right],\qquad(9)$$

– Q = mass flow of pollutant emitted (mg/s);
– u = wind speed, (m/s).

5. Calculation of averaged concentration:

$$\overline{\psi}_{j,k} = \frac{1}{n_i} \sum_i^{n_i} c_{i,j,k};\qquad(10)$$

$$\overline{\psi}_{i,j} = \frac{1}{n_k} \sum_k^{n_k} c_{i,j,k},\qquad(11)$$

– i, j, k = current index on x, y, z direction for receptor;
– $c_{i,j,k}$ = pollutant concentration in space at receptor, on wind direction, (mg/m³);
– $c_{j,k}, c_{i,j}$ = averaged concentration on one direction, x, k, respectively.

Table 1. Weather condition classes assigning; correlation of weather conditions defined by cases I – VI, with wind speed

Case	Wind speed (m/s)				
	<2	between 2 and 3	between 3 and 5	between 5 and 6	>6
I	1	1;2	2	3	3
II	1;2	2	2;3	3;4	4
III	2	3	3	4	4
IV	4	4	4	4	4
V	5	5	4	4	4
VI	6	6	5	4	4

2. Method

2.1. The experimental part. Case study

Computer simulation of powders dispersion from multiple sources, using the *Pol 15sm* model, whose basic elements have been previously presented, was applied to 14 point sources, which belong to a number of 5 economical agents in the agroindustrial area of Vaslui city. It was necessary to correctly evaluate the total contribution of a specific pollutant, emitted by the group of sources with this characteristic and to establish the contribution of every source generating a certain type of pollutant, to "show" the total impact of that pollutants concentration at immission, determined in precisely set receiver points.

Encoding of the 14 point sources has been done as follows:

- S1–S3 for SC Stemar SRL Vaslui;
- S4–S9 for SC Termica SA Vaslui;
- S10 for Brick Factory SRL Vaslui;
- S11–S13 for SC Ulerom SA Vaslui; and
- S14 for SC Vascar SA Vaslui.

Six points were set – *initial receivers* in conjunction with existing fixed point for air quality monitoring from the network of Environmental Protection Agency Vaslui, so we can establish the territorial correspondence below:

- R1 – Point "APM headquarters";
- R2 – Point "Station Vaslui 1 – Public Finance Department Vaslui – urban background station";
- R3 – Point "Vaslui County Hospital";
- R4 – Point "Watewater Treatment Plant";
- R5 – Point "SC Termica SA Vaslui";
- R6 – Point "SC AMC Badotherm SA Vaslui".

As an additional element absolutely necessary for such a model, it is required to enter, in a certain form, the coordinates of emission sources, on West-East and South-North directions, that are *cxs* and *cys*, set to a fixed reference point, chosen by user, as well as the coordinates for receptors on West-East and South-North directions, that are *cxr* and *cyr*, set to the same reference point, called the "reference".

The input data entered in the modelling program in this particular case, to characterize the emissions of pollutants, the point sources and weather conditions are those presented in Table 2 and Table 3, specifying that the classification in stability classes was made considering the instructions in the presentation of *Pol 15sm* model, and the

Table 2. Input data – emission sources for *Pol 15sm* model

Emission sources	Geographical coordinates latitude/longitude	Q [mg/s]	Q [m³/s]	T_f [°C]	h [m]	cxs [m]	cys [m]	Source encoding
SC Stemar SA Vaslui – chimney 1	46° 37' 8.352" 27° 43' 61.092"	0.3138	0.0707	110	12	0	0	S1
SC Stemar SA Vaslui – chimney 2	46° 37' 8.352" 27° 43' 61.092"	3.5357	0.2898	160	12	0	0	S2
SC Stemar SA Vaslui – chimney 3	46° 37' 8.352" 27° 43' 61.092"	3.3494	0.8588	100	25	0	0	S3
SC Termica SA Vaslui – chimney 1	46° 37' 35.4" 27° 43' 38.87	45.62	37.39	200	80	820	−495	S4
SC TermicaSA Vaslui – chimney 2	46° 37' 35.4" 27° 43' 38.87	4.82	4.15	160	60	820	−495	S5
SC Termica SA Vaslui – chimney 3	46° 37' 35.4" 27° 43' 38.87	199.3	4.91	95	16.5	820	−495	S6
SC Termica SA Vaslui – chimney 4	46° 37' 35.4" 27° 43' 38.87	164.4	4.91	95	16.5	820	−495	S7
SC Termica SA Vaslui – chimney 5	46° 37' 35.4" 27° 43' 38.87	118.4	1.76	95	16.5	820	−495	S8
SC Termica SA Vaslui – chimney 6	46° 37' 35.4" 27° 43' 38.87	328.9	4.91	95	16.5	820	−495	S9
SC Fabrica de caramizi SRL – chimney 1	46° 37' 46.25" 27° 43' 13.84"	156.0065	2.1488	95	12	1150	−1050	S10
Sc Ulerom SA Vaslui – chimney 1	46° 38' 41.52" 27° 43' 22.24"	0.5089	0.2827	174	25	1300	350	S11
Sc Ulerom SA Vaslui – chimney 2	46° 38' 41.52" 27° 43' 22.24"	0.5089	0.2827	174	25	1300	350	S12
Sc Ulerom SA Vaslui – chimney 3	46° 38' 41.52" 27° 43' 22.24"	43.1027	0.3848	160	218	1300	350	S13
Sc Vascar SA Vaslui – chimney 1	46° 37' 38.22" 27° 43' 65.59"	2.4504	0.6126	160	45	1200	100	S14

Table 3. Input data – receptors for *Pol 15sm* model

Receptors	Geographical coordinates latitude/longitude	h [m]	cxr [m]	cyr [m]	Terrain conditions regarding the source (rural/urban)	Receptor encoding
APM Vaslui	46° 38 '19.827" 27° 43'21.788"	1.5	2170	1000	urban	R1
Stația Vaslui 1	46° 37'55.777" 27° 43'51.307"	1.5	1450	800	urban	R2
Vaslui County Hospital	46° 38' 18" 27° 45' 61"	1.5	2250	2700	urban	R3
Statia de Epurare Vaslui	46° 37' 25" 27° 13' 65"	1.5	−515	−1050	urban	R4
SC TermicaA SA Vaslui	46° 37' 29.8 27° 43'61.092"	0	650	0	urban	R5
SC AMC Badotherm SA Vaslui	46° 38' 25.84" 27° 42' 51.404"	0	2450	−1500	urban	R6

simulation was realized taking into account two categories of weather conditions: unstable and highly stable.

The geographical coordinates of the emission point sources considered (S1–S14), as well as the coordinates of receptors R1–R6 have been determined directly on field, using a GPS system. To facilitate calculation and subsequent graphic representations, the reference, (fixed or zero point) with the coordinates: 46° 37' 8,352" / 27° 43' 61,092", was established in the same place as the sources S1, S2, S3, the *Pol 15sm* model not introducing any interdiction or restriction in this regard.

Figure 1 shows the scheme for territorial arrangement of multiple sources and industrial receptors – points, by reporting to a fixed reference point, chosen by the user as input data for the program – reference point with the coordinates of SC Stemar SA Vaslui (S1–S3).

For every receptor, the output data provide by the model as strings of data that indicate:

– local concentration of pollutants due to every emission source;

– total local concentration of the pollutant due to emission of all sources considered.

Data were then selected and processed in table and graphic form, to become more relevant and accessible for analysis and to formulate correct and coherent conclusions.

3. Results and discussions

Using the processing facilities of this model tables and graphic representations for 10 different situations depending on weather conditions, have been made. Linking the data and information regarding the emission sources and the receptors with data and information regarding the direction and speed of air currents in Vaslui area, it is observed that:

– prevailing wind in the agroindustrial area of Vaslui is North-West, with an average of 18.5% out of total per an and a higher share (over 50% of the time) in the hot season;

– second prevailing wind – with a share of only 17.1% – is the wind from the North;

– in this evaluation of air currents characteristics in the agroindustrial area, a special place is occupied by the atmospheric calm with a share of over 24% a year's total.

In these conditions, returning to placement scheme of sources and receptors, some assessing guidelines to analyze the modelling results obtained using the simulation program for multiple sources *Pol 15sm* can be stated:

– with the prevailing wind direction from N-V. R5r-eceptors are partially influenced by emissions of sources S4. S5. S6. S7. S8. S9 and S10 and not at all – influenced by emissions of sources S1. S2. S3. S11. S12 and S13;

– considering the next direction as a share, respectively the wind from N. In this case the receptors R5 and R4 are influenced by the emissions generated

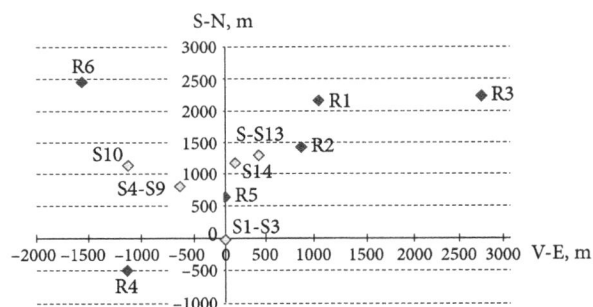

S1-S3 – SC SETMAR SRL
S4-S9 – SC TERMICA SRL
S10– SC bricks factory SRL
S11-S13 – SC ULEROM SA
S14 – SC VASCAR SA

R1 – APM headquarter point
R2 – Vaslui 1 station – public finances direction point
R3 – Vaslui county hospital point
R4 – wastewater treatment plant point
R5 – SC TERMICA SA Vaslui point
R6 – SC AMC badotherm SA Vaslui point

Fig. 1. Scheme for territorial arrangement of multiple sources and industrial receptor – points, by reporting to a fixed reference point

by sources S4, S5, S6, S7, S8, S9, S10, S11, S12, S13 and S14;

– in situations of atmospheric calm, characterized by wind speeds under 1 m/s and without a definite prevalence of air currents direction (situations frequently accompanied by temperature inversion phenomena), an estimation of the influence shared of each source on the considered receptors cannot be a sufficiently exact. In this case, dispersion conditions must be taken into account, in approximately equal proportions. The direction of air currents in the studied area was also considered in connection with the effect of pollution source on each receptor, respectively – in the order to find the average frequency recorded in one year:

– North-West wind – sources influence on receptors R5;

– South-East wind – sources influence on receptors R6 ;

– North-East wind – sources influence on receptors R4. R5;

– North wind – sources influence on receptors R4. R5;

– South-West wind – sources influence on receptors R1. R2. R3. R5.

Each of the four dominant wind directions has been considered for two atmospheric stability categories, selected as indicated by *Pol 15sm* model and according to weather characteristics of the investigated area, namely unstable and stable atmosphere.

Further, analyzing graphics and tables for modelling the dispersion of suspended powders from the multiple sources for ten of these cases, it is found that:

A. If atmosphere is unstable:

For wind direction from W to E:

– in receptors R1. R2. R3 and R6 there is no contribution of any powders generating sources. reason for which the total concentration of powders at emission considered in these receiver-points is zero;

– at R4 receptor dust emissions arrive from sources S4. S6. S7. S8. S9. S10;

– at R5 receptor dust emissions arrive from sources S12. S13. S14.

For wind direction from N-W to S-E:

– in receptors R1. R2. R3. R4 and R6 there is no contribution of any powders generating sources. reason for which the total concentration of powders at emission considered in these receiver-points is zero;

– at R5 receptor dust emissions arrive from sources S4. S5. S6. S7. S8. S9. S10.

For wind direction from S-E to N-W:

– in receptors R1. R2. R3. R4 and R5 there is no

contribution of any powders generating sources. Reason for which the total concentration of powders at emission considered in these receiver-points is zero;

– at R5 receptor dust emissions arrive from S4. S6. S7. S8. S9. S10 and S13.

For wind direction from N-E to S-W:

– at R1 receptor dust emissions arrive from sources S5. S6. S7. S8. S9. S10. S13 and S14;

– at R2 receptor dust emissions arrive from S4. S5. S6. S7. S8. S9. S10 and S14;

– at R3 receptor dust emissions arrive from S6. S7. S8. S9 and S14;

– in receptors R4, R5, R6 there is no contribution of any powders generating sources to the total concentration of powders at emissions, this being considered null.

Figure 2 is a data combination of Table and diagram and summarizes the distribution of powders emissions from the 14 sources in "less unstable" atmospheric conditions.

Reviewing all these observations and correlating them with the results of air quality monitoring, obtained by the surveillance network of the Environmental Protection Agency Vaslui, especially for period of 2000–2009, it is understandable why the annual frequency of powders exceedance registered in monitoring points is so small (not exceeding a maximum of 7% per year). Although, in this time, all said sources have almost continuously emitted powders with concentrations at emission that do not exceed with a lot the maximum allowed, though it happens.

Basically, the location of network points has partially captured the individual and cumulative influence of selected sources, on one wind direction, that is not the prevailing one (from N-E to S-W).

B. If stable atmosphere:

For wind direction from S-W to N-E:

– in receptors R1. R2. R3. R5 and R6 there is no contribution of any powders generating sources to the total concentration of powders at emissions. this being considered null;

– at R4 receptor dust emissions arrive from S11.

For wind direction from S to N:

– in receptors R1. R2. R3. R4 R5. R6 there is no contribution of any powders generating sources to the total concentration of powders at emissions, this being considered null.

For wind direction from E to V:

– in receptors R1. R2. R3 and R4 there is no contribution of any powders generating sources to the total concentration of powders at emissions, this being considered null;

– at receptor R5 dust emissions arrive from S1. S2. S3;

Powders concentrations generated by the 14 sources to receptors R1–R6 in certain weather conditions											
Receptor No.	Weather conditions				Powders concentration [µg/m³]						
	Wind			T							
	Speed [m/s]	θ [degrees]	Pasquill stability classes	[°C]	Source	R1	R2	R3	R4	R5	R6
R1	5	45	1	10	S1	0.000	0.000	0.000	0.000	0.000	0.000
R2	5	45	1	10	S2	0.000	0.000	0.000	0.000	0.000	0.000
R3	5	45	1	10	S3	0.000	0.000	0.000	0.000	0.000	0.000
R4	5	45	1	10	S4	0.000	0.004	0.000	0.000	0.000	0.000
R5	5	45	1	10	S5	0.001	0.001	0.000	0.000	0.000	0.000
R6	5	45	1	10	S6	0.025	0.025	0.002	0.000	0.000	0.000
					S7	0.021	0.021	0.002	0.000	0.000	0.000
					S8	0.016	0.015	0.001	0.000	0.000	0.000
					S9	0.042	0.042	0.003	0.000	0.000	0.000
					S10	0.007	0.001	0.000	0.000	0.000	0.000
					S11	0.000	0.000	0.000	0.000	0.000	0.000
					S12	0.000	0.000	0.000	0.000	0.000	0.000
					S13	0.011	0.000	0.000	0.000	0.000	0.000
					S14	0.001	0.006	0.001	0.000	0.000	0.000
					Total concentrations for receptor	0.125	0.118	0.010	0.000	0.000	0.000

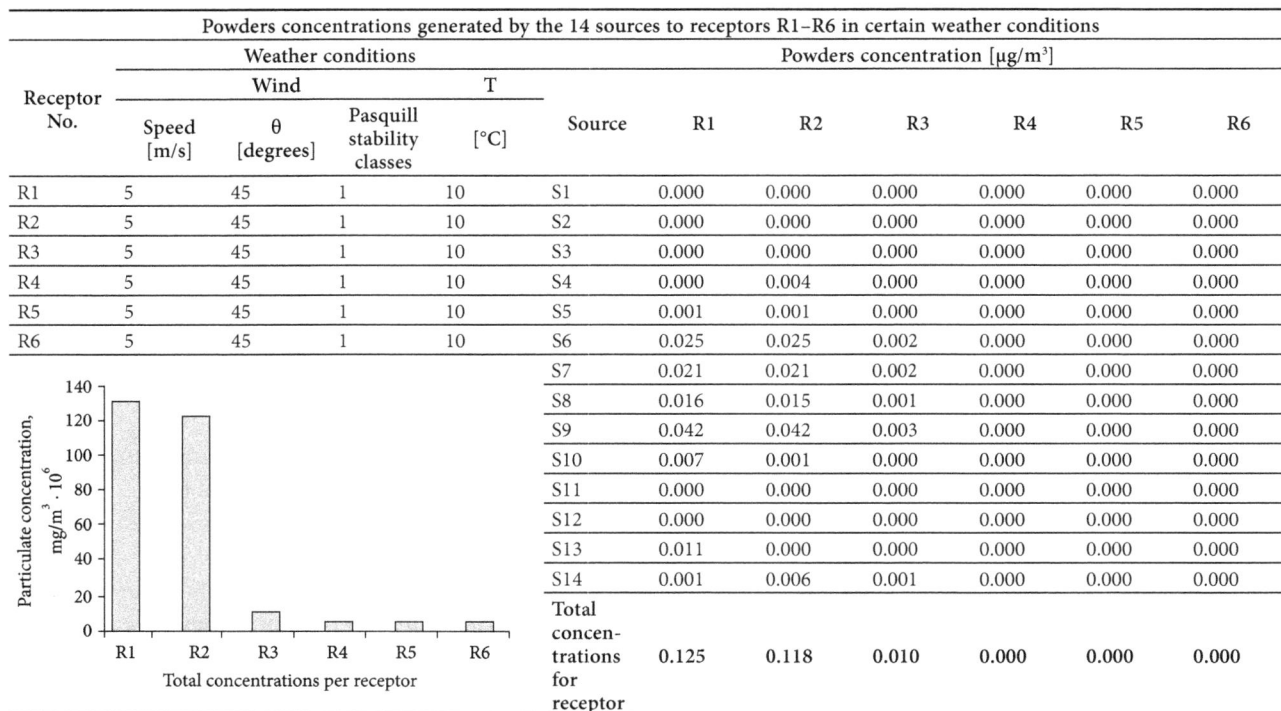

Fig. 2. Data and graphical representation for distribution of powders emissions from the 14 sources in "highly unstable" atmospheric conditions

Concentrations of powders generated by the 14 sources at receptors R1–R6 for certain atmospheric conditions											
Receptor No.	Weather conditions				Concentration of powders [µg/m³]						
	Wind			T							
	Speed [m/s]	E [degrees]	Pasquill stability class	[°C]	Source	R1	R2	R3	R4	R5	R6
R1	1	0	5	15	S1	0.000	0.000	0.000	0.000	0.000	0.000
R2	1	0	5	15	S2	0.000	0.000	0.000	0.000	0.000	0.000
R3	1	0	5	15	S3	0.000	0.000	0.000	0.000	0.000	0.000
R4	1	0	5	15	S4	0.000	0.000	0.000	0.000	0.000	0.000
R5	1	0	5	15	S5	0.000	0.000	0.000	0.000	0.000	0.000
R6	1	0	5	15	S6	0.000	0.000	0.000	0.000	0.014	0.000
					S7	0.000	0.000	0.000	0.000	0.000	0.000
					S8	0.000	0.000	0.000	0.000	0.000	0.000
					S9	0.000	0.001	0.000	0.000	0.000	0.000
					S10	0.000	0.324	0.000	0.000	0.000	0.000
					S11	0.000	0.000	0.000	0.000	0.000	0.000
					S12	0.000	0.000	0.000	0.000	0.000	0.000
					S13	0.000	0.000	0.000	0.000	0.000	0.000
					S14	0.000	0.000	0.000	0.000	0.000	0.000
					Total concentration for receptor	0.000	0.325	0.000	0.000	0.062	0.000

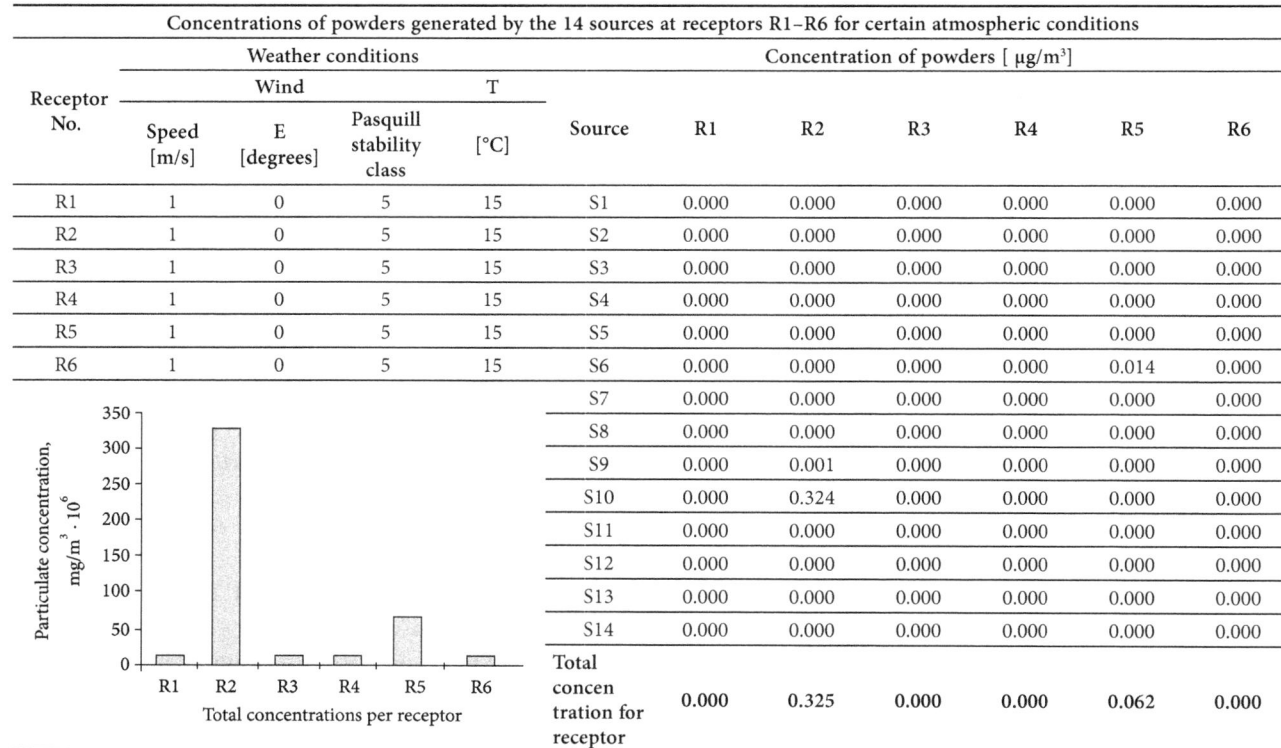

Fig. 3. Data and graphic representation for distribution of powders emissions from 14 sources in atmospheric condition "highly stable"

– at receptor R6 dust emissions arrive from S9. S10 and S11.

For wind direction from N to S (Fig. 3):

– in receptors R1. R3. R4. R6 there is no contribution of any powders generating sources to the total concentration of powders at emissions, this being considered null;

– at receptor R2 dust emissions arrive from S9 and S10;

– at receptor R5 dust emissions arrive from S6. S7. S8. S9.

Figure 3 is a combination of Table and diagram that summarizes the distribution of powders emissions from the 14 sources in "highly stable" atmospheric conditions.

In terms of influence areas and the intensity of impact made by every point source, it can be concluded that:

– total concentrations on receptors from all 14 sources, calculated with the simulation program *Pol 15sm* for multiple emission sources, are well below maximum admitted limit, and are between 0.004 and 0.076 micrograms/cubic meter;

– the calculations were done for the majority of weather conditions encountered in Vaslui city – from atmospheric calm to strong wind – for both models. In none of the situations maximum allowed concentrations have been reached;

– in any case, maximum concentrations at emission are, without exception, well below maximum allowed limits for suspended powders;

– results of dispersion of suspended powders that contain heavy metals using the simulation program *Pol 15sm* are consistent with those determined experimentally by the Environmental Protection Agency Vaslui over time and are a justification for reduction of monitoring points for air quality.

Conclusions

1. Dispersion modelling has been studied for 14 point sources of emission, from 5 economic agents: SC Stemar SRL Vaslui, SC Termica SA Vaslui, Fabrica de cărămizi SRL Vaslui, SC Ulerom SA Vaslui and SC Vascar SA Vaslui, with an original mathematical model of pollutants dispersion, which is the *Pol 15sm* model. Based on this model, a simulation program for a chosen pollutant's dispersion, emitted simultaneously by multiple sources has been made. As a result, pollutant concentrations at emission for many set receptors has been obtained.

2. The maximum number of usable sources in the same time (15) by the simulation program, as well as the maximum number of receptors that can be set (15) covers the needed investigations for industrial platforms. This, together with the fact that the model implies a plain input data, easy to introduce and accomplish, recommends the use of the model for very simple flat-land scenarios.

3. The POL15sm model application, as a tool for analyzing the problems of air pollution, provides a significant reduction in both cost and time than the application for other models (e.g. AERMOD) making possible the estimation of human activities' impact on the environment cheaper and faster.

References:

Ani, E.-C.; Cristea, V. M.; Agachi, P. S. 2012. Mathematical models to support pollution counteraction in case of accidents, *Environmental Engineering and Management Journal* 11: 13–20.

Baldauf, R.; Watkins, N.; Heist, D.; Bailey, C.; Rowley, P.; Shores, R. 2009. Near-road air quality monitoring: factors affecting network design and interpretation of data, *Air Quality, Atmosphere & Health* 2(1): 1–9. http://dx.doi.org/10.1007/s11869-009-0028-0

Branquinho, C.; Gaio-Oliveira, G.; Augusto, S.; Pinho, P.; Máguas, C.; Correia O. 2008. Biomonitoring spatial and temporal impact of atmospheric dust from a cement industry, *Enviromental Pollution* 151(2): 292–299. http://dx.doi.org/10.1016/j.envpol.2007.06.014

Budianu, M.; Macoveanu, M. 2010. Modeling of dispersions of powder emissions from the industrial area of Vaslui, *Bulletin of the Polytechnic Institute of Iasi. Section of Chemistry and Chemical Engineering* LVI(LX): 129–138.

Candelieri, A.; Giordani, I.; Testa, P.; Arosio, G.; Archetti, F. 2012. A Markov-based model to forecast emergency hospital admissions due to air pollution: the Lenvis project approach, *Environmental Engineering and Management Journal* 11: 999–1008.

Carpentieri, M.; Kumar, P.; Robins, A. 2011. An overview of experimental results and dispersion modelling of nanoparticles in the wake of moving vehicles, *Environmental Pollution* 159: 685–693. http://dx.doi.org/10.1016/j.envpol.2010.11.041

Gaba, A. 2010. Air pollution reduction by using of low NOx burners for furnaces and boilers, *Environmental Engineering and Management Journal* 9: 165–170.

Hanna, S. R. 1981. Applications in air pollution modelling, in FTM; van Dop, H. (Eds.). *Atmospheric turbulence and air pollution modelling*. Nieuwstadt: D. Reidel Publishing company, 275–310.

Huang, Z.; Jin, G.; Wang, X. 2011. Numerical simulation applied for unsymmetrical dimethylhydrazine propellant gas dispersion based on CFD technology, *Environmental Engineering and Management Journal* 10: 971–974.

Mezdrea-Cojocareanu-Pata, I.; Pata, S. M.; Macoveanu, M. 2010. Biomonitoring of atmospheric pollution of Fe and Zn using native epigeic mosses, *Environmental Engineering and Management Journal* 9: 1217–1225.

Miclaus, C.; Dezsy, S.; Nicu, M. 2006. Passive samplers – checking method in the air quality control network, *Environmental Engineering and Management Journal* 5: 1333–1340.

Moreira, D. M.; Vilhena, M. T.; Tirabassi, T.; Buske, D.; Pinto da Costa, D. 2010. Comparison between analytical models to simulate pollutant dispersion in the atmosphere, *International Journal of Environment and Waste Management* 6: 327–344. http://dx.doi.org/10.1504/IJEWM.2010.035066

Nagacevschi, V.; Macoveanu, M. 1994. Modeling of pollutants dispersion into the atmosphere, in *Conference: Progress in Chemistry and Chemical Technology*, 27–29 October, 1994, Polytechnic Institute of Iasi, Iasi, Romania, 41–47.

Nagacevschi, V.; Macoveanu, M. 1995. Modeling of dispersion for pollutants emitted into the atmosphere from multiple sources, in *The 5th National Colloquium for Atmosphere Protection*, 15–17 November, 1995, SOROPA, Bucureşti, Romania, 64–70.

Nagacevschi, V.; Macoveanu, M. 1996. Dynamic simulation of air pollution, in *National Symposium: Technologies With Low Environmental Impact for the Tanning of Hides and Skins*, 6–8 June, 1996, Technical University of Iaşi, Iasi, Romania, 39–46.

Nagacevschi, V.; Macoveanu, M.; Peiu, N. 1997. Model for the air dispersion of a pollutant, *Bulletin of the Polytechnic Institute of Iasi*. Technical University "Gheorghe Asachi" Iaşi, LI, 23–31.

Nagacevschi, V.; Macoveanu, M. 2002. Contract: the study of dispersion for pollutants emitted into the atmosphere at the site of Bread Factory No.1. Galaţi belonging to S.C. GALMO-PAN. Galaţi. *Execution – S.C. REDICOM S.R.L. Iaşi*. 2002, 6–12.

Nisulescu, C.; Călinoiu, D.; Timofte, A.; Boscornea, A.; Talianu, C. 2011. Diurnal variation of particulate matter in the proximity of Rovinari fossil-fuel power plant, *Environmental Engineering and Management Journal* 10: 99–105.

Pasquill, F. 1961. The estimation of the dispersion of windborn material, *Meteorological Magazine* 90: 33–49.

Popescu, F.; Ionel, I.; Talianu, C. 2011. Evaluation of air quality in airport areas by numerical simulation, *Environmental Engineering and Management Journal* 10: 115–120.

Radulescu, C.; Prisecaru, T.; Mihaescu, L.; Pisa, I.; Lazaroiu, G.; Zamfir, S.; Vairenu. D.; Popa, E. 2010. Researches on the negative effects assessment (slugging, clogging, ash deposits) developed at the biomass-coal co-firing, *Environmental Engineering and Management Journal* 10: 17–26.

San Jose, R. 1997. *Sensitivity study of dry deposition fluxes in ANA air quality model over Madrid Mesoscale area – measurements and modelling in environmental pollution*. Southampton, Boston. 119–129.

Smaranda, C.; Gavrilescu, M. 2008. Migration and fate of persistent organic pollutants in the atmosphere – a modelling approach, *Environmental Engineering and Management Journal* 7: 743–761.

Stătescu, F.; Cotiusca-Zauca, D. 2006. Heavy metal soil contamination, *Environmental Engineering and Management Journal* 5: 1205–1213.

Ştefan, S.; Radu, C.; Belegante, L. 2013. Analysis of air quality in two sites with different local conditions, *Environmental Engineering and Management Journal* 12: 371–379.

Vătăşescu, M. M.; Diaconescu, D.; Duţă, A.; Burduhos, B. G. 2011. Atmospheric pollution evaluation in Brasov Romania based on turbidity factor analysis, *Environmental Engineering and Management Journal* 5: 251–256.

Mihaela BUDIANU. PhD student in Department of Environmental Engineering and Management, Faculty of Chemical Engineering, Technical University of Iasi. Publications: author/co-author of 10 scientific papers. Research interest: environmental engineering.

Valeriu NAGACEVSCHI. Professor at the Faculty of Chemical Engineering, Technical University of Iasi (retired).

Matei MACOVEANU is a Consultant Professor at the Faculty of Chemical Engineering and Environmental Protection Iasi. In the 44 years of teaching and scientific activity has received numerous awards for his inventions (31 patents obtained in the country and abroad). He is the Founder of the international academic journal *Environmental Engineering and Management Journal*, initiator of the PhD establishment in Environmental Engineering, from the Technical University "Gheorghe Asachi" of Iasi. Publications: author/co-author of 23 books published in central publishing, over 200 scientific papers published in specialized periodicals in the country and abroad.

3

GREEN ALGAE CHLORELLA VULGARIS CULTIVATION IN MUNICIPAL WASTEWATER AND BIOMASS COMPOSITION

Petras VENCKUS[a] Jolanta KOSTKEVIČIENĖ[a], Vida BENDIKIENĖ[b]

[a] Department of Botany and Genetics, Faculty of Natural Sciences, Vilnius University,
M. K. Ciurlionio 21/27, LT-03101 Vilnius, Lithuania
[b] Department of Biochemistry and Molecular Biology, Faculty of Natural Sciences, Vilnius University,
M. K. Ciurlionio 21/27, LT-03101 Vilnius, Lithuania

Abstract. This paper deals with the accumulation of lipids, carbohydrates and proteins in the biomass of the green algae *Chlorella vulgaris* that is cultivated in the municipal wastewater of Vilnius City. The growth rate of the culture on different chemical compositions of media was investigated. Dependence of lipid, carbohydrate and protein content on total phosphorus and nitrogen initial concentrations in wastewater and removal of nutrients was investigated. Data showed that the higher amount of total nitrogen is the main factor leading to a higher rate of biomass increase. The study showed that *Chlorella vulgaris* is capable of very efficient nutrient removal from wastewater (up to 86% of total nitrogen and 87% phosphorus was removed). Data showed that there is strong correlation between the initial concentration of nitrogen, and in some cases phosphorus, in the media and content of proteins and carbohydrates in the biomass. A higher amount of nitrogen in the starting media leads to a higher amount of proteins and a lower amount of carbohydrate in the biomass. There was no correlation found between the initial nitrogen or phosphorus concentration in the media and content of lipids in the biomass.

Keywords: *Chlorella vulgaris*, wastewater management, biomass, lipids, carbohydrates, proteins.

Introduction

There is a plentiful supply of fossil fuels at a reasonably low cost, but a rising use of fossil fuels is unlikely to be sustainable in the longer term principally due to the attributed increase in greenhouse gas (GHG) emissions from using these fuels and the environmental impact of these emissions on global warming (Hill *et al.* 2006; Pittman *et al.* 2011). The potential of microalgae as a source of lipids, proteins and carbohydrates has received considerable interest. The growth and photosynthetic activity of algae have been extensively studied under different environmental conditions over the last few decades (Burlew 1953; Moraine *et al.* 1979; Soeder *et al.* 1985; Cromar *et al.* 1996; Torzillo *et al.* 2003; Park, Craggs 2010; Zheng *et al.* 2013). For sustainability and economic viability, production optimization of mass culture conditions is needed. The growth of microalgae is faster than other photosynthetic plants, which indicates a high productivity per area. Microalgae can generally double their biomass within 24 h, and the lipid content in several microalgae species exceed over 30–80% of the dry biomass weight (Chisti 2007; Cho *et al.* 2011; Pittman *et al.* 2011). However, there remains a major price gap between microalgae-derived biofuels and fossil fuels despite the tremendous efforts to reduce the costs of algae production and processing (Pittman *et al.* 2011).

Wastewater derived from municipal water treatment plants can provide a low cost and sustainable means to grow algae for biofuels. The municipal wastewater stream from primary and secondary settling tanks is rich in nutrients including phosphorus, ammonia and organic nitrogen (Zhou *et al.* 2011). In addition, algae could be used in the aerobic treatment of waste in the secondary treatment process and in the removal of nitrogen and phosphorus from the wastewater (Hameed, Ebrahim 2007).

Green algae are one of the most used algae in wastewater treatment experiments and pilot plants for algae cultivation (Hameed, Ebrahim 2007). Different studies have shown that *Chlorella vulgaris* is capable of treating various wastewaters including textile wastewater (Lim *et al.* 2010; Chu *et al.* 2009), piggery wastewaters (Godos *et al.*

Corresponding author: Petras Venckus
E-mail: petras.venckus@gf.vu.lt

2009; Ji *et al.* 2013), industrial effluent (Valderrama *et al.* 2002) and municipal wastewater (Cho *et al.* 2011; Zhou *et al.* 2011). Research on strain differences and the methods to acclimate strains to the wastewater environment are limited. In order to effectively couple the wastewater treatment with algae cultivation in the northern climate, selection and establishment of an adequate microalgae pool from local habitats becomes particularly important and necessary (Zhou *et al.* 2011). There are some limiting factors for wastewater usage as the source of microalgae cultivation media. A proper pre-treatment method to remove algae-feeding microorganisms and microorganisms competing for nutrient should be applied to increase algae biomass production (Cho *et al.* 2011). However, comparatively little effort is being conducted in northern regions due to strong seasonal effects on climate and solar insolation levels, which are likely to negatively affect feasibility and production costs. However, these disadvantages can be potentially mitigated by careful selection of strains that can be grown in wastewater at lower temperatures (Park *et al.* 2015).

The aim of this study is to investigate and establish mathematical models describing the rate of *C. vulgaris* biomass growth and accumulation of lipids, proteins and carbohydrates in biomass in dependence on total nitrogen and phosphorus concentration in wastewater and efficiency of removal these nutrient.

1. Materials and methods

Microalgae *Chlorella vulgaris*, in combination with the bacteria Flavimonas oryzihabitans was investigated in the present study. Flavimonas oryzihabitans (previously known as Pseudomonas oryzihabitans) is a gram-negative organism that survives in moist environments. There are a few studies that show the existence of interactions between microalgae and bacteria (Fukami *et al.* 1997; Krustok *et al.* 2015a; Krustok *et al.* 2015b). These studies postulate that the bacteria encourage microalgae growth due to the production of various growth factors, and that the microalgae produce simultaneously organic substances that encourage bacterial growth (Verschuere *et al.* 2000;

Riquelme, Avendaño-Herrera 2003; Krustok *et al.* 2015a, 2015b). Microalgae was cultivated in wastewaters taken from Vilnius City Municipal Wastewater Treatment Plant. Wastewater samples were taken at two different treatment stages: after mechanical treatment (after sedimentation of solid particles) and biological treatment, on 30/06/2011. The concentration of nitrates (NO_3^-), ammonium (NH_4^-), total nitrogen (TN), phosphates (PO_4^{2-}), and total phosphorus (TP) on wastewater was determined at the Joint-Stock Company "Vilniaus Vandenys" laboratory. Several dilutions of wastewater (2–20 times) in order to get the maximum effect for the algae biomass growth and quality conducted (Table 1). Samples of wastewater were autoclaved 30 min at 120° C to avoid contamination of other algae or bacteria. Distilled water was used for wastewater dilutions. The volume of one sample was 165 mL (150 mL of wastewater (diluted or not diluted) and 15 ml of algae culture which was maintained in Basal media). Flasks were kept in daylight at room temperature (21–24 °C) for 19 days. Flasks were mixed once a day before measurements of biomass concentration. The optical density of cultures was determined with a spectrophotometer at a wavelength of 680 nm. To estimate algal concentration standard routines include direct cell counts, absorbance or turbidity and chlorophyll content measurement. Typically 680 nm is associated with chlorophyll absorption. Rodrigues and co-authors published that when spectrophotometrical absorbance is the chosen method, a reading wavelength of 680 nm and 687 nm have also been used (Rodrigues *et al.* 2011). Analyses were performed in triplicate and then means were calculated. After 19 days of growth the algal biomass was centrifuged, frozen, lyophilized and used for the determination of the biomass biochemical composition. The supernatant in every flask was analyzed by the Joint-Stock Company "Vilniaus Vandenys" laboratory for determination of chemical composition. The LAND 58:2003, LAND 84:2006 and LAND 38:2000 were used to measure total phosphorus, total nitrogen and ammonium nitrogen, respectively.

The biochemical composition of *Chlorella vulgaris* (in combination with the bacteria Flavimonas oryzihabitans)

Table 1. Chemical composition of Vilnius City municipal wastewaters of different clarification stages used in the experiment

Wastewater sample type	Percentage of raw wastewater in the media	Parameter					
		BOD/COD	TN mg N L^{-1}	NO_3^-, mg L^{-1}	NH_4^- mg L^{-1}	TP mg P L^{-1}	PO_4^{2-} mg L^{-1}
After mechanical treatment	100% (Undiluted)	312/455	85	0.26	34.5	8.37	7.28
After biological treatment	100% (Undiluted)	11/55	35	0.21	29.1	0.14	0.062

biomass was determined at the laboratory of the Institute of Ecosystem Study (Florence, Italy). The total carbohydrate content of the biomass was checked using the phenol-sulphuric acid method (Dubois et al. 1956). The total protein content was determined using Lowry's method (Lowry et al. 1951). The total lipid content was determined spectrophotometrically (Blight, Dyer 1959) after carbonization of the material was extracted with a 2:1 methanol/chloroform solution (Marsh, Weinstein 1966). Glyceryl tripalmitate (tripalmitin) (Sigma-Aldrich, Milan, Italy) was used as a standard for lipid extraction (Holland, Gabbott 1971). Analyses were performed in triplicate. All data presented in mg of compound (lipids, proteins, carbohydrates) in 1g of dry algae biomass (mg g^{-1} d. wt.).

For statistical analysis multiple linear regressions (Quinn, Keough 2002) were used. Analyses were conducted with Microsoft Excel 2013 data analysis tool pack (Microsoft. Microsoft Excel. Redmond, Washington: Microsoft, 2013. Computer Software). The significance level (α) is 0.05.

2. Results and discussion

2.1. Consumption of nitrogen and phosphorus

Algae cells can absorb nitrogen and phosphorus from culture medium (Georgianna, Mayfield 2012). Nutrient (N and P) removal is directly linked to photosynthetic activity and in microalgae-based systems they are converted into algal biomass that can be sustainably recycled (Singh et al. 2011). In this experiment the initial N:P ratio in mechanically treated wastewater was 10:1 and in biologically treated – 10:0.04. It is known, that such ratio is suitable for photosynthetic microorganisms' cultivation (Lee et al. 2000; Vezie et al. 2002). Our results shows, that the highest consumption of total N and total P was observed in mechanically treated

undiluted wastewater (73 mg N L^{-1} and 7,26 mg P L^{-1}) (Fig. 1). By lowering the amount of N and P in wastewater the consumption of elements lowers too. Another study (Lau et al. 1996) showed that most of the nitrogen (>80%) was removed in the five days of cultivation and most of the phosphorus was removed in one to two days of cultivation. Our experiment showed that during the 19 day cultivation period 87% of initial nitrogen and 86% of initial phosphorus was removed from the wastewater. Results of this study demonstrated that Chlorella vulgaris is capable of efficient nitrogen and phosphorus removal from wastewater. After the cultivation of C. vulgaris in the wastewater the nitrogen concentration lowers significantly, but never lowered below 1.4 mg N L^{-1}. Phosphorus was consumed too, but the concentration did not drop below 0.2 mg P L^{-1}. The remaining amount of nitrogen and phosphorus possibly have not been absorbed by algae due to physiological limitations of algae or chemical properties of compounds left in the wastewaters. Growing C. vulgaris in different diluted wastewater the amount of remaining nitrogen and phosphorus after cultivation was quite similar and varied from 1.4 to 3.4 mg N L^{-1} and from 0.2 to 0.37 mg P L^{-1} Exception was undiluted wastewater after mechanical treatment. The concentration of nitrogen and phosphors remaining after cultivation was 12.0 mg N L^{-1} and 1.11 mg P L^{-1}. This phenomenon could be explained by the high concentrations of nutrients in the medium. In that case the limiting factors for algae growth could be not chemical but physical, for example light intensity or density of culture.

The significant increase of initial phosphorus concentrations in Figure 1B compared to Table 1 is due to addition of phosphorus from phosphorus-rich growth media which was used preparing the inoculate of Chlorella vulgaris culture.

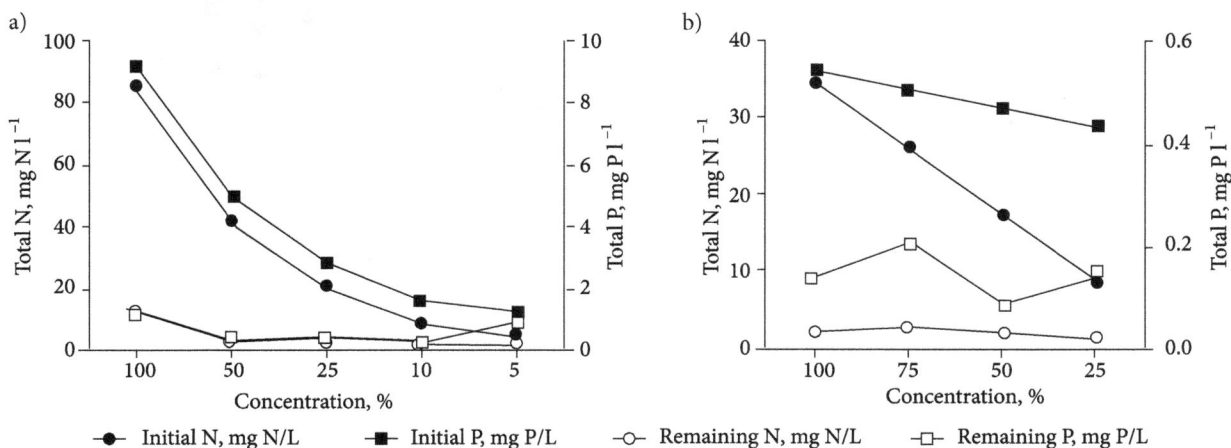

Fig. 1. Consumption of nitrogen and phosphorus by Chlorella vulgaris in different type and concentration of wastewaters of Vilnius city municipal wastewater treatment plant: (a) – after mechanical treatment; b) – after biological treatment). "%" shows the percentage of raw wastewater in the culture medium. Note the different scales in each panel

2.2. Growth dynamics

The growth rate of *C. vulgaris* was determined spectro-photometrically by measuring the optical density of the culture every day. During the first two days of cultivation a lag phase was observed in all cultures and in some cultures, optical density decreased on the second day (Fig. 2). It is known, that the changing environmental conditions could cause stress to algae cells and lead to a decrease of pigment content, especially chlorophyll (Boussiba *et al.* 1999). From the second to the fifth day an increase of growth rate was observed. From day five until the end of the experiment, growth stabilized and the growth rates of the cultures with different initial nitrogen and phosphorus concentrations began to differ. The highest growth rate (+0.078 and +0.076 OD per day) was measured in the culture with the highest amount of nitrogen and phosphorus. It were in the mechanically treated undiluted and double

diluted wastewater. The lowest growth rate (+0.041 OD per day) was measured in the culture with the lowest amount of nitrogen (but not phosphorus). The lowest growth rate was in the mechanically treated wastewater that was diluted 20 times.

The best-fit trend line shows that the growth rate depends logarithmically on the amount of nitrogen from 4.25 to 85.0 mg L^{-1} (Fig. 3; Eq. (1)).

$$y = 0.0163\ln(x) + 0.0141; \ R^2 = 0.847, \quad (1)$$

where: y is the growth rate and x is the concentration of nitrogen in the starting medium.

Equation (1) indicates that the growth rate is 85% determined by the concentration of nitrogen in the starting medium. Since this dependence is logarithmic, the addition of nitrogen has a lower impact to the growth rate as the concentration of nitrogen in the starting medium getting higher.

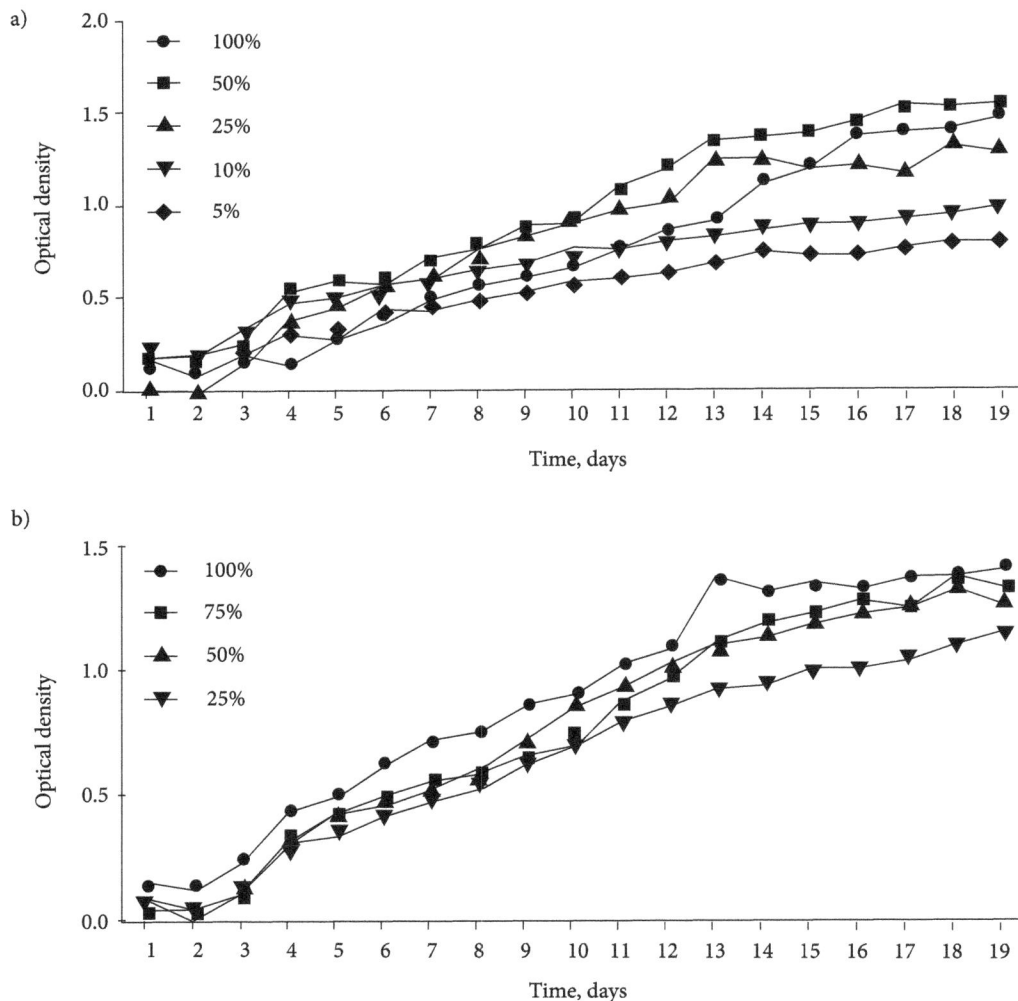

Fig. 2. Growth dynamics of *Chlorella vulgaris* (in combination with bacteria Flavimonas oryzihabitans) in different type and concentration of wastewater of Vilnius city municipal wastewater treatment plant (a) – wastewater after mechanical treatment; b) – wastewater after biological treatment). "%" shows the percentage of raw sewage water in the medium

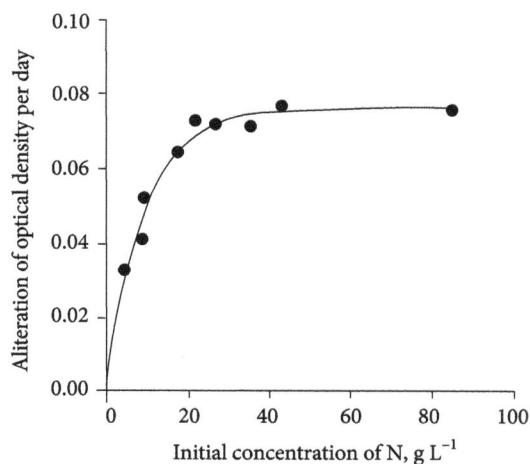

Fig. 3. Variation of the optical density with the initial nitrogen concentration

2.3. Biochemical composition of the biomass

There were obvious differences in biochemical composition in *Chlorella vulgaris* biomass treated under different conditions. It is known, that proteins comprise a large fraction of the biomass of actively growing microalgae and in certain species protein content can reach values as high as 50–60% by dry weight biomass (Gonzalez-Lopez *et al.* 2010). Our data shows that algae cultivated under nitrogen-rich conditions accumulated higher amounts of proteins and lower amounts of carbohydrates in their biomass. Lower initial concentrations of nitrogen led to a lower amount of proteins and higher amount of carbohydrates in the algae cells. The highest protein content (432 mg g^{-1} d. wt.) was detected in the biomass cultivated in undiluted mechanically treated wastewaters. The highest carbohydrate content (577 mg g^{-1} d. wt.) was in the biomass cultivated in 10% mechanically treated wastewaters. A step wise multiple linear regression analysis was performed in order to predict the biochemical biomass composition under different initial concentrations of nitrogen and phosphorus in the medium. Lipids and carbohydrates represent the main energy storage molecules in algae. Nutrient stress has been the method for increasing lipid and starch accumulation in green algae and diatoms. Under nitrogen deplete conditions, some green algae accumulate high levels of lipids as triacylglycerols (TAG), (Hu *et al.* 2008). Tables 2 and 3 summarize the results of the multiple regression analysis in which the amounts of proteins and carbohydrates were calculated as a function of variables of initial nitrogen concentration and initial phosphorus concentration. A multiple regression model for protein analysis was modified using backward elimination of P from the initial model due to low significance of this variable (p > 0.05). A multiple linear regression model was unsuitable for describing the dependence of lipid content and the initial nitrogen and phosphorus

concentrations in the medium (No significant ($p < 0.05$) R^2 value was found).

Table 2. Multiple linear regression for content of proteins prediction. Nine cases were included in the analysis

Source	df	Sum of squares	Mean square	F-ratio	Significance F
Regression	2	45900.82	22950.41	69.30	7,14E-05
Residual	6	1987.17	331.19		
Total	8	47888			

Variable	Coefficients	Standard Error	t	P-value
Constant	204,14	11,61	17,57	4,75E-07
N	2,98	0,32	9,32	3,38E-05

Note: Dependent variable was content of proteins. Independent variable was initial concentration of nitrogen (*N*).

$$C = 204.139 + 2.983N , \qquad (2)$$

where: *C* – Content of proteins in the biomass (mg g^{-1} d.wt.), *N* – initial concentration of nitrogen (mg ^{1} L^{-1}).

Table 3. Multiple linear regression for content of carbohydrates prediction. Nine cases were included in the analysis

Source	df	Sum of squares	Mean square	F-ratio	Significance F
Regression	2	134579.37	67289.68	490.26	2.024E-07
Residual	6	823.51	137.25		
Total	8	135402.88			

Variable	Coefficients	Standard Error	t	p-value
Constant	604.16	6.83	88.45	1.40E-10
N	–6.81	0.36	18.46	1.62E-06
P	16.66	3.26	5.10	2.21E-03

Note: Dependent variable was content of carbohydrates. Independent variables were initial concentration of nitrogen (N) and initial concentration of phosphorus (P).

$$C = 604.16 + (-6.81N) + (16.66P) , \qquad (3)$$

where: *C* – Content of carbohydrates in the biomass (mg g^{-1} d.wt.), *N* – initial concentration of nitrogen (mg N L^{-1}), *P* – initial concentration of phosphorus (mg P L^{-1}).

Figure 4 and the regression analysis (Tables 2–3 and Equations 2–3) show that the initial concentration of nitrogen has a significant influence over the biochemical composition of algae biomass. A higher concentration of initial nitrogen concentration leads to the accumulation of proteins in the biomass (Fig. 4 A); however, a shortage of both nitrogen and phosphorus leads to the accumulation of carbohydrates (Fig. 4 A, B). Mutlu *et al.* (2011) described this dependence too, but they noticed

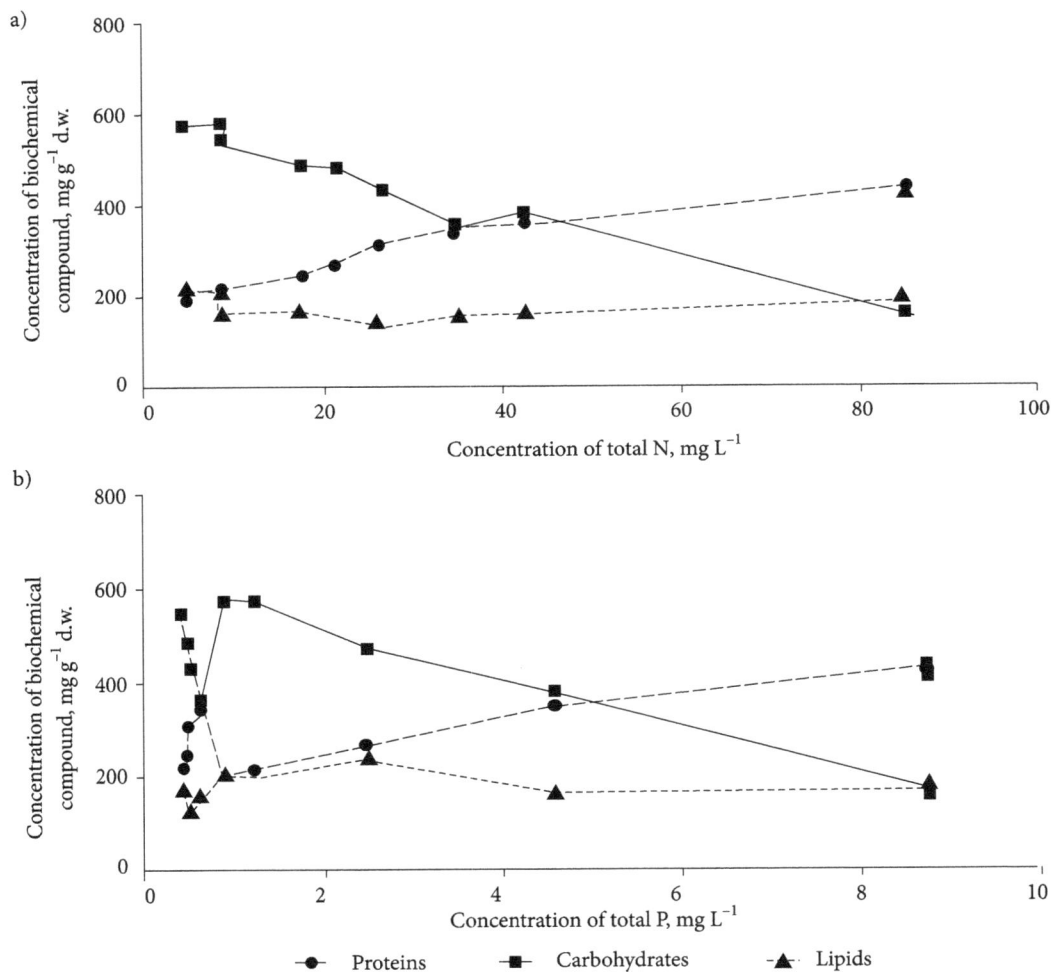

Fig. 4. Dependence of concentration of lipids, proteins and carbohydrates in *Chlorella vulgaris* (grown with bacteria Flavimonas oryzihabitans) biomass after cultivation period in different initial concentration of total nitrogen (a) and total phosphorus (a) in cultivation media

the accumulation of lipids due to the shortage of nitrogen. However Rodolfi *et al.* (2009) observed an increase in both lipid content and areal lipid productivity through nutrient starvation. Several authors (Perez-Garcia *et al.* 2011) have proposed that lipid accumulation may not be dependent on nitrogen starvation but on excess carbon in the culture medium. In our study no significant correlation between concentration of nitrogen (Fig. 4a) or phosphorus (Fig. 4b) in the initial media and the content of lipids in the biomass was observed.

Conclusions

Chlorella vulgaris is capable of efficient nutrient (nitrogen and phosphorus) removal from Vilnius City wastewater. During the experiment 87% to 93% of total nitrogen and up to 87% of total phosphorus were removed from the municipal wastewaters of Vilnius City. The growth rate and the concentration of proteins in the *C. vulgaris* biomass are directly dependent on the starting concentration of nitrogen

in the medium. The highest growth rate and the concentration of proteins (432 mg g^{-1}) were reached in the wastewater with the concentration of total nitrogen in the medium from 42.5 to 85.0 mg N L^{-1}. The lower concentration of total nitrogen and the higher concentration of total phosphorus lead to a higher concentration of carbohydrates in the biomass of *C. vulgaris*. The highest concentration of carbohydrates in the biomass (577 mg g^{-1}) was reached by cultivating *C. vulgaris* in the medium containing 4.0 mg N L^{-1}and 0.4 mg P L^{-1}. The remediation of wastewater using *C. vulgaris* provides an environmentally acceptable and effective option for wastewater remediation, which not only recycles valuable nutrients but also improves water quality. Therefore *C. vulgaris* may be regarded as efficient nutrient remover.

Acknowledgements

We are grateful for the dr. Pietro Carlozzi research group for the possibilities to make biochemical analyses of the *Chlorella vulgaris* biomass.

References

Blight, E. G.; Dyer, W. J. 1959. A rapid method of total lipid extraction and purification, *Canadian Journal of Biochemistry and Physiology* 37(8): 911–917. https://doi.org/10.1139/o59-099

Boussiba, S.; Bing, W.; Yuan, J. P.; Zarka1, A.; Chen, F. 1999. Changes in pigments profile in the green alga *Haeamtococcus pluvialis* exposed to environmental stresses, *Biotechnology Letters* 21(7): 601–604. https://doi.org/10.1023/A:1005507514694

Burlew, J. S. 1953. *Algae culture: from laboratory to pilot plant.* Carnegie Institution of Washington, Washington DC.

Chisti, Y. 2007. Biodiesel from microalgae, *Biotechnology Advances* 25(3): 294–306. https://doi.org/10.1016/j.biotechadv.2007.02.001

Cho, S.; Luong, T. T.; Lee, D; Oh, Y.; Lee, T. 2011. Reuse of effluent water from a municipal wastewater treatment plant in microalgae cultivation for biofuel production, *Bioresource Technology* 102(18): 8639–8645. https://doi.org/10.1016/j.biortech.2011.03.037

Chu, W. L.; See, Y. C.; Phang, S. M. 2009. Use of immobilised *Chlorella vulgaris* for the removal of colour from textile dyes, *Journal of Applied Phycology* 21: 641–648. https://doi.org/10.1007/s10811-008-9396-3

Cromar, N. J.; Fallowfield, H. J.; Martin, N. J. 1996. Influence of environmental parameters on biomass production and nutrient removal in a high rate algal pond operated by continuous culture, *Water Science and Technology* 34(11): 133–140. https://doi.org/10.1016/S0273-1223(96)00830-X

Dubois, M.; Gilles, K. A.; Hamilton, J. K.; Rebers, P. A.; Smith, F. 1956. Colorimetric method for determination of sugars and related substances, *Analalytical Chemistry* 28(3): 350–356. https://doi.org/10.1021/ac60111a017

Fukami, K.; Nishijima, T.; Ishida, Y. 1997. Stimulative and inhibitory effects of bacteria on the growth of microalgae, *Hydrobiologia* 358: 185–191. https://doi.org/10.1023/A:1003139402315

Georgianna, D. R.; Mayfield, S. P. 2012. Exploiting diversity and synthetic biology for the production of algal biofuels, *Nature* 488: 329–335. https://doi.org/10.1038/nature11479

Godos, E.; Blanco, S.; Garcia-Encina, P. A.; Becares, E.; Munoz, R. 2009. Long-term operation of high rate algal ponds for the bioremediation of piggery wastewaters at high loading rates, *Bioresource Technology* 100: 4332–4339. https://doi.org/10.1016/j.biortech.2009.04.016

Gonzalez-Lopez, C.V.; Ceron-Garcia, M. C.; Acien-Fernandez, F. G.; Segovia-Bustos, C.; Chisti, Y.; Fernandez-Sevilla, J. M. 2010. Protein measurements of microalgal and cyanobacterial biomass, *Bioresource Technology* 101(19): 7587–7591. https://doi.org/10.1016/j.biortech.2010.04.077

Hameed, M. S. A.; Ebrahim, O. H. 2007. Biotechnological potential uses of immobilized algae, *International Journal of Agriculture and Biology* 9(1): 183–192.

Hill, J.; Nelson, E.; Tilman, D.; Polasky, Tiffany, D. 2006. Environmental, economic, and energetic costs and benefits of biodiesel and ethanol biofuels, *Proceedings of the Nation Academy of Science of the United States of America* 103(30): 11210–11210. https://doi.org/10.1073/pnas.0604600103

Holland, D.; Gabbott, P. 1971. A micro – analytical scheme for determination of protein, carbohydrate, lipid and RNA levels in marine invertebrate larvae, *Journal of the Marine Biological Association of the UK* 59(1): 95–101. https://doi.org/10.1017/S0025315400015034

Hu, Q.; Sommerfeld, M.; Jarvis, E.; Ghirardi, M.; Posewitz, M.; Seibert, M.; Darzins, A. 2008. Microalgal triacylglycerols as feedstocks for biofuel production: perspectives and advances, *Plant Journal* 54(4): 621–639. https://doi.org/10.1111/j.1365-313X.2008.03492.x

Ji, M.; Kim, H.; Sapireddy, V. R.; Yun, H.; Abou-Shanab, R. A. I.; Choi, J.; Lee, W.; Timmes, T. C.; Inamuddin, J. B. 2013. Simultaneous nutrient removal and lipid production from pretreated piggery wastewater by *Chlorella vulgaris* YSW-04, *Applied Microbiology and Biotechnology* 97(6): 2701–2710. https://doi.org/10.1007/s00253-012-4097-x

Krustok, I.; Truu, J.; Odlare, M.; Truu, M.; Ligi, T.; Tiirik, K.; Nehrenheim, E. 2015b. Effect of lake water on algal biomass and microbial community structure in municipal wastewater-based lab-scale photobioreactors, *Applied Microbiology and Biotechnology* 99(15): 6537–6549. https://doi.org/10.1007/s00253-015-6580-7

Krustok, I.; Odlare, M.; Shabiimam, M. A.; Truu, J.; Truu, M.; Ligi, T.; Nehrenheim, E. 2015a. Characterization of algal and microbial community growth in a wastewater treating batch photo-bioreactor inoculated with a lake water, *Algal Research* 11: 421–427. https://doi.org/10.1016/j.algal.2015.02.005

LAND 38:2000. *Vandens kokybė. Amonio kiekio nustatymas. Rankinis spektrotometrinis metodas* [Water quality – determination of ammonia – Manual spectrofotometrical method]. Lithuanian standard.

LAND 58:2003. *Vandens kokybė. Fosforo nustatymas* [Water quality – Determination of phosphorus]. Lithuanian standard.

LAND 84:2006. *Vandens kokybė. Kjeldalio azoto nustatymas. Mineralizavimo selenu metodas* [Water quality – Determination of Kjeldahl nitrogen – Selenium mineralization method]. Lithuanian standard.

Lau, P. S.; Tam, N. F. Y.; Wong, Y. S. 1996. Wastewater nutrients removal by *Chlorella vulgaris*: Optimization through acclimation, *Environmental Technology* 17(2): 183–189. https://doi.org/10.1080/09593331708616375

Lee, S. J.; Jang, M. H.; Kim, H. S.; Yoon, B. D.; Oh, H. M. 2000. Variation of microcystin content of *Microcystis aeruginosa* relative to medium N:P ration and growth stage, *Journal of Applied Microbiology* 89: 323–329. https://doi.org/10.1046/j.1365-2672.2000.01112.x

Lim, S. L.; Chu, W. L.; Phang, S. M. 2010. Use of *Chlorella vulgaris* for bioremediation of textile wastewater, *Bioresource Technology* 101(19): 7314–7322. https://doi.org/10.1016/j.biortech.2010.04.092

Lowry, O. H.; Rosebrough, N. J.; Farra, L.; Randall, R. J. 1951. Protein measurement with the Folin's phenol reagent, *Journal of Biological Chemistry* 193(1): 263–275.

Marsh, J.; Weinstein, D. 1966. Simple charring method of determination of lipids, *Journal of Lipid Research* 7: 574–576.

Moraine, R.; Shelef, G.; Meydan, A.; Levi, A. 1979. Algal single cell protein from wastewater treatment and renovation process, *Biotechnology and Bioengeneering* 21: 1191–1207. https://doi.org/10.1002/bit.260210709

Mutlu, Y. B.; Isik, O.; Uslu, L.; Koc, K.;. Durmaz, Y. 2011. The effects of nitrogen and phosphorus deficiencies and nitrite addition on the lipid content of *Chlorella vulgaris*

(*Chlorophyceae*), *African Journal of Biotechnology* 10: 453–456.

Park, J. B. K.; Craggs, R. J. 2010. Wastewater treatment and algal production in high rate algal ponds with carbon dioxide addition, *Water Science and Technology* 61(3): 633–639. https://doi.org/10.2166/wst.2010.951

Park, K. C.; Whitney, C. E. G.; Kozera, C.; O'Leary, S. J. B.; McGinn, P. J. 2015. Seasonal isolation of microalgae from municipal wastewater for remediation and biofuel applications, *Journal of Applied Microbiology* 119(1): 76–87. https://doi.org/10.1111/jam.12818

Perez-Garcia, O.; Escalante, F. M. E.; de-Bashan, L. E.; Bashan, Y. 2011. Heterotrophic cultures of microalgae: metabolism and potential products, *Water Research* 45(1): 11–36. https://doi.org/10.1016/j.watres.2010.08.037

Pittman, J. K.; Dean, A. P.; Osundeko, O. 2011. The potential of sustainable algal biofuel production using wastewater resources, *Bioresource Technology* 102(1): 17–25. https://doi.org/10.1016/j.biortech.2010.06.035

Quinn, G.; Keough, M. 2002. *Experimental design and data analysis for biologists*. Cambridge University Press. https://doi.org/10.1017/CBO9780511806384

Riquelme, C. E.; Avendaño-Herrera, R. E. 2003. Microalgae and bacteria interaction in the aquatic environment and their potential use in aquaculture, *Revista Chilena de Historia Natural* 76(1): 725–736.

Rodolfi, L.; Zitelli, G. C.; Bassi, N.; Padovani, G.; Biondi, N.; Bonini, G.; Tredici, M. R. 2009. Microalgae for oil: strain selection, induction of lipid synthesis and outdoor mass cultivation in a low-cost photobioreactor, *Biotechnology and Bioengineering* 102(1): 100–112. https://doi.org/10.1002/bit.22033

Rodrigues, L. R.; Arenzon, A.; Raya-Rodriguez, M. T.; Fontoura, N. F. 2011. Algal density assessed by spectrophotometry: a calibration curve for the unicellular algae *Pseudokirchneriella subcapitata*, *Journal of Environmental Chemistry and Ecotoxicology* 3(8): 225–228.

Singh, M.; Reynolds, D. L.; Das, K. C. 2011. Microalgal system for treatment of effluent from poultry litter anaerobic digestion, *Bioresource Technology* 102(23): 10841–10848. https://doi.org/10.1016/j.biortech.2011.09.037

Soeder, C. J.; Hegewald, E.; Fiolitakis, E.; Grobbelaar, J. U. 1985. Temperature dependence of population growth in a green microalga: thermodynamic characteristics of growth intensity and the influence of cell concentration, *Zeitschrift Natur* 40: 227–233.

Torzillo, G.; Goksan, T.; Faraloni, C.; Kopecky J.; Masojidek J. 2003. Interplay between photochemical activities and pigment composition in an outdoor culture of *Haematococcus pluvialis* during the shift from the green to red stage, *Journal of Applied Phycology* 15: 127–136. https://doi.org/10.1023/A:1023854904163

Valderrama, L. T.; Del Campo, C. M.; Rodriguez, C. M.; Bashan, L. E.; Bashan, Y. 2002. Treatment of recalcitrant wastewater from ethanol and citric acid production using the microalga *Chlorella vulgaris* and the macrophyte *Lemna minuscula*, *Water Research* 36(17): 4155–4192. https://doi.org/10.1016/S0043-1354(02)00143-4

Verschuere, L.; Rombaut, G.; Sorgeloos, P.; Verstraete, W. 2000. Probiotic bacteria as biological control agents in aquaculture, *Microbiology and Molecular Biology Reviews* 64(4): 655–671. https://doi.org/10.1128/MMBR.64.4.655-671.2000

Vezie, C.; Rapala, J.; Vaitomaa, J.; Seitsonen, J.; Sivonen, K. 2002. Effect of nitrogen and phosphorus on growth of toxic and nontoxic *Microcystis* strains and on intracelular microcystin concentrations, *Microbial Ecology* 43(1): 443–454. https://doi.org/10.1007/s00248-001-0041-9

Zheng, Y.; Li, T.; Yu, X.; Bates, P. D.; Dong, T.; Chen, S. 2013. High-density fed-batch culture of a thermo tolerant microalga *Chlorella sorokiniana* for biofuel production, *Applied Energy* 108: 281–287. https://doi.org/10.1016/j.apenergy.2013.02.059

Zhou, W.; Li, Y.; Min, M.; Hu, B.; Chen, P.; Ruan, R. 2011. Local bioprospecting for high-lipid producing microalgal strains to be grown on concentrated municipal wastewater for biofuel production, *Bioresource Technology* 102(13): 6909–6919. https://doi.org/10.1016/j.biortech.2011.04.038

Petras VENCKUS. PhD student (biology). Lecturer at Department of Botany and Genetics, Vilnius University (VU). Research interests: algae biotechnology, wastewater treatment, various high value biochemical compounds obtained from algae, development of technologies to optimise the growth of algae for renewable fuels.

Jolanta KOSTKEVIČIENĖ. Dr. Assistant Professor, Vilnius University (VU). Doctor of Biomedical Sciences, Vilnius University (VU), 2001. Publications: author of 2 textbooks, over 20 research papers. Scientific supervisor of research work at PhD, B. Sc., and M. Sc. levels. Research interest: algae taxonomy, biology, ecology, and biotechnology (optimization of algae cultivation for the biomass rich in different bioactive compounds; treatment of different type of wastewater).

Vida BENDIKIENĖ. Dr, Habilitation Procedure (HP), Chief Research Fellow of Department Biochemistry and Molecular Biology, Vilnius University (VU). Doctor of Science Habilitation Procedure (biochemistry), VU, 2008. Doctor of Science (biochemistry), Byelorussian State University (Minsk, Byelorussia), 1980. Publications: author of 1 study-guide, 1 patent, over 40 research papers and 12 inventions. Scientific supervisor of PhD projects and diploma projects of BSc and MSc levels. Research interests: biotechnology, applied enzymology, enzymes-catalyzed biotechnologically important processes optimization methods; development of technologies to optimize the growth of algae for renewable fuels, specialty chemicals and value-added chemical compounds.

4

EVALUATION OF NOISE EXPOSURE BEFORE AND AFTER NOISE BARRIERS, A SIMULATION STUDY IN ISTANBUL

Nuri İLGÜREL[a], Neşe YÜĞRÜK AKDAĞ[b], Ali AKDAĞ[c]

[a, b] Department of Architecture, Yıldız Technical University, D-107, Beşiktaş, İstanbul, 34349 Turkey
[c] Hidrotek Architecture and Engineering Ltd., İstanbul, Turkey

Abstract. İstanbul is the most prominent Turkish city in terms of the gradually worsening noise problems associated with the rapid increase in population. This study aims to investigate the noise exposure in the settlements around the link roads connecting the Bosporus Bridge to the European side of the city by the aid of simulations and noise mapping, in the frame of action planning studies performed in İstanbul. Noise maps were generated for L_{den} and L_n noise indicators with the help of a noise mapping software. Since a considerable part of the settlements is exposed to high noise levels, a noise barrier alongside the link road was proposed as a control measure by the aid of the acoustic simulation. Simulations with the noise barriers suggest that for the L_{den} time interval the noise affected area over 55 dBA would reduce by 10%, the number of dwellings by 26% and the number of inhabitants by 25%, whereas for the L_n time interval the noise affected area over 45 dBA would reduce by 5%, the number of dwellings by 20% and the number of inhabitants by 20%.

Keywords: Bosporus Bridge, noise pollution, noise mapping, noise mitigation, noise barriers, noise indicators.

Introduction

Noise is an important environmental pollution factor that should be dealt seriously in accordance with the noise control strategies in order to mitigate it to the acceptable levels in the noise sensitive urban areas. Noise mitigation studies should be initially considered at the urban planning stage. Noise sources with different levels and characteristics and their location regarding the noise sensitive areas in the city are important decision factors at this stage.

In order of importance, measures for noise control should be taken first of all at the source, then between the source and the receiver, and finally at the receiver. While noise control measures are most effective and economic at the source, when this is not possible or sufficient, it may become important to implement other measures between the source and the receiver. In mitigation of urban noise, control of the noise generated by traffic flow on roads becomes crucial, since roads cover the city as a pervasive web and tend to affect noise sensitive urban areas. Thus, measures taken on the path of noise propagation, such as the use of noise barriers and improvement of building envelopes against noise are needed in addition to the

arrangement of the roads and the traffic flow in a scientific manner. Today, the method of benefiting from noise maps for supply of statistical data about areas which are exposed to road noise is frequently applied in determination of necessary noise measures and development of effective solutions against noise. Noise maps, which offer the possibility to make predictions regarding the future, are considered a highly useful tool in planning urban development zones and determining the correct routes and locations for roads and other urban noise sources (Ibbeken, Krüger 2013).

Strategic noise maps are now required in the European Union for all population centres of more than 250,000 inhabitants, as well as for major roads, railways, and airports, and are becoming required for urban areas with over 100,000 people (END 2002). Debates surrounding environmental noise pollution with a particular focus on the European Union are gaining importance. Environmental noise pollution is an emerging public policy and environmental concern and is considered to be one of the most important environmental stressors affecting public health throughout the world. (Licitra 2013; Murphy, King 2014) In respect of noise mapping and action plans, there

Corresponding author: Nuri İlgürel
E-mail: milgurel@yildiz.edu.tr

are many studies which may be taken as a model for the implementation of the Environmental Noise Directive in European countries and the utilization of noise maps (Blanes Guardia, Nugent 2013; Braunstein 2013; Ramírez, Domínguez 2013; Tracz, Wozniak 2013; Seong et al. 2011; King, Rice 2009).

In European countries, where related works still continue, 2nd phase of the studies were completed in 2012. As a European Union candidate country, Turkey made legal arrangements on the issue as well, and the national legislation for environmental noise management were revised. The Regulation on Environmental Noise Assessment and Management (RENAM 2010) which was prepared in harmony with the relevant European Directive (Directive 2002/49/EC), was published in the Official Journal of Turkey and put into force in July of 2005 and revised in 2010. The European Directive and the corresponding Turkish Regulation require noise mapping studies to be conducted for the evaluation of noise environment in urban areas. As the initial stage, RENAM stipulated the generation of noise maps until 31th of December, 2016 and development of related action plans until 31th of December, 2017 for settlements of more than 250,000 inhabitants, for major roads with annual traffic volume exceeding 6,000,000 motor-vehicles, for major railroads with annual traffic volume exceeding 60,000 trains and for major airports having more than 50,000 flights a year.

The European Commission has assigned a twinning project concerning the implementation of the Directive 2002/49/EC in Turkey to the Turkish Ministry of Environment and Forestry and the German Federal Ministry of Environment, Nature Conservation and Nuclear safety. Accordingly, noise maps and action plans have been developed for five pilot areas located in five major Turkish cities (Adana, Ankara, Bursa, İstanbul and İzmir) (Irmer et al. 2007). In another EU-supported project (Technical Assistance Project for Implementation Capacity for the Environmental Noise Directive), which addresses various noise sources in fourteen Turkish municipalities, noise mapping studies have been realized (Ministry of Environment and Urban Planning of Turkey 2015). The main purpose of the project is to strengthen the capacity of the stakeholders (Ministry of Environment and Urbanization, other ministries and relevant offices of municipalities) taking part in the field of noise management. As a leading research institution of Turkey, TUBITAK (Scientific and Technological Research Council of Turkey) has completed the generation of city noise maps for several cities. The institution also prepared noise maps for 13 Turkish airports and carries on related studies for 26 additional airports (TÜBİTAK MAM 2015; Özkurt et al. 2014; Sarı et al. 2014).

According to the European Union Directive 2002/49/EC, acceptable noise levels for settlements are 55 dBA for the whole day average including day, evening and night time intervals (L_{den}), and 45 dBA for the night time interval (L_{night}). It usually requires special measures to meet these acceptable levels in areas located around busy major roads. The purpose of this study is to exemplify the noise reduction potential of noise barriers, which may play an important role within the scope of environmental noise action plans, through a practical work carried out on access roads of the Bosporus Bridge. As a vital link between the European and Asian sides of Istanbul, the Bosporus Bridge has experienced an enormous increase of traffic density since its construction in 1973 (Fig. 1). In the practical study conducted for the area around the viaduct which serves as the European side connection of the bridge, noise maps were generated for the current situation by use of a simulation software and accuracy of the maps were validated through comparisons with the results of actual noise measurements. Afterwards, noise level reductions that can be obtained with the implementation of noise barriers were modelled on the noise maps and the results were evaluated by making comparisons with the current situation.

1. Methodology

The study was conducted in the following steps:
- Collection of data about the study area the road traffic.
- Generation of grid noise maps and statistical noise maps for the L_{den} (day-evening-night) and L_n (night) (23.00–07.00) time intervals, for the current situation.
- Validation of the accuracy of noise maps through comparisons with the results of actual noise measurements.
- Generation of grid noise maps and statistical noise maps for the L_{den} and L_n time intervals, this time with the addition of noise barriers to the model.
- Comparative evaluation of the data for the current situation and the model with noise barriers.

Fig. 1. General view of settlements around of the Bosporus Bridge in European side of İstanbul (wallpaperup 2014)

1.1. Collection of data about the study area and the road traffic

Noise maps were generated in the SoundPLAN 7.3 simulation software (Braustain+Berndt GMBH 2006) for an area of approximately 3.5 km² in order to reveal the noise problem experienced in the area in detail. For being able to generate accurate noise maps in the software, topography of the study area is required to be modeled as a beginning. Ready-to-use actual maps of the study area at the scale of 1/1000 were provided by the Metropolitan Municipality of Istanbul in the digital environment. Digital maps which were prepared in the UTM coordinate system and contained necessary data regarding elevation (x, y, z [height]) and the building information system (building use, number of floors, etc.) were loaded in the SoundPLAN 7.3 software. In order for the buildings in the area to be defined in the software, various formal and dimensional information including building height, number of floors and floor height, and other information including number of habitants were entered in the software for each building. Since the actual maps did not contain population data, total population of the area was equally distributed between the buildings, based on the data provided by the Turkish Statistical Institute (TUIK 2014).

For the purposes of simulation, prevailing and secondary winds observed in Istanbul were defined to be NNE and SSW respectively (Windfinder 2014); average temperature for the whole year was defined to be 15 °C and average relative humidity was defined to be 70%. The sound absorption value (G) defined for the green fields in the study area was 0.6 as specified in the ISO 9613 standard (ISO 9613-2 1996).

Trucks, semi-trailers and other similar heavy vehicles are not allowed to use the Bosporus Bridge. Thus, heavy vehicle use of the bridge is limited to urban mass transportation means including busses and metro-busses. Latest data related to traffic volume on the bridge were released in 2012. However, 2012 was the year when the heavy vehicle traffic between two sides of the city was directed to the Bosporus Bridge during the evening and night time intervals, because the second bridge which is normally handling heavy vehicle traffic was temporarily put under maintenance works. For this reason, data for the year 2011 were preferred to be used in the study instead of the year 2012, and necessary updates were made on the data based on actual observations. Total number of vehicles passages on the bridge throughout the year 2011 was approximately 68,500,000 (Turkish Administration Privatization 2011). Table 1 shows the results of the observations which were carried out to provide up-to-date data regarding the number of vehicles using the bridge during certain hours of the day, evening and night time intervals in 2014. Since the traffic volume on the bridge does not differ significantly for monthly or daily variations, results of the observation in May 2014 were considered suitable for the purposes of the study. In addition, average traffic speed data were determined by performing experimental drives on the bridge. Traffic data of other primary and secondary roads in the area were also taken into account during the generation of noise maps.

In the preparation of noise maps, the NMPB Routes 96 (Guide de Bruit) standard was applied as the method of calculation, as recommended in the European Union Directive. Calculations were performed at a height of 4 m above ground and at grid dimensions of 10×10 m as specified in the same Directive.

Table 1. Traffic data for the Bosporus Bridge

Time interval	Hours	Light vehicles per hour	Heavy vehicles per hour	Light vehicle speed (km/s)	Heavy vehicle speed (km/s)	Proportion of heavy vehicles (%)
Day	17.00–18.00	9475	225	50	50	2.4
Evening	19.00–20.00	11 500	240	50	50	2.0
Night	04.00–05.00	3500	45	80	70	1.3

2. Determination of noise exposure in the current situation

In Figure 2 and Figure 3, the grid noise map and statistical noise map for the L_{den} noise indicator are given for the current situation experienced in the area around the Bosporus Bridge link road. When the maps are reviewed, three main areas of dense settlements, which mainly consist of dwellings, are seen to be significantly affected by traffic noise. Level of noise affecting those areas are found to exceed 80 dBA for L_{den} and 70 dBA for L_n. The areas mostly affected by traffic noise are the settlements located on the ridges, as a result of the hilly terrain. In addition, dwellings around the Beşiktaş junction are seen to be under the effect of high noise levels. When it is remembered that the acceptable noise level is 45 dBA for the night time and 55 dBA for the day time, the whole population living in the area is understood to be exposed to unacceptable levels of noise. According to the statistical data given in Table 2 and based on the daily average figures, the number of inhabitants being subject to a noise level over 55 dBA during the day time is 16,550, while the number of inhabitants being subject to a noise level over 45 dBA during the night time is 22,110.

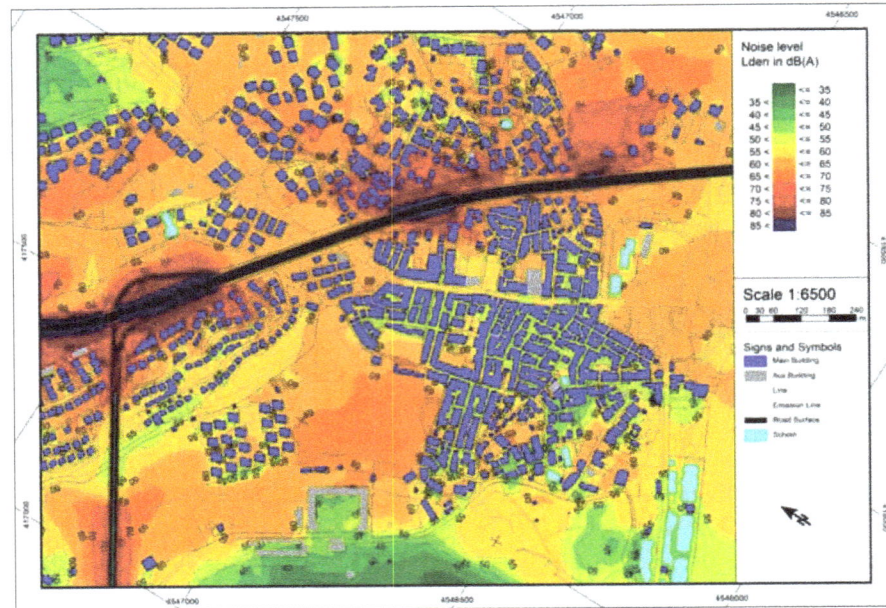

Fig. 2. Current situation noise map of the Bosporus Bridge link road for the day-evening-night average (L_{den})

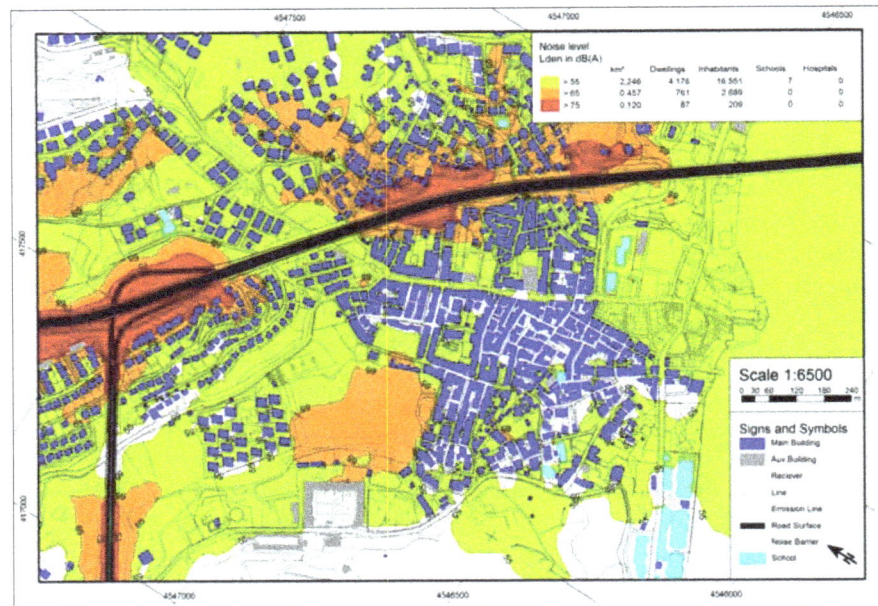

Fig. 3. Current situation statistical noise map for the day-evening-night average (L_{den})

Table 2. Statistical values of exceeded noise levels in L_{den} and L_n for the current situation

Time interval	L_{den} (day, evening, night average)			L_n (night)		
Noise level	>55 dBA	>65 dBA	>75 dBA	>45 dBA	>55 dBA	>65 dBA
Affected area (km²)	2.246	0.457	0.120	2.501	0.845	0.161
Affected number of dwellings	4200	761	87	5900	1410	210
Affected number of inhabitants	16 550	2689	209	22 110	5755	500

3. Noise measurements and validation of the model

For realistic generation of noise maps, various data including topography of the study area, formal and dimensional properties of buildings, type of ground cover and properties of noise sources (parameters such as vehicle density and percentage of heavy vehicles and variation of traffic density in time for road noise) must be defined in the most accurate way possible in the simulation software. The most suitable method for validating the accuracy of a noise map is to carry out actual noise measurements and compare the actual results with the values calculated by the simulation. In our study, noise level measurements were performed at ten different points during both day and night time intervals for this purpose. Since the level of noise tends to decrease with higher densities of traffic due to the decrease in average vehicle speed, the measurements were carried out for day and night time intervals between the hours 11:00–13:00 and 24:00–02:00 respectively, when the measured noise levels rise to peak points

for the whole day. Measurement points of measurements are seen in Figure 4. All measurements were performed in accordance with the ISO 1996-1 standard (ISO 1996-1 2003) in A-weighting by use of Brüel&Kjær Type 2236 and Type 2258 sound level meters. Devices were placed to a height of 1.5 m above ground, at least 2 m away from any surface in order to eliminate the possibility of reflection, and windscreens were installed to the microphones. The results obtained through 15 minutes measurements and the data regarding number of vehicles are given in Table 3. On the other hand, spectral distribution of the noise measured for day time interval can be found in Figure 5 for the eight octave bands between 63 Hz and 8000 Hz.

According to the observations of Maruyama et al., when the number of vehicles exceeds 170, uncertainty indication $\Delta LAeq_T$ must be defined to be ± 1 dB, independent from the time of measurement. In our study, number of vehicles passing was higher than 170 at all measurement points, hence the level of noise emitted from the

Fig. 4. Google Earth view of the European side link road of the Bosporus Bridge (2015) and noise measurement points

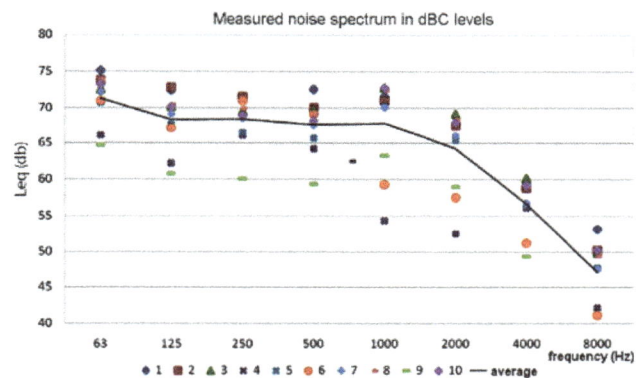

Fig. 5. Octave band noise levels measured on the European side link road of the Bosporus Bridge

Table 3. Noise measurement results and number of vehicles

Measurement time	Statistical evaluation	Measurement points									
		1	2	3	4	5	6	7	8	9	10
Day time (for 15 minutes)	L_{Aeq}	75.5	74.5	68.7	63.7	70.2	67.5	73.1	70.5	60.8	72.7
	dBC	80.4	79.7	80.0	71.2	75.6	75.1	77.6	78.3	68.5	78.1
	$L_C - L_A$	4.9	5.2	7.9	7.5	5.4	7.6	4.5	7.8	7.7	5.4
	Number of vehicles	2521	2451	2425	2356	2468	2471	2418	2436	2423	2476
	Heavy vehicles	45	44	64	69	50	61	48	54	58	49
	Heavy vehicle (%)	1.75	1.76	2.57	2.85	1.99	2.41	1.95	2.17	2.34	1.94
Night time (for 15 minutes)	L_{Aeq}	71.3	69.6	65.7	58.7	63.2	64.5	69.2	61.5	57.8	64.5
	dBC	73.4	72.2	70.1	62.6	66.4	69.0	71.4	66.3	63.6	67.4
	$L_C - L_A$	2.1	2.6	4.4	3.9	3.2	4.5	2.2	4.8	5.8	2.9
	Number of vehicles	915	904	887	856	882	876	918	845	847	897
	Heavy vehicles	6	8	12	15	14	16	7	9	21	8
	Heavy vehicle (%)	0.65	0.88	1.33	1.72	1.56	1.79	0.76	1.05	2.42	0.88

road can be accepted to be between 60.8 and 75.5 dBA with an approximation of ±1 dB (Maruyama *et al.* 2012; Garg, Maji 2014).

Viaducts and bridges are known to be subject to vibration during the passage of vehicles because of their separate decks sitting on piers, which may cause significant increases on the level of low frequency noise. Vehicle number counting studies reveal a heavy vehicle proportion of 1.3–2.4% on the Bosporus Bridge. However, this figure does not consist of heavy trucks of semi-trailers, which may cause high amplitude vibration of the bridge deck and create dominant low frequency noise as a result, but of busses, metro-busses and smaller trucks. As can be seen in Table 3, L_c – L_A values vary between 4.5–7.9 dB for L_{den} and between 2.1–5.8 dB for L_n time intervals. Since the L_c – L_A value is lower than 8 dB at all measurement points, it can be foreseen that any significant disturbance will not be experienced in the area in terms of low frequencies, when noise barriers are installed on the bridge (Ascari *et al.* 2015).

For the purpose of validation of the noise maps generated, single point receiver calculations were performed by the software for the L_d and L_n time intervals for the actual measurement points, at an elevation of 1.5 m above ground and with different distances to the highway. Results of the calculations made by the SoundPLAN 7.3 software are given in Table 4, together with the results of actual measurements. For day time interval the measured noise levels vary between 60.8–75.5 dBA with a standard deviation of 4.70, while for night time interval between 57.8–71.3 dBA with a standard deviation of 4.53. The differences between the calculated and measured levels vary for day time interval between (–0.3)–(1.4) dB according to the measurement point with a standard deviation of 1.08, while for night time interval between (–0.3)–(1.5) dB with a standard deviation of 0.93. The calculated levels are in good agreement with the measured levels for both time intervals, since the differences between measured and calculated levels are sufficiently small.

According to the Good Practice Guide for Strategic Noise Mapping and the Production of Associated Data on Noise Exposure (WG-AEN 2006), the difference between actual measurement results and the figures presented by the model must not exceed 1 dB in a distance of 300 m from the source, 3 dB in a distance of 600 m from the source and 10 dB in a distance of 2,000–3,000 m from the source. As can be seen in Table 4, highest difference between measured and calculated noise levels is 1.5 dBA, which easily validates the reliability of the results presented by the simulation.

4. Determination of noise exposure after the addition of noise barriers to the model

In order to propose a solution for the negative effects of noise on the settlements located around the Bosporus Bridge link road, noise barriers were virtually designed for both sides of the road and additional noise maps were generated for the new scenario with noise barriers. The attributes of the designed noise barriers are as follows:

- According to the simulations the length of the noise barriers should be 900 m. from the joining point of the bridge on the European side to the junction point of the roads in order to protect the settlement zones from noise.
- Noise level decrease for 2 m, 3 m and 4 m high barriers have been calculated, nevertheless 4 m high barrier was not significantly efficient as compared to others with very small gains in noise level decrease. Thus, 3 m high barrier was chosen for noise mapping process.
- In order to minimize the over-height visual effect of the barrier and increase the barrier effectiveness, the barriers were designed as; the lower 2 m vertical part made of opaque materials and the upper 1 m transparent part slopped 10° to the road side.
- In order to minimize the reflections the lower part of the barrier facing the road side was designed sound absorbent (load bearing structure + sound

Table 4. Noise measurement data and calculation statistics

Measurement time	Statistical evaluations	Measured and calculated levels										Standard deviation
	Measurement points	1	2	3	4	5	6	7	8	9	10	
	approx. distance to the highway (m)	11	14	18	60	12	22	2	2	70	8	
Day 11:00 – 13:00	Measured LAeq	75.5	74.5	68.7	63.7	70.2	67.5	73.1	70.5	60.8	72.7	4.70
	Calculated LAeq	74.8	74.1	70.0	62.8	71.3	68.5	72.0	69.1	59.7	73.0	4.86
	Difference	0.7	0.4	–1.3	0.9	–0.9	–1.0	1.1	1.4	1.1	–0.3	1.08
Night 24:00 – 02:00	Measured LAeq	71.3	69.6	65.7	58.7	63.2	64.2	69.2	61.5	57.8	64.5	4.53
	Calculated LAeq	71.8	68.8	65.0	57.5	63.5	62.7	70.3	62.3	57.2	64.0	4.91
	Difference	–0.5	0.8	0.7	1.2	–0.3	1.5	–1.1	–0.8	0.6	0.5	0.93

absorbing materials + perforated metal plates) and the upper part slopped.

– The mass of the lower part of the barrier should be nearly 24 kg/m² for sufficient high sound transmission loss.

Figure 6 and Figure 7 show the grid noise map and statistical noise map generated for L_{den} noise indicator after the implementation of the noise barrier in the software.

Table 5 shows the values of noise exposure for the day-evening-night and night time intervals in terms of

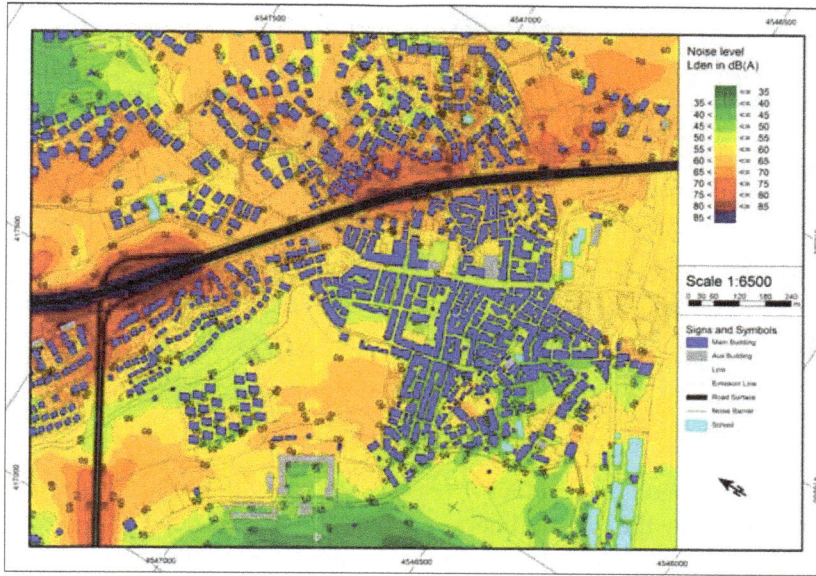

Fig. 6. Bosporus Bridge link road noise map for the day-evening-night average (L_{den}) with noise barriers

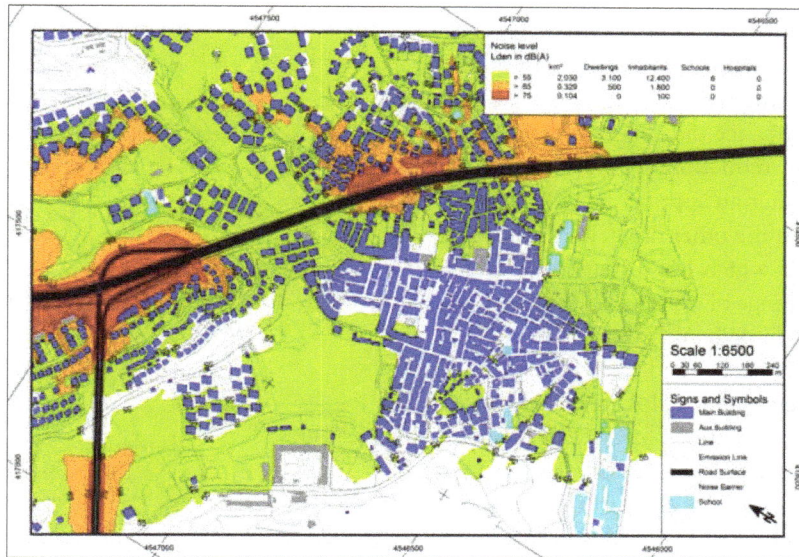

Fig. 7. Statistical noise map for the day-evening-night average (L_{den}) with noise barriers

Table 5. Noise exposure values for L_{den} and L_n time intervals with noise barrier

Time interval	L_{den} (day, evening, night average)			L_n (night)		
Noise level	>55 dBA	>65 dBA	>75 dBA	>45 dBA	>55 dBA	>65 dBA
Affected area (km²)	2.030	0.329	0.104	2.377	0.587	0.133
Affected number of dwellings	3087	466	35	4708	905	112
Affected number of inhabitants	12 450	1770	121	18 116	3705	205

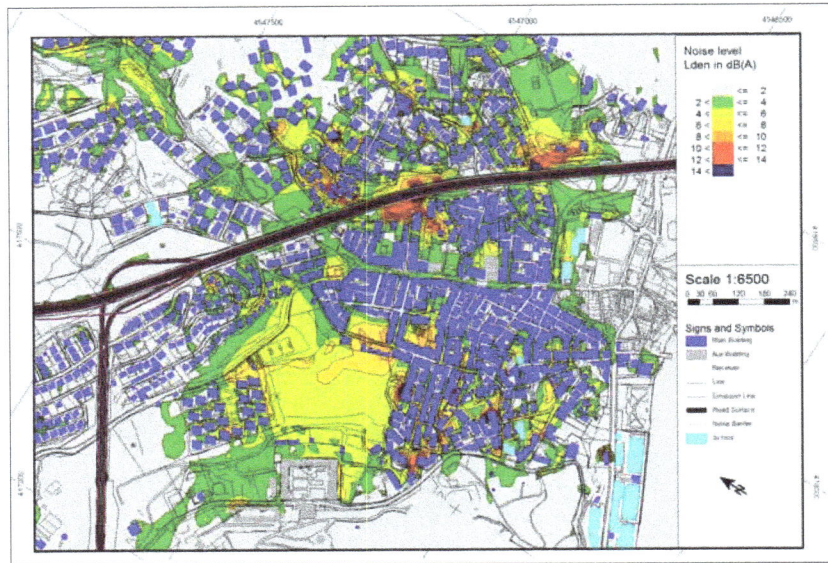

Fig. 8. Reduction in the level of exposure obtained with the addition of noise barriers to the model

affected area, number of dwellings and number of inhabitants, based on the data provided by the statistical noise maps generated with noise barriers.

Conclusions

1. Figure 8 shows the noise reduction obtained with the addition of noise barriers to the model, in comparison with the current situation without any barrier. Lower elevation of the section on the south-western side of the bridge is understood to enhance the effectiveness of barriers. In this section, the reduction provided by noise barriers falls below 2 dB after an approximate distance of 500 m from the bridge. On the other hand, on the north-eastern side of the bridge, which has a higher elevation than the bridge, effective range of noise barriers is seen to be limited to around 250 m. When the area is evaluated as a whole, many buildings are found to benefit from a reduction of noise between 2 and 4 dBA, certain regions are seen to obtain higher reductions as a result of their favourable topographic and settlement conditions, and the level of noise effecting a small plot close to the bridge is seen to be up to 14 dBA lower than the current situation without barriers.

2. Figure 9 and 10 reveal exposure conditions of the area in terms of affected area, dwellings and inhabitants in the L_{den} and L_n time intervals, with and without noise barriers. These two graphs indicate changes in the affected parameters in terms of the area, dwelling and population. Since the affected noise levels before and after barrier states can be assessed by using these different types of parameters, they have been shown on individual graphs for the L_{den} and L_n noise indicators in order to easy the comparison between them.

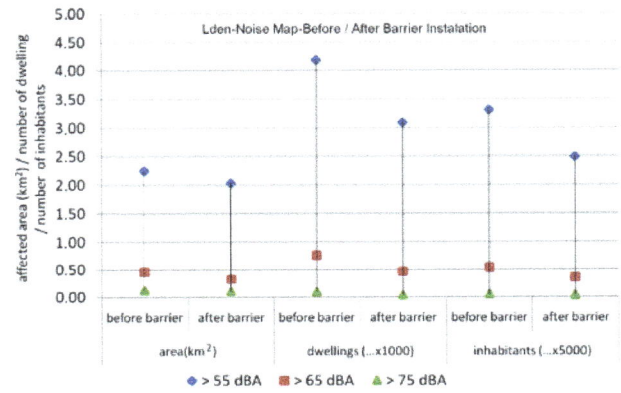

Fig. 9. Changes in figures of affected area-dwellings-habitants in the day-evening-night (L_{den}) time interval with the addition of noise barriers to the model. (e.g. number of dwellings exposed to noise before barrier over 65 dBA is calculated approx. by 0.8*1000 = 800)

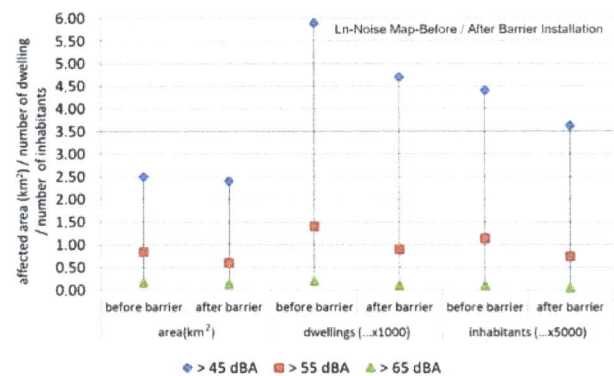

Fig. 10. Changes in numbers of affected area-dwellings-habitants in the night (L_n) time interval with the addition of noise barriers to the model. (e.g. number of dwellings exposed to noise before barrier over 55 dBA is calculated approx. by 1.4*1000 = 1400)

When the graphs in Figure 9 and 10 are reviewed, it is seen that construction of the proposed noise barrier can provide a certain amount of improvement in terms of the amount of area, number of dwellings and inhabitants being exposed to excessive noise levels. When a threshold exposure value for Lden is considered as 55 dBA, number of inhabitants being exposed to excessive noise levels can be reduced by 25%, from 16,550 to 12,450, by the aid of the noise barrier. When the same comparison is made for a threshold exposure value of 65 dBA, number of inhabitants affected can be reduced by 35%, from 2,689 to 1,770. For the threshold value of 75 dBA, reduction reaches to 42%, with the number of inhabitants affected decreasing from 209 to 121. At the night time interval, the number of dwellings being exposed to noise can be reduced by 19% for the 45 dBA threshold, 35% for the 55 dBA threshold and 59% for the 65 dBA threshold.

3. Simulations with the noise barriers suggest that the number of people in the acceptable noise exposure area would increase as to cover 4100 people more than before barrier; considering the acceptable noise levels for L_{den} and L_n indicators determined as 55 dBA and 45 dBA respectively.

4. According to the simulations made it is obvious that the barrier efficiency, which is determined by the degree of the noise level decrease, is significantly influenced by the topography and vegetation; for instance at the upper parts of the sloped hills the decrease in noise levels due to the barrier gets smaller in accordance with the diminishing acoustic shadow effect of the barrier.

5. It is well known that low frequency components of noise enhance with the addition of the noise barriers on the edges of the viaduct, especially vibrations caused by heavy vehicles are effective. In order to avoid this undesirable effect the noise barriers to be placed on the viaduct should be constructed by using structure components not too heavy but rigid (Monazzam, Nassiri 2009; Monazzam 2009). That is one of the reasons why the noise barrier was not designed homogenous in this study, instead from two different components; one is opaque and the other is transparent.

6. As seen in most cities of the world, inhabitants in Turkey spend most of their time under exposure to unacceptable levels of noise. For being able to create desirable acoustic environments, many further studies similar to the one exemplified in this paper must be conducted at higher levels of detail, and action plans must be carried into practice based on the results.

7. Many Turkish cities do not have maps in GIS format, while a significant proportion of existing maps can only provide inadequate or outdated data. Inadequacy of data in terms of noise sources is another important problem encountered in the course of noise mapping studies.

8. Accordingly, municipalities being in the first place, many institutions and organizations in Turkey continue their efforts towards filling the gap of available data. It is thought that noise mapping and action planning works can be performed much more rapidly when these important priorities are met.

References

Ascari, E.; Licitra G.; Teti, L.; Cerchiai, M. 2015. Low frequency noise impact from road traffic according to different noise prediction methods, *Science of the Total Environment* 505: 658–669. http://dx.doi.org/10.1016/j.scitotenv.2014.10.052

Blanes Guardia, N.; Nugent, C. 2013. Overview of the current state of the environmental noise directive implementation in Europe and exploitation of results, in *Proceedings of Inter-Noise 2013*, 15–18 September 2013, Innsbruck, Austria.

Braustain+Berndt GMBH. 2006. *SoundPLAN Manual* V 7.3. Backnang: Braustain+Berndt GMBH.

Braunstein, G. 2013. Suitable tools for the optimization of the modelling large noise maps and a discussion about the selection of appropriate input data, in *Proceedings of Inter-Noise 2013*, 15–18 September 2013, Innsbruck, Austria.

END. 2002. ENC Directive 2002/49/EC relating to the assessment and management of environmental noise, *Official Journal of European Communities*, L189. 14 p. [online], [cited 04 November 2015]. Available from Internet: http://www.noiseproject.gen.tr/uploads/files/Ek_VI_Noise%20EU%20Directive.pdf

Garg, N.; Maji, S. 2014. A critical review of principal traffic noise models: strategies and implications, *Environmental Impact Assessment Review* 46: 68–81. http://dx.doi.org/10.1016/j.eiar.2014.02.001

Ibbeken, S.; Krüger, M. 2013. Noise mapping of major roads for the state of Baden-Wuerttemberg according to the EU Environmental Noise Directive, in *Proceedings of Inter-Noise 2013*, 15–18 September 2013, Innsbruck, Austria.

Irmer, V.; Öncüer, B.; Saral, S.; Yükcel, H. 2007. Noise mapping in Turkey-implementation of the environmental noise directive, in *Proceedings of Inter-Noise 2007*, 28–31 August 2007, İstanbul, Turkey.

ISO 9613-2:1996. Acoustics – Attenuation of sound during propagation outdoors – part 2: general method of calculation. International Organization for Standartization, Geneva.

ISO 1996-1:2003. Acoustics – Description, measurement and assessment of environmental noise – part 1: basic quantities and assessment procedures. International Organization for Standartization, Geneva.

King, E. A.; Rice, H. J. 2009. The development of a practical framework for strategic noise mapping, *Applied Acoustics* 70(8): 1116–1127. http://dx.doi.org/10.1016/j.apacoust.2009.01.005

Licitra, G.; 2013. *Noise mapping in the EU: models and procedures.* USA: CRC Press. http://dx.doi.org/10.1201/b12885

Maruyama, M.; Kuna, K.; Sone, T. 2012. The minimum measurement time for estimating L_{AeqT} of road traffic noise from the number of vehicles pass-bys, *Applied Acoustics* 74(3): 317–24. http://dx.doi.org/10.1016/j.apacoust.2012.08.005

Ministry of Environment and Urban Planning of Turkey. 2015. *Technical Assistance Project for Implementation Capacity for the Environmental Noise Directive*. EuropeAid/131352/D/SER/TR [online], [cited 04 November 2015]. Available from Internet: http://www.noiseproject.gen.tr/en/about-project/5

Monazzam, M. R.; Nassiri, P. 2009. Performance of profiled vertical reflective parallel noise barriers with quadratic residue diffusers, *International Journal of Environmental Research* 3(1): 69–84.

Monazzam, M. R. 2009. Optimization of profiled diffuser barrier using the new multi-impedance discontinuities model, *International Journal of Environmental Research* 3(3): 327–334.

Murphy, E.; King; E. A. 2014. *Environmental noise pollution*. Noise Mapping, Public Health and Policy. USA, Elsevier Inc. http://dx.doi.org/10.1016/B978-0-12-411595-8.00001-X

Özkurt, N.; Sarı, D.; Akdağ, A.; Kütükoğlu, M.; Gürarslan, A. 2014. Modelling of noise pollution and estimated human exposure around İstanbul Atatürk Airport in Turkey, *Science of the Total Environment* 482–483: 486–492. http://dx.doi.org/10.1016/j.scitotenv.2013.08.017

Ramirez, A.; Dominguez, E. 2013. Modelling urban traffic noise with stochastic and deterministic traffic models, *Applied Acoustics* 74(4): 614–621. http://dx.doi.org/10.1016/j.apacoust.2012.08.001

RENAM. 2010. Regulation of Environmental Noise Assessment and Management, *The Official Gazette* No. 25862. Turkey (in Turkish).

Sarı, D.; Özkurt, N.; Hamamcı, S. F.; Ece, M.; Yalçındağ, N.; Akdağ, A.; Akdağ, N. Y. 2014. November. Assessment of noise pollution sourced from entertainment places in Antalya, Turkey, in *Proceedings of Inter-Noise 2014*, 16–19 November 2014, Melbourne, Australia.

Seong, J. C.; Park, T. H.; Ko, J. H.; Chang, S. I.; Kim, M.; Holt, J. B.; Mehdi, M. R. 2011. Modelling of road traffic noise and estimated human exposure in Fulton Country, Georgia, USA, *Environmental International* 37: 1336–1341. http://dx.doi.org/10.1016/j.envint.2011.05.019

Tracz, M.; Wozniak, K. 2013. Use of noise maps in designing of bypass vertical alignment in relation to housing location, in *Proceedings of Inter-Noise 2013*, 15–18 September 2013, Innsbruck, Austria.

TUIK. 2014. *Document of regional accounts statistics* [online]. Turkish Statistical Institute [cited 10 April 2014]. Available from Internet: http://www.turkstat.gov.tr/PreTabloarama.do

Turkish Administration Privatization. 2012. *Presentation document of highways and bridges privatization* [online], [cited 10 October 2015]. Available from Internet: http://www.oib.gov.tr/2011/dosyalar/otoyollar _tanitim_Ocak_2011.pdf

TÜBİTAK MAM. 2015. *Projects* [online], [cited 1 February 2015]. Available from Internet: http://ctue.mam.tubitak.gov.tr/en/ arastirma-alanlari/projects

wallpaperup [online]. 2014 [cited 10 August 2014]. Available from Internet: http://www.wallpaperup.com/349001/ Istanbul citysea_of_marmara_night_ bosphorus_bridge_turkey.html

WG-AEN. 2006. *Good practice guide for strategic noise mapping and the production of associated data on noise exposure* [online], European Commission Working Group Assessment of Exposure to Noise [cited February 2015]. Available from Internet: http://ec.europa.eu/environment/noise/pdf/wg_aen.pdf

Windfinder [online]. 2014 [cited 10 February 2014]. Available from Internet: http://www.windfinder.com

Nuri İLGÜREL graduated in 2000 from the Department of Architecture at Balıkesir University. He attained his MSc degree in 2004 and PhD degree in 2010 in "Building Physics" at the Department of Architecture at the Yıldız Technical University (YTU). He became Assoc. Professor in 2014 and he continues his research studies at the Building Physics Unit of the Department of Architecture at YTU. He has profession in Noise Control and Architectural Acoustics. He has many national and international publications in his field. He is a member of the Turkish Acoustical Society.

Neşe YÜĞRÜK AKDAĞ studied at Yildiz Technical University and obtained a degree in Architecture in 1984, an MSc degree in "Building Physics" in 1987, and received her PhD in "Architectural Acoustics" from Yildiz Technical University in 1995. She is full professor since 2011 in Building Physics Unit of the Faculty of Architecture in Yildiz Technical University. She has range of articles, papers, books and applications on room and building acoustics. She has also several researches on noise mapping. She prepared and organized some national projects and contributed to several national and international projects. She is a member of the Turkish Acoustical Society.

Ali AKDAĞ graduated in 1985 in Electronics and Communication Engineering at Istanbul Technical University. He received his MSc degree in 1989. He is owner of Hidrotek Mimarlik Muh. Tic. Ltd. He is providing training services for the preparation of noise maps and projecting noise prevention measures. He has involved several noise mapping projects like noise mapping of 31 airports in cooperation with TUBITAK Marmara Research Center, Environment and Cleaner Production Institute and strategic noise mapping of several cities in Turkey.

5

REMOVAL OF HEAVY METALS USING BACTERIAL BIO-FLOCCULANTS *OF BACILLUS SP.* AND *PSEUDOMONAS SP.*

Ahmed M. AZZAM[a], Ahmed TAWFIK[b]

[a]*Environmental Researches Department, Theodor Bilharz Research Institute, 30-12411, Imbaba, Giza, Egypt*
[b]*Department of Environmental Engineering, Egypt-Japan University of Science and Technology (E-JUST), New Borg El Arab City, 21934, Alexandria, Egypt*

Abstract. Bio-flocculants produced by *Bacillus sp.* and *Pseudomonas sp.* were evaluated as flocculating agents for the removal of Cu (II), Pb (II) and Cd (II) from chemical and textile wastewater industries. Both bio-flocculants were very effective for removal of heavy metals at a dosage not exceeding 0.1 mg/ml. However, the removal efficiency of heavy metals was dependant on initial concentration and type of bio-flocculants. 84.0% of Cu^{2+} and 99.5% of Pb^{2+} were removed from industrial wastewater using *Bacillus sp.* Bio-flocculant resulting residual values of 28.5 and 1.13μg/l respectively in the treated effluent. Lower removal efficiencies of 70.4% for Cu^{2+} and 97.8% for Pb^{2+} occurred using *Pseudomonas sp.* bioflocculant. Nevertheless, *Pseudomonas sp.* bio-flocculant achieved a substantially higher removal efficiency of Cd^{2+} (93.5%) as compared to 72.9% using *Bacillus sp.* Based on these results bio-flocculants are considered as a viable alternative for the treatment of industrial wastewater containing heavy metals.

Keywords: heavy metals, wastewater, bio-flocculants, *Bacillus sp.*, *Pseudomonas sp.*

Introduction

Heavy metal pollution is one of the most serious environmental problems that threaten a large number of people. Heavy metals are not biodegradable and tend to be accumulated in organisms and cause numerous diseases and disorders. Extensive damage to human organs, such as liver, kidney, digestion system, and nervous system can be caused by uptake of excess heavy metals (Ozer, Pirincci 2006; He *et al.* 2013). The discharge of heavy metals into the environment due to agricultural, industrial, and military operations and the effect of this pollution on the ecosystem and human health are growing concerns. Recent research in the area of heavy metals removal from wastewaters and sediments has focused on the development of materials with increased affinity, capacity, and selectivity for target metals (Gadd 1990; Gadd, White, 1993; Totura 1996). The use of microorganisms to sequester, precipitate, or alter the oxidation state of various heavy metals has been extensively studied (Gadd 1990; Gadd, White, 1993; Macaskiey 1990; Rittle *et al.* 1995; Shen, Wang 1993). Industrial wastes contain toxic and hazardous substances, most of which are detrimental to human health (Gana

et al. 2008; Ogunfowokan et al. 2005; Rajaram, Ashutost 2008). Heavy metals from industrial processes are of special concern because they produce water or chronic poisoning in aquatic animals (Ellis 1989). More researchers are focusing their attention on industrial wastewater treatments (El-Sheekh *et al.* 2005; Gao *et al.* 2009; Liu *et al.* 2009; Lu *et al.* 2005; Wang *et al.* 2007) and on drinking water treatment (Liu *et al.* 2009) using microbial bio-flocculants. The use of bio-flocculants in wastewater treatment seems to be an economical alternative to physical and chemical means (Vijayalakshmi, Raichur 2003).

Biological materials or dead bacterial cells (Zheng *et al.* 2008), algae (El-Sheekh *et al.* 2005), protozoans (Rehman *et al.* 2008), yeasts, fungi (Guangyu, Thiruvenkatachari 2003) and plants (Heredia, Martin 2009) have been shown to play significant roles in heavy metal removal and recovery. Amongst bio-remediants, bio-flocculants have gained increasing attention since they are environmentally-friendly, biodegradable and non-toxic (Shih *et al.* 2001). Bio-flocculants contain various organic groups, which are responsible for binding metals, such as uronic acids (containing a carbonyl and a carboxylic

Corresponding author: Ahmed Tawfik
E-mails: ahmed.tawfik@ejust.edu.eg, drahmedazzam@hotmail.com

acid function) (Aguilera *et al.* 2008; Lu *et al.* 2005; Wu,Ye 2007), glutamic and aspartic acid in the protein component (Dignac *et al.* 1998) or galacturonic acid and glucuronic acid in the polysaccharide component (Bender *et al.* 1994). The fact that these bio-flocculants have higher efficiencies at low metal concentrations makes them very attractive for the removal of heavy metals from industrial effluents/wastewaters (Kotrba *et al.* 1999). Feng *et al.* (2013) used microbial flocculant GA1 (MBFGA1) to remove Pb^{2+} ions from aqueous solution. The removal efficiency of Pb^{2+} reached up to 99.85%. Moreover, Lin, Harichund (2011b) recorded that up to 90% of Pb^{2+} was removed by bio-flocculant produced by *Pseudomonas sp.* CH8. Bio-flocculants produced by *Pseudomonas* sp. CH6 and *Herbaspirillium* sp. CH13 flocculated 78% of Hg^{2+} and 66% of Cd^{2+} from the metal solutions respectively.

In order to mitigate the metal pollution in Egyptian industrial wastewater, metal-tolerant bacterial species that are capable of producing bio-flocculants have been isolated from industrial sludge. The removal efficiency of heavy metals (Cu^{2+}, Pb^{2+} and Cd^{2+}) present in industrial wastewater using bacterial bio-flocculants of *Bacillus sp.* and *Pseudomonas sp.* was investigated at different initial concentrations.

1. Materials and methods

1.1. Wastewater characteristics

Industrial wastewater containing heavy metals was collected from chemical and textile manufacturing companies (Shoubra El-Kheima – Qaluobyia Governorate, Egypt). The samples were collected in sterilized plastic bottles. Copper (Cu), lead (Pb) and cadmium (Cd) concentrations were determined by GBC Atomic Adsorption Spectroscopy (AAS) (Savanta AA). pH was measured using HANNA HI 9024. pH values of the wastewater ranged between 6.4–7.9. The concentration of Cu, Pb and Cd are highly fluctuated and varied (47 ± 1–482.6 ± 2.5 µg/l), (85 ± 1.5–304.3 ± 1.5 µg/l) and (34.4 ± 0.66–321 ± 6 µg/l) respectively.

1.2. Isolation and identification of bacteria

Bacillus sp. and Pseudomonas sp. bacteria species resistant to heavy metals were isolated from industrial sludge harvested from pipe effluent of chemical and textile industry. *Bacillus sp.* and *Pseudomonas sp.* were selected for preparation of bacterial bio-flocculants (Duguid *et al.* 1975; Nanda *et al.* 2011). Serial dilution and pour plating method using nutrient agar, blood agar and MacConky agar (Himedia Company) were used. Strains were maintained in agar slants containing nutrient agar. The colony morphology, physiological and biochemical characteristics of different strains were assessed. Every week, isolated strains were transferred to a new medium in order to keep

metabolic activity and check the purity by microscopic examination process. The growth of *Bacillus sp.* took place on blood agar with the following biochemical characteristics, large Gram-positive rod, given positive results with indole, urease, nitrate and negative with catalase, citrate and oxidase. A *Pseudomonas sp.* bacterium is Gram negative rod gave positive results with motility, ornithine, catalase, citrate and oxidase.

1.3. Production of bio-flocculants

The selected bacterial species (*Bacillus sp.* and *Pseudomonas sp.*) were cultivated in a 250 ml Erlenmeyer flask containing 30 ml YMPG medium (0.3% yeast extract, 0.3% malt extract, 0.5% polypeptone, 1% glucose and 2% agar at pH 7) at 28°C, 220 rpm for 20 h. A portion amounting to 0.7 ml of the cultivated bacterial strains was inoculated into 70 ml of production medium (0.5% yeast extract, 0.5% polypeptone, 2% ethanol, 1% glycerol, 0.05% K_2HPO_4, 0.05% $MgSO_4 \cdot 7H_2O$, 0.2% NaCl, and 0.2% $CaCO_3$) at the above conditions for 72 h (Nakata, Kurane 1999). Bio-flocculants were recovered from the supernatant after centrifugation (4000 rpm) for 15 min and precipitated by adding 2 volumes of ethanol at 4 °C overnight. The pellet was centrifuged at 4000 rpm for 15 min and dried in desiccators containing anhydrous cobalt chloride at room temperature under reduced pressure (Kurane *et al.* 1994; Lin, Harichund 2011a).

1.4. Experimental set-up

Erlenmeyer flasks contained 150 mL of wastewater with different concentrations of heavy metals (Cu^{2+}, Pb^{2+} and Cd^{2+} ions) and 15.0 mg of bio-flocculants (0.1 mg/ml). The flasks were kept, under constant agitation (120 rpm), at room temperature for 24 hrs. Bio-flocculants were separated from the medium after 24 hrs and residual metal concentrations in the inlet and outlet were determined by GBC Atomic Adsorption Spectroscopy (AAS) (Savanta AA). The percentage of removed heavy metals (Cu^{2+}, Pb^{2+} and Cd^{2+}) was calculated as follows,

$$Removal\% = \left\langle \frac{C_0 - C_f}{C_0} \, x \, 100 \right\rangle,$$

where C_0 and C_f are the initial and final concentrations of metal ion in wastewater in µg/L.

1.5. Statistical analysis

All experiments were performed in triplicate and the results were expressed as the means±SD. T-tests were used to examine the statistically significant differences ($p < 0.001$) for heavy-metal removal after bio-flocculants treatment using SPSS version 17. The results of each experiment were assumed to be independent with different variance.

2. Results and discussion

2.1. Lead (Pb²⁺) removal

The results presented in Table 1 and Figure 1 show the bio-flocculants of *Bacillus sp.* and *Pseudomonas sp.* is very effective for removal of lead (Pb²⁺) from industrial wastewater. However, the removal efficiency of lead (Pb²⁺) using *Pseudomonas sp.* was significantly affected at increasing the initial concentration from 85±1.5 to 304.3±1.5 µg/l. The residual values of Pb²⁺ in the treated water were increased from 0.32±0.82 to 9.2±0.29 µg/l at increasing the initial concentration from 85±1.5 to 304.3±1.5 µg/l respectively. This was not the case for bio-flocculants of *Bacillus sp.* where the removal efficiency of Pb²⁺ was not largely affected at increasing the initial concentration from 85±1.5 to 304.3±1.5 µg/l. The removal efficiency slightly changed from 99.4% to 99.7% at increasing the initial concentration from 85±1.5 to 304.3±1.5 µg/l respectively. This indicates that bio-flocculants of *Bacillus sp.* has a higher capability of adsorption of Pb²⁺ as compared to *Pseudomonas sp.* Similar findings are recorded by Lin, Harichund (2011a) who found that bio-flocculant produced by *Bacillus sp.* CH15 was capable of removing Pb²⁺ (87%) and Cr²⁺ (86%). The main mechanism removal of Pb²⁺ by bio-flocculants is mainly due to charge neutralization and adsorption bridging. Guo, Yu (2014) found that the removal efficiency of Pb²⁺ reached 94.7 % using MBFR10543. Fourier transform infrared spectra analysis indicated that functional groups, such as –OH, C = O, and C – N, were existed in MBFR10543 molecular chains, which had strong capacity for removing Pb²⁺. The maximum bio-floccution activity (95%) was recorded by *Achromobacter sp.* TL-3 bacteria for Pb²⁺ removal (Batta *et al.* 2013). Moreover, Feng *et al.* (2013) showed that the results of Pb²⁺ adsorption by microbial flocculant GA1 could be described by the Langmuir adsorption model, and being the monolayer capacity negatively affected with an increase in temperature and the adsorption process could be described by pseudo-second-order kinetic model.

2.2. Cupper (Cu²⁺) removal

Similar trends were observed for removal of Cupper (Cu²⁺) ions from industrial wastewater using bio-flocculants of *Bacillus sp.* and *Pseudomonas sp.* as shown in Table 2 and Figure 2. The reduction in Cu²⁺ concentration was substantially higher using *Bacillus sp.* bio-flocculant as compared to *Pseudomonas sp.* Nevertheless, the removal efficiency of Cu²⁺ was highly deteriorated at increasing the initial concentration from 47±1 to 482.6±2.5 µg/l. Bio-flocculant of *Bacillus sp.* had the highest copper removal percentage (87.2%) than *Pseudomonas sp.* (68.1%) at initial concentration of47±1µg/l. The removal efficiency of Cu²⁺ was significantly increased from 68.1% to 75.8% at increasing the initial concentration from 47±1 to

Table 1. The efficiency of bio-flocculants of *Bacillus sp.* versus *Pseudomonas sp.* for removal of Lead (Pb²⁺)

Bioflocculants	*Bacillus sp.*		*Pseudomonas sp.*	
Metal initial concentration (µg/l)	Residual values (µg/l)	%R	Residual values (µg/l)	%R
85±1.5	0.47±0.06	99.4	0.32±0.82	99.6
142.7±2	0.47±0.15	99.7	2.5±0.35	98.2
286±1	1.5±1.5	99.5	5.2±0.87	98.2
287±1	1.4±0.15	99.5	6.4±0.4	97.8
304.3±1.5	1.8±15	99.4	9.2±0.29	97.0

Note: %R*: percentage removal

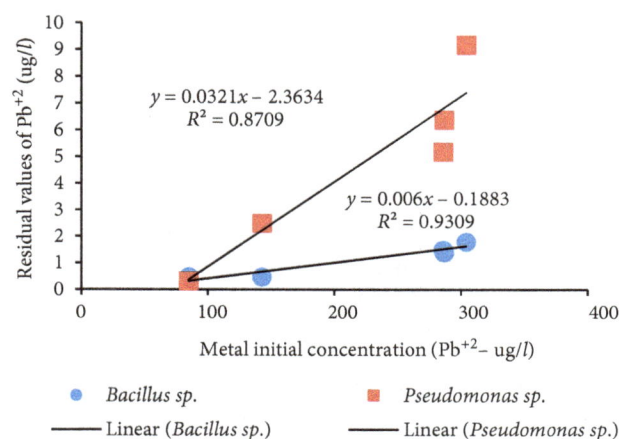

Fig. 1. Relationship between the initial concentration and residual values of Lead (Pb²⁺)

Table 2. The efficiency of bio-flocculants of *Bacillus sp.* versus *Pseudomonas sp.* for removal of Cupper (Cu²⁺)

Bioflocculants	*Bacillus sp.*		*Pseudomonas sp.*	
Metal initial concentration (µg/l)	Residual values (µg/l)	%R*	Residual values (µg/l)	%R*
47±1	6±1	87.2	15.4±0.7	68.1
58.9±1.5	6.7±1.5	88.6	17.2±0.6	71.2
113.7±1.5	17.3±1.5	84.8	43.3±0.9	62.6
191±1.5	36±1	81.2	71.6±1.4	62.5
482.6±2.5	76.7±1.5	84.1	117.1±1.5	75.8

Note: %R*: percentage removal

482.6±2.5 µg/l using *Pseudomonas sp.* This indicates that the efficiency of *Pseudomonas sp.* is initial concentration dependant. On the contrary, the removal efficiency of Cu²⁺ was slightly affected at increasing the initial concentration from 47±1 to 482.6±2.5 µg/l where the removal efficiency dropped from 87.2% to 84.1% respectively. The bio-flocculant adsorption/participates is mainly depends on the available hydroxyl, carbonyl and carboxyl groups which induces very high binding capacity. The negative charge groups could react with the positively charged site of heavy metals present in the wastewater, in this case, the metals can approach sufficiently close to each other so that attractive forces become effective. Chemical groups in the bio-flocculants act like a bridging agent of metals complexes and reduce inter-metals distances through the

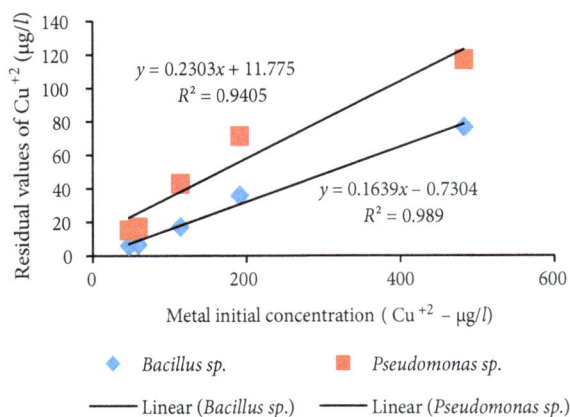

Fig. 2. Relationship between the initial concentration and residual values of Cupper (Cu^{2+})

Table 3. The efficiency of bio-flocculants of *Bacillus sp.* versus *Pseudomonas sp.* for removal of Cadmium (Cd^{2+})

Bio-flocculants	*Bacillus sp.*		*Pseudomonas sp.*	
Metal initial concentration (µg/l)	Residual values (µg/l)	%R*	Residual values (µg/l)	%R*
34.4±0.66	6.4±0.53	81.4	1.8±0.15	94.8
38.3±0.64	10.3±0.7	73.1	1.4±0.08	96.3
67.2±0.91	21.2±1	68.5	4.1±0.95	93.9
272±2.1	78.5±1.8	71.1	19.2±0.92	92.9
321±6	81.9±0.65	74.5	21.2±1.2	93.4

Note: %R*: percentage removal

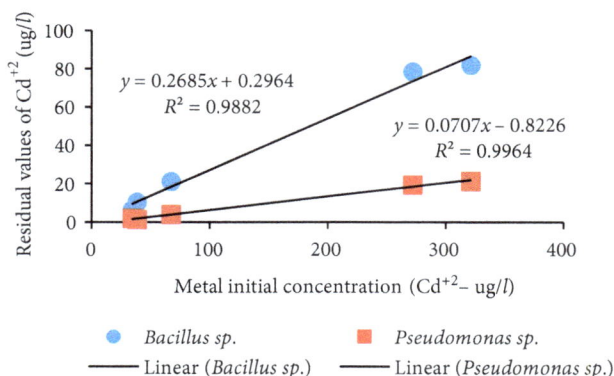

Fig. 3. Relationship between the initial concentration and residual values of Cadmium (Cd^{2+}).

ionic bonds mechanism, metals adsorbed onto one bio-flocculant molecular chain, and they could be adsorbed simultaneously by other chains, leading to the formation of three-dimensional flocs, which were capable of rapid biosorption process. Shuhonga *et al.* (2014) found that exopolysaccharide (EPS) from *Arthrobacter* ps-5 have strong biosorption capability, up to 169.15 mg/g of Cu^{2+}, 216.09 mg/g of Pb^{2+} and 84.47 mg/g of Cr^{6+}, respectively.

2.3. Cadmium (Cd^{2+}) removal

The results for Cadmium (Cd^{2+}) revealed that *Pseudomonas sp.* achieved higher removal efficiency as compared to *Bacillus sp.*, bio-flocculants at the same initial concentrations (Table 3 and Fig. 3). Moreover, the removal efficiency of Cd^{2+} using *Pseudomonas sp.* bio-flocculants was slightly affected (96.3–93.4%) at largely fluctuated the initial concentration from 38.3±0.64 to 321±6 µg/l. likely, the removal efficiency of Cd^{2+} using *Bacillus sp.* bio-flocculants remained unaffected at a level of 74% at increasing the initial concentration from 38.3±0.64 to 321±6 µg/l. This indicates that both *Bacillus sp.*, and *Pseudomonas sp.* bio-flocculants are very efficient for removal of Cd^{2+} from industrial wastewater. Nevertheless, the residual values of Cd^{2+} in the treated effluent were quite low using *Pseudomonas sp.* as compared to *Bacillus sp.* (Table 3). This metal biosorption potential of *Bacillus sp.*, and *Pseudomonas sp.* could probably be attributed to the role of bio-flocculant produced by it. Ion uptake by the range of bacterial component, such as cell wall component and extracellular polysaccharide, plays important roles in controlling heavy metal pollution in the treatment processes. Bio-flocculant producing bacteria such as *B. subtilis* WD 90, *B. subtilis* SM 29, and *E. agglomerans* SM 38 were reported to absorb the heavy metal by the cell component and the biopolymer (Saithong, Poonsuk 2002). Shamim, Rehman (2012) reported that *Klebsiella pneumonia* CBL-1 that isolated from heavy metal laden industrial wastewater was capable to remove 54% and 82% cadmium from the industrial effluents after 4 and 8 days of incubation at room temperature.

2.4. Comparison between the efficiency of *Bacillus sp.* and *Pseudomonas sp.* bio-flocculants for removal of Pb^{2+}, Cu^{2+} and Cd^{2+} at the same initial concentration

The results in Figure 4 show comparison between the efficiency of *Bacillus sp.* and *Pseudomonas sp.* bio-flocculants for removal of Pb^{2+}, Cu^{2+} and Cd^{2+} at the same initial concentration. Available data revealed that *Bacillus sp.* was very effective for removal of Cu^{2+} and Pb^{2+} from industrial wastewater resulting residual values of 28.5 and 1.13 µg/l in the treated effluent. Lower removal efficiencies of 70.4% for Cu^{2+} and 97.8% for Pb^{2+} were occurred using *Pseudomonas sp.* flocculants. Nevertheless *Pseudomonas sp.*

bio-flocculants achieved a substantially higher removal efficiency of Cd^{2+} (93.5%) as compared to 72.9% using *Bacillus sp.* bio-flocculants. Bio-flocculants cause aggregation of particles and cells by bridging or charge neutralization, colloid entrapment and double layer compression (Salehi-zadeh, Shojaosadati 2001). Shuhonga *et al.* (2014) found that exopolysaccharide (EPS) from *Arthrobacter* ps-5 have strong biosorption capability, where infrared spectrometry analysis demonstrated that the groups of O──H, C══O, C──O──C and C══O──C of the EPS involved in metal biosorption process and were the main functional groups for binding metal ions. Initial pH of the culture medium is known to play a key role in determining the electric charge of the cells together with the oxidation potential that affects the nutrient absorption and enzymatic action (Xia *et al.* 2008).

Conclusions

- Bio-flocculants produced by bacteria isolated from industrial sludge are capable of removing heavy metals simultaneously and effectively however, the removal efficiencies were initial concentration dependant.

- Bio-flocculants produced from *Bacillus sp.* and *Pseudomonas sp.* with a concentration 0.1 mg/ml provided a maximum removal efficiency of 99.5% and 97.9% for Pb^{2+} & 83.8% and 68% for Cu^{2+} respectively.

- The removal efficiency of Cd^{2+} by the bio-flocculants originated from *Pseudomonas sp.* was significantly higher (93.5%) than those achieved (72.9%) using *Bacillus sp.*

- These results demonstrate that bio-flocculants *Bacillus sp.* and *Pseudomonas sp.* would serve as a potential candidate for bioremediation of industrial wastewater containing heavy metals.

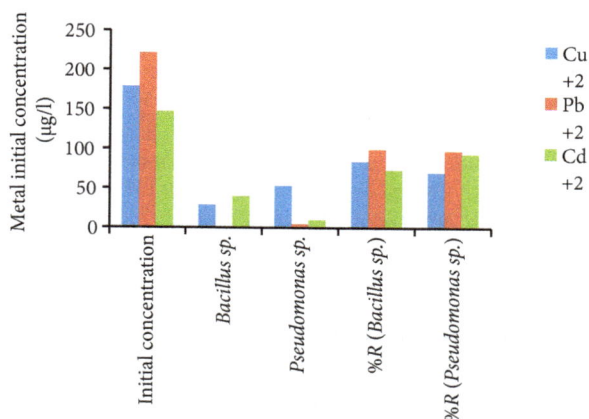

Fig. 4. Comparison between the efficiency of *Bacillus sp.* and *Pseudomonas sp.* bio-flocculants for removal of Pb^{2+}, Cu^{2+} and Cd^{2+} at the same initial concentration

References

Aguilera, M.; Quesada, M. T.; Aguila, V. G. D.; Morillo, J. A.; Rivadeneyra, M. A.; Cormenzana, A. R.; Sanchez, M. M. 2008. Characterization of *Paenibacillus jamilae* strains that produce exopolysaccharide during growth on and detoxification of olive mill wastewaters, *Biouresource Technology* 99: 5640–5644. http://dx.doi.org/10.1016/j.biortech.2007.10.032

Batta, N.; Subudhi, S.; Lal, B.; Devi, A. 2013. Isolation of a lead tolerant noval bacterial species, *Achromobacter sp.* TL-3: assessment of bio-flocculant activity, *Indian Journal of Experimental Biology* 51: 1004–1011.

Bender, J.; Rodriguez, E. S.; Ekanemesang, U. M.; Phillips, P. 1994. Characterization of metal-binding bio-flocculants produced by the cyanobacterial component of mixed microbial mats, *Applied and Environmental Microbiology* 60 (7): 2311–2315.

Dignac, M. F.; Urbain, V.; Rybacki, D.; Bruchet, A.; Snidaro, D.; Scribe, P. 1998. Chemical description of extracellular polymers: implication on activated sludge flock structure, *Water Science and Technology* 38(3): 9–46. http://dx.doi.org/10.1016/s0273-1223(98)00676-3

Duguid, J. P.; Marmion, B. P.; Swain, R. H. A. 1975. *Medical microbiology: the practices of medical microbiology.* 12th ed. Vol. 11. Edinburgh, London and New York: Churchill Livingston, Medical Division of Longman Group Limited, 170–188.

Ellis, K. V. 1989. *Surface water pollution and its control.* Basingstoke: Macmillan Press Ltd. http://dx.doi.org/10.1007/978-1-349-09071-6

El-Sheekh, M. M.; El-Shouny, W. A.; Osman, M. E. H.; El-Gammal, E. W. E. 2005. Growth and heavy metals removal efficiency of *Nostoc muscorum* and *Anabaena subcylindrica* in sewage and industrial wastewater effluents, *Environmental Toxicology Pharmacology* 19(2): 357–365. http://dx.doi.org/10.1016/j.etap.2004.09.005

Feng, J.; Yang, Z.; Zeng, G.; Huang, J.; Xu, H.; Zhang, Y.; Wei, S.; Wang, L. 2013. The adsorption behaviour and mechanism investigation of Pb(II) removal by flocculation using microbial flocculant GA1, *Bioresource Technology* 148: 414–421. http://dx.doi.org/10.1016/j.biortech.2013.09.011

Gadd, G. M. 1990. Metal tolerance, in C. Edwards (Ed.). *Microbiology of extreme environments.* New York: McGraw-Hill, 178–210.

Gadd, G. M.; White, C. 1993. Microbial treatment of metal pollution – a working biotechnology, *Trends of Biochemistry Technology* 11: 353–359. http://dx.doi.org/10.1016/0167-7799(93)90158-6

Gana, J. M.; Ordóñez, R.; Zampini, C.; Hidalgo, M.; Meoni, S.; Isla, M. I. 2008. Industrial effluents and surface waters genotoxicity and mutagenicity evaluation of a river of Tucuman, Argentina, *Journal of Hazardous Materials* 155(3): 403–406. http://dx.doi.org/10.1016/j.jhazmat.2007.11.080

Gao, Q.; Zhu, X. H.; Mu, J.; Zhang, Y.; Dong, X. W. 2009. Using *Ruditapes philippinarum* conglutination mud to produce bio-flocculant and its applications in wastewater treatment, *Bioresource Technology* 100: 4996–5001. http://dx.doi.org/10.1016/j.biortech.2009.05.035

Guangyu, Y.; Thiruvenkatachari, V. 2003. Heavy metals removal from aqueous solution by fungus *Mucor rouxii*, *Water Research* 37(18): 4486–4496. http://dx.doi.org/10.1016/S0043-1354(03)00409-3

Guo, J.; Yu, J. 2014. Sorption characteristics and mechanisms of Pb (II) from aqueous solution by using bio-flocculant MBFR10543, *Applied Microbiology and Biotechnology* 98(14): 6431–6441. http://dx.doi.org/10.1007/s00253-014-5681-z

He, B.; Yun, Z.; Shi, J.; Jiang, G. 2013. Research progress of heavy metal pollution in China: Sources, analytical methods, status, and toxicity, *Chinese Science Bulletin* 58(2): 134–140. http://dx.doi.org/10.1007/s11434-012-5541-0

Heredia, J. B.; Martín, J. S. 2009. Removing heavy metals from polluted surface water with a tannin-based flocculant agent, *Journal of Hazardous Materials* 165: 1215–1218. http://dx.doi.org/10.1016/j.jhazmat.2008.09.104

Kotrba, P.; Doleková, L.; De Lorenzo, V.; Ruml, T. 1999. Enhanced bioaccumulation of heavy metal ions by bacterial cells due to surface display of short metal binding peptides, *Applied and Environmental Microbiology* 65: 1092–1098.

Kurane, R.; Hatamochi, K.; Kakuno, T.; Kiyohara, M.; Kawaguchi, K.; Mizono, Y.; Hirano, M.; Taniguchi, Y. 1994. Production of a bio-flocculant by *Rhodococcus ertythropolis* S-l grown on alcohols, *Bioscience, Biotechnology and Biochemistry* 58: 428–429. http://dx.doi.org/10.1271/bbb.58.428

Lin, J.; Harichund, C. 2011a. Industrial effluent treatments using heavy-metal removing bacterial bioflocculants, *Water Research Commission* 37(2): 265–270.

Lin, J.; Harichund, C. 2011b. Isolation and characterization of heavy metal removing bacterial bio-flocculants, *African Journal of Microbiology Research* 18 (5–6): 599–607.

Liu, W. J.; Yuan, H. L.; Yang, J. S.; Li, B. Z. 2009. Characterization of bioflocculants from biologically aerated filter backwashed sludge and its application in dying wastewater treatment, *Bioresource Technology* 100: 2629–2632. http://dx.doi.org/10.1016/j.biortech.2008.12.017

Lu, W. Y.; Zhang, T.; Zhang, D. Y.; Li, C. H.; Wen, J. P.; Du, L. X. 2005. A novel bio-flocculant produced by *Enterobacter aerogenes* and its use in defecating the trona suspension, *Biochemical Engineering Journal* 27: 1–7. http://dx.doi.org/10.1016/j.bej.2005.04.026

Macaskiey, L. E. 1990. An immobilized cell bioprocess for the removal of heavy metals from aqueous flows, *Journal of Chemical Technology and Biotechnology* 49: 357–379. http://dx.doi.org/10.1002/jctb.280490408

Nakata, K.; Kurane, R. 1999. Production of an extracellular polysaccaride bio-flocculant by *Klebsiella pneumonia, Bioscience, Biotechnology and Biochemistry* 63: 2064–2068. http://dx.doi.org/10.1271/bbb.63.2064

Nanda, M.; Sharma, D.; Kumar, A. 2011. Removal of heavy metals from industrial effluent using bacteria, *International Journal of Environmental Science* 2(2): 781–787.

Ogunfowokan, A. O.; Okoh, E. K.; Adenuga, A. A.; Asubiojo, O. I. 2005. An assessment of the impact of point source pollution from a university sewage treatment oxidation pond on a receiving stream – a preliminary study, *Journal of Applied Science* 5(1): 36–43. http://dx.doi.org/10.3923/jas.2005.36.43

Ozer, A.; Pirincci, H. B. 2006. The adsorption of Cd (II) ions on Sulphuric acid treated wheat bran, *Journal of Hazardous Materials* 13(2): 849–855. http://dx.doi.org/10.1016/j.jhazmat.2006.03.009

Rajaram, T.; Ashutost, D. 2008. Water pollution by industrial effluents in India: discharge scenario and case for participatory ecosystem specific local regulation, *Environmental Journal* 40(1): 56–69.

Rehman, A.; Shakoori, F. R.; Shakoori, A. R. 2008. Heavy metal resistant freshwater ciliate, *Euplotes mutabilis*, isolated from industrial effluents has potential to decontaminate wastewater of toxic metals, *Bioresource Technology* 99: 3890–3895. http://dx.doi.org/10.1016/j.biortech.2007.08.007

Rittle, K. A.; Drever, J. L.; Colberg, P. J. S. 1995. Precipitation of arsenic during bacterial sulfate reduction, *Geo-Microbiology Journal* 13: 1–11. http://dx.doi.org/10.1080/01490459509378000

Saithong, K.; Poonsuk, P. 2002. Biosorption of heavy metal by thermotolerant polymer producing bacterial cells and the bioflocculant, *Journal of Science and Technology* 24: 421–430.

Salehizadeh, H.; Shojaosadati, S. A. 2001. Extracellular biopolymeric flocculants-recent trends and biotechnological importance, *Biotechnology advances* 19(5): 371–385. http://dx.doi.org/10.1016/S0734-9750(01)00071-4

Shamim, S.; Rehman, A. 2012. Cadmium resistance and accumulation potential of *Klebsiella Pneumoniae* strain CBL-1 isolated from industrial wastewater, *Pakistan Journal of Zoology* 44(1): 203–208.

Shen, H.; Wang, Y. T. 1993. Characterization of enzymatic reduction of hexavalent chromium by *Escherichia coli* ATCC 33456, *Applied and Environmental Microbiology* 59: 3771–3777.

Shih, I. L.; Van, Y. T.; Yeh, L. C.; Lin, H. G.; Chang, Y. N. 2001. Production of a biopolymer flocculant from *Bacillus licheniformis* and its flocculation properties, *Bioresource Technology* 78(3): 267–272. http://dx.doi.org/10.1016/S0960-8524(01)00027-X

Shuhonga, Y.; Meipingb, Z.; Honga, Y.; Hana, W.; Shana, X.; Yanb, L.; Jihuia, W. 2014. Biosorption of Cu^{2+}, Pb^{2+} and Cr^{6+} by a novel exopolysaccharide from *Arthrobacter* ps-5, *Carbohydrate Polymers* 101(30): 50–56. http://dx.doi.org/10.1016/j.carbpol.2013.09.021

Totura, G. 1996. Innovative uses of specialty ion exchange resins provide new cost-effective options for metals removal, *Environmental Progress* 15(3): 208–212. http://dx.doi.org/10.1002/ep.670150322

Vijayalakshmi, S. P.; Raichur, A. M. 2003. The utility of *Bacillus subtilis* as a bio-flocculant for fine coal, *Colloids Surfaces* 29: 265–275. http://dx.doi.org/10.1016/S0927-7765(03)00005-5

Wang, S. G.; Gong, W. X.; Liu, X. W.; Tian, L.; Yue, Q. Y.; Gao, B. Y. 2007. Production of a novel bio-flocculant by culture of *Klebsiella mobilis* using dairy wastewater, *Biochemical Engineering Journal* 36: 81–86. http://dx.doi.org/10.1016/j.bej.2007.02.003

Wu, J. Y.; Ye, H. F. 2007. Characterization and flocculating properties of an extracellular biopolymer produced from a *Bacillus subtilis* DYU1 isolate, *Process Biochemistry* 42: 1114–1123. http://dx.doi.org/10.1016/j.procbio.2007.05.006

Xia, S.; Zhang, Z.; Wang, X.; Yang, A.; Chen, L.; Zhao, J.; Leonard, D.; Jaffrezic-Renault, N. 2008. Production and characterization of bio-flocculant by Proteus mirabilis TJ1, *Bioresource Technology* 99: 6520–6527. http://dx.doi.org/10.1016/j.biortech.2007.11.031

Zheng, L.; Tian, Y.; Ding, A. Z.; Wang, J. S. 2008. Adsorption of Cd(II), Zn(II) by extracellular polymeric substances extracted from waste activated sludge, *Water Science and Technology* 58(1): 195–200. http://dx.doi.org/10.2166/wst.2008.646

Ahmed AZZAM, Dr, is an associate professor working at Environmental Researches Department, Theodor Bilharz Research Institute, P.O. Box 30-2411, Imbaba, Giza, Egypt. His research of interest lies in bioremediation, wastewater treatment technologies. He has published twelve papers in international journals and attended twenty two conferences.

Ahmed TAWFIK, Professor. He is working at Department of Environmental Engineering, Egypt-Japan University of Science and Technology (E-JUST); Egypt. His research of interests is wastewater treatment technologies and energy production from waste materials. He has published 63 papers in international peer reviewed journals and presented his research activities in 120 international conferences. He was awarded with several prizes in Environmental Science and Technology in 2014; 2013; 2012; 2010; 2007 and 2006.

6

AGROECOSYSTEMS TO DECREASE DIFFUSE NITROGEN POLLUTION IN NORTHERN LITHUANIA

Laura Masilionytė, Stanislava Maikštėnienė, Aleksandras Velykis, Antanas Satkus

Joniškėlis Experimental Station, Lithuanian Research Centre for Agriculture and Forestry,
39301 Joniškėlis, Lithuania

Abstract. The paper presents the research conducted at the Joniškėlis Experimental Station of the Lithuanian Research Centre for Agriculture and Forestry on a clay loam *Gleyic Cambisol* during the period of 2006–2010. The research investigated the changes of mineral nitrogen in soil growing catch crops during the winter wheat post-harvest period and incorporating their biomass into the soil for green manure. Green manure implications for environmental sustainability were assessed. The studies were carried out in the soil with a low (1.90–2.00%) and moderate (2.10–2.40%) humus content in organic and sustainable cropping systems. The crop rotation, expanded in time and space, consisted of red clover (*Trifolium pretense* L.) → winter wheat (*Triticum aestivum* L.) → field pea (*Pisum sativum* L.) → spring barley (*Hordeum vulgare* L.) with undersown red clover. Investigations of mineral nitrogen migration were assessed in the crop rotation sequence: winter wheat + catch crops → field pea. Higher organic matter and nitrogen content in the biomass of catch crops were accumulated when *Brassisaceae* (white mustard, *Sinapis alba* L.) was grown in a mixture with buckwheat (*Fagopyrum esculentum* Moench.) or as a sole crop, compared with oilseed radish (*Raphanus sativus* var. *Oleiferus* Metzg.) grown with the long-day legume plants blue lupine (*Lupinus angustifolius* L.). Mineral nitrogen concentration in soil depended on soil humus status, cropping system and catch crop characteristics. In late autumn there was significantly higher mineral nitrogen concentration in the soil with moderate humus content, compared with soil with low humus content. The lowest mineral nitrogen concentration in late autumn in the 0–40 cm soil layer and lower risk of leaching into deeper layers was measured using organic cropping systems with catch crops. The highest mineral nitrogen concentration was recorded in the sustainable cropping system when mineral nitrogen fertilizer (N_{30}) was applied for winter wheat straw decomposition. In the organic cropping system, the incorporation of catch crop biomass into soil resulted in higher mineral nitrogen reserves in soil in spring than in the sustainable cropping system, (mineral nitrogen fertilizer (N_{30}) applied for straw decomposition in autumn and no catch crop grown). Applying organic cropping systems with catch crops is an efficient tool to promote environmental sustainability.

Keywords: environmental sustainability, low and moderate soil humus content, organic and sustainable cropping systems, catch crops, straw, biomass, mineral nitrogen.

Introduction

In intensive agriculture, mineral fertilization intended to meet plant nutritional needs poses a threat to ecological balance. Much research has been completed and recommendations have been presented about the best-suited plant fertilizer forms and rates, application timing and methods. However, there is little research into the effects of technologies on nutrients, especially nitrogen (N) immobilization in soil and leaching after harvesting of the main crops, when the soil during the post-harvest period stays bare for a prolonged period. Intensive fertilization results in significant increases in mineral N in soil in autumn, which poses a threat of groundwater contamination (Shevtsova *et al.* 2003). The largest quantities of N are leached from light soils. However, in heavy soils, conditions are created for N migration into deeper layers and groundwater contamination (Arlauskienė *et al.* 2011). Contributory factors include increased stickiness resulting in the appearance of vertical rills during crop growing season and due to low soil permeability resulting in higher runoff from ploughed soil. Nutrient leaching is also promoted by uneven distribution of rainfall during the growing season and more frequent downpours fuelled by global warming. Mineral nitrogen accumulation and dynamics in soil depend on soil texture and humus content, weather during the growing season and soil and crop management technologies, especially the fertilizers used (mineral and organic), their rates and application methods (Tonitto *et al.* 2006; Diacono, Montemurro 2010). However, even organic agriculture poses some risk of nitrate leaching, due to abundant accumulation of organic matter and mineralization (Loges *et al.* 2006; Torstensson *et al.* 2006).

Seeking to alleviate environmental pollution, it is important that appropriate preventive measures are chosen for inclusion of nutrients which are not utilized by plants into biological cycling (Di *et al.* 2002). For this, technologies involving catch crops that accumulate nutrients remaining in the soil during the most intensive leaching period in the end of summer-autumn (winter) and prevents them from leaching, are widely used in West Europe (Dawson *et al.* 2008; Torstensson *et al.* 2006; Komatsuzaki, Ohta 2007; Möller *et al.* 2008). Moreover, cover crop management, manure application and no-tillage practices during the cereal post-harvesting period reduce evaporation and CO_2 and N_2O emission from humus-rich clay soils (Komatsuzaki, Ohta 2007). Rapid and constant cover of soils with crops during the growing season and autumn protects it from the negative effects of direct atmospheric phenomena (Tonitto *et al.* 2006; Fullen *et al.* 2011).

It is suggested that under favourable weather conditions the biomass of catch crops incorporated into the soil starts to decompose in autumn (Lahti, Kuikman 2003). The higher the decomposition rate, the more mineral N is accumulated in soil (Vinther *et al.* 2004; Tripolskaja 2005; Stadler *et al.* 2005). If it is not bound into soil organic compounds or not utilized by growing plants, losses may occur during the winter-early spring period (Hasegawa, Denison 2005). The rate of mineralization of organic matter incorporated into the soil and N losses can be reduced by delaying the incorporation of catch crops biomass, by cultivating winter catch crops (whose choice is rather limited today) or by leaving catch crops as a frozen mulch over winter and incorporating it in spring (Crandall *et al.* 2005; Larsson *et al.* 2005). Researchers have reported that with straw incorporated together with N-rich catch crop biomass binds excess mineral N in soil, restores soil humus, increases the amount of stable humic substances and improves aeration (Arlauskienė *et al.* 2010). The technique may also improve soil physical state. Incorporation results in improved structure and increased amount of water stable soil aggregates, which is especially relevant for soils containing high silt contents (Velykis, Satkus 2008).

There are contrasting opinions that, in the case of continuous cultivation of winter cereals, during an average five-year period, the content of nitrates in the soil solution was found to be lower compared with the treatments grown with cereals alternating in growing season with catch crops (Catt *et al.* 1998). Research at the Lithuanian Institute of Agriculture's Joniškėlis Experimental Station on a fine-textured soil established that having incorporated N-rich biomass of lucerne and red clover as green manure, early in spring the soil 0–40 cm layer had a low mineral N (30–40 kg ha^{-1}) content. However, a significantly higher cereal grain yield was obtained (Arlauskienė, Maikštėnienė 2010; Arlauskienė *et al.* 2011), which suggests slower mineralization of organic fertilizers incorporated into fine-textured soils.

When making the shift from intensive cropping to alternative systems and replacement of fertilization systems it is important to quantify changes in soil productivity parameters. In alternative cropping systems, plant demand for major nutrients is compensated by soil reserves and nutrients released from organic matter. Thus, the present research was designed to ascertain the effects of catch crops, cultivated in the winter wheat post-harvest period and the impact of their biomass incorporated for green manure on mineral N dynamics in autumn and spring in clay loam *Cambisol* in organic and sustainable cropping systems. The impacts of these changes on environmental sustainability were assessed.

1. Material and methods

Experimental site and soil. Field experiments were conducted at the Joniškėlis Experimental Station of the Lithuanian Research Centre for Agriculture and Forestry (LRCAF). The station is situated in the northern part of Central Lithuania's lowland (40–60 m above the sea level; latitude: 56°12" N; longitude: 24° 20"). Northern Lithuania has a climate mid-way between maritime and continental. The climate is changeable, with mild, wet summers and cold winters. Annual precipitation is 500–600 mm. The soil of the experimental site is *Endocalcari–Endohypogleyic Cambisol*, whose texture is clay loam on silty clay with deeper lying sandy loam. The parent material is glacial lacustrine clay, which at 70–80 cm depth transits into morainic loam. Clay (<0.002 mm) in the Ap horizon (0–30 cm) account for 27.0%, in the B$_w$ horizon (31–51 cm) 59.6%, in the B$_k$ horizon (52–76 cm) 51.6%, in the C$_1$ horizon (77–105 cm) 10.7%, in the C$_2$ horizon (106–135 cm) 11.0%. Soil bulk density in the plough layer (0–25 cm) is 1.3–1.4 Mg m^{-3}, total porosity is 40–45%, and air-filled porosity is 8–10%. Investigations were performed in soil with the following chemical characteristics in the plough layer: pH$_{KCl}$ 6.6, available P$_2$O$_5$ in the soil low in humus was 75–101, in the soil moderate in humus 111–134 mg kg^{-1}. The corresponding values for available K$_2$O were 207–235 and 221–240 mg kg^{-1}, respectively.

Field experiment. The field experiment was arranged according to the following design:

Soil humus content – *factor A*:
1) *low* (1.90–2.01%); 2) *moderate* (2.10–2.40%).

Cropping systems – *factor B*:

Organic I (O I): red clover biomass as green manure was applied for winter wheat, winter wheat straw was applied as manure and the catch crop (blue lupine, *Lupinus angustifolius* L.) in a mixture with oilseed radish (*Raphanus sativus* var. *Oleiferus* Metzg.) was cultivated as green manure for field pea.

Organic II (O II): red clover biomass as green manure and farmyard manure (40 Mg ha^{-1}) was applied for winter wheat, winter wheat straw was applied as manure and white mustard (*Sinapis alba* L.) was cultivated as green manure for field pea.

Sustainable I (S I): farmyard manure (40 Mg ha^{-1}) was applied for winter wheat, winter wheat straw was applied as manure + mineral N fertilizer (N$_{30}$) in the form of ammonium nitrate for straw decomposition in autumn and white mustard in a mixture with buckwheat (*Fagopyrum esculentum* Moench.) was cultivated as green manure for field pea.

Sustainable II (S II): red clover biomass as green manure + mineral fertilizer N$_{30}$P$_{60}$K$_{60}$ was applied for winter wheat, winter wheat straw as manure + mineral N fertilizer (N$_{30}$) for straw decomposition in autumn and mineral fertilizer N$_{10}$P$_{40}$K$_{60}$ in spring was applied for field pea.

The field experiment was arranged as a randomized single row design in four replicates. The crop rotation, expanded in time and space, and consisted of red clover (*Trifolium pretense* L.) cv. 'Vyliai' → winter wheat (*Triticum aestivum* L.) cv. 'Ada' → field pea (*Pisum sativum* L.) cv. 'Pinochio' → spring barley (*Hordeum vulgare* L.) cv. 'Luoke' with undersown red clover. The investigated cropping systems were assessed in winter wheat + catch crops (for green manure) → field pea.

In the sustainable II cropping system, in autumn, before sowing mineral fertilizer P$_{60}$K$_{60}$ was applied for winter wheat, and in spring after resumption of winter wheat growth, mineral N fertilizer (N$_{30}$) was applied in the form of ammonium nitrate. After winter wheat harvesting, the stubble was broken by a combined breaker and chopped straw was simultaneously incorporated into the soil. In the sustainable system I and sustainable system II, winter wheat straw decomposition was promoted with the addition of mineral N fertilizer (N$_{30}$) in the form of ammonium nitrate. Catch crops were sown: a mixture of blue lupine (cv. 'Boruta') and oilseed radish (cv. 'Rufus') at a seed rate of 150 and 20 kg ha^{-1}, respectively; white mustard (cv. 'Sinus') at 22 kg ha^{-1}, and when grown in mixture with buckwheat (cv 'Smuglianka') at 15 and 80 kg ha^{-1}, respectively.

Plant analyses. Biomass of catch crops was established by weighing. Dry matter (DM) content in the biomass was determined by drying samples at 105 °C to constant weight. The yield of catch crop biomass is presented in absolutely dry matter Mg ha^{-1}. Analyses of N content were performed on samples taken from the catch crop biomass before incorporation into the soil. The biomass of catch crops was analysed for the concentration (%) of N using the standard macro-Kjeldahl procedure (Koutroubas *et al.* 2008). N content was recalculated in kg ha^{-1}. Chemical analyses of plants were made at the Agrochemical Research Laboratory of the LRCAF.

Soil chemical properties. Soil samples for the determination of soil chemical characteristics of the plough layer (0–25 cm) were collected before the establishment of the experiment. Available P$_2$O$_5$ and K$_2$O in the soil were determined by ammonium lactate extraction Egner-Riem-Domingo (A-L), and pH$_{KCl}$ by electro-potensiometric methods. The effect of the investigated measures on mineral N dynamics was estimated in autumn before catch crops biomass incorporation by ploughing and in the spring of the following year before the sowing of field pea. Twenty soil samples were used for the determination of changes in mineral N within each plot. Each of the 20 samples was taken from different places within the plot from the 0–40 cm depth and composite samples were formed. Mineral N was determined: N-NH$_4$ by the spectrophotometric method and N-NO$_3$ by the ionometric method. Soil chemical analyses were made at the Chemical Research Laboratory of the Institute of Agriculture of LRCAF.

Meteorological conditions. The mean daily air temperature and precipitation are presented for more important experimental periods (Fig. 1). In 2006, during the growing season of the main crops (May–July), when cereals grow intensively and utilize nutrients from the soil, there was also a moisture deficit. The rainfall that fell during that period accounted for only 38.9% of the long-term mean (Table 1). Meanwhile, the mean daily temperature in June and July was higher by 0.7 and 3.7 °C, respectively. The yield of the main crops was low. However, the growing season of catch crops (August–October 2006) was the most favourable for catch crop growth compared with the other growing seasons of catch crops. Minimum daily temperature dropped <10 °C only in the second half of September. Moreover, in August, September and October the mean daily air temperature exceeded the long-term mean by 1.5, 2.9 and 3.1 °C, respectively. During the growing season of catch cops, precipitation exceeded the long-term mean by 85.5 mm, and there was ample rainfall in August and October. The period from November 2006 to April 2007 was warmer than usual, except for February. The mean daily temperature of the months was >0 °C. Precipitation amount accorded with the long-term mean. Only in

January there was relatively more precipitation. Early in spring (March and April) there was little precipitation.

After the dry year of 2006, the growing season of the main crops (May–July 2007) was relatively warm and wet. Thus, conditions were especially favourable for cereal grows. Catch crops sown after cereal harvesting grew poorly. In August, September, October the mean daily temperature was higher than the long-term mean, the minimum daily air temperature dropped <10 °C by late August and persisted all through September, although the days were relatively warm. Poor plant emergence and establishment were determined by the droughty first half of August, and heavy rain only fell on August 20. The period from November 2007 to April 2008 was one of the warmest and wettest. The winter months were especially warmer (by 4.2–6.9 °C compared with the long-term mean). Spring months (March and April) were distinguished by higher precipitation amounts (when the amount of precipitation exceeded the long-term mean by 16.8 and 17.2 mm, respectively).

In 2008, the mean daily air temperature during the main crop growing season differed little from the long-term mean. However, this period was one of the driest recorded. Plants were already short of moisture at the early growth stages (May and June). This impeded plant nutrient uptake from the soil. The growing season of catch crops was rather wet. However, precipitation was distributed very unevenly. In August rainfall exceeded the long-term mean by 48.6 mm, and in October by 29.3 mm. September was extremely dry (rainfall was only 6.5 mm).

The period from November 2008 to April 2009 was 1.8 °C warmer than the long-term mean. However, it was wet. Much precipitation fell in December and March (63.1 and 51.7 mm, respectively). Dry weather in April and May 2009 suppressed the germination and establishment of the main crops. However, June and July were very wet. October was cooler, with rainfall 27.1 mm more that the long-term mean. In 2009, November and December were wet with rainfalls of 11.2 and 23.8 mm, respectively, higher that the long-term mean.

Winter 2010 was cold. In January, air temperature was 6.6 °C lower and precipitation was 14.7 mm lower that

Table 1. Precipitation and long-term (30 years) mean air temperature

Months	April	May	June	July	August	September	October
Temperature °C	6.2	12.3	15.6	17.1	17.1	12.0	6.3
Rainfall mm	37.4	45.6	59.4	69.2	67.9	57.9	45.5

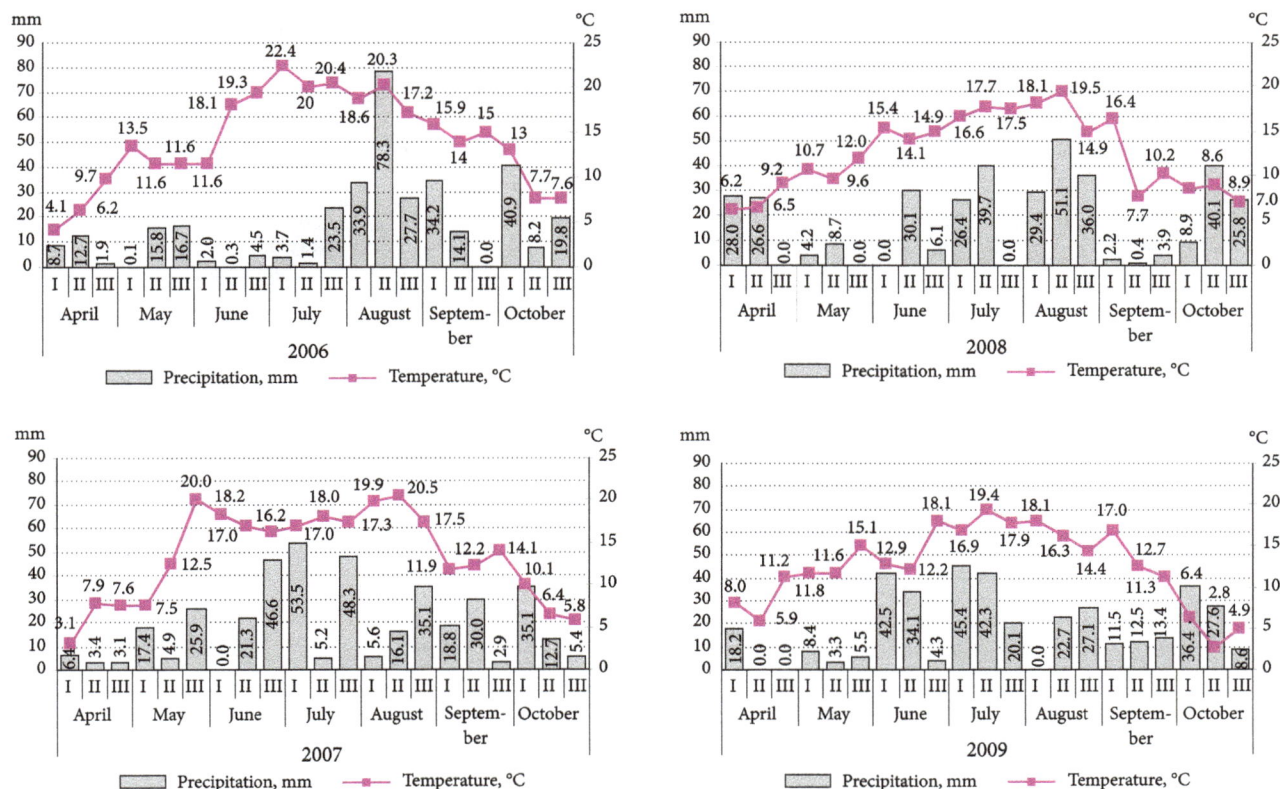

Fig. 1. Precipitation and mean air temperature during the experimental periods

the long-term mean. In spring, the mean daily temperature was close to the long-term mean, but rainfall was higher in March and May (by 2.5 and 23.7 mm, respectively). The high amount of rainfall had negative affects throughout the 2010 growing season.

Statistical analysis. The research data were statistically processed by a two-factor analysis of variance and correlation-regression analysis using the program package Selekcija (Tarakanovas, Raudonius 2003), as well as "R" statistical software after Crawley (2007).

2. Results and discussion

Dry matter yield of catch crops. In the experiment established in 2006, there was a moisture deficit during the growing season of the main crop (winter wheat). Thus, most nutrients in soil were not utilized by the crop. After cereal harvesting, during the warm and wet autumn, catch crops grew rapidly by effectively utilizing nutrients

left in the soil. In the soil with low and moderate humus content, when the conditions for catch crop growth were favourable, the largest dry matter yield was produced in the organic cropping system by short-vegetation plants (white mustard). The dry matter yield was 16.7 and 76.5% more, respectively, than the longer-vegetation blue lupine combined with oilseed radish and 5.6 and 11.8% more, respectively, than in the sustainable cropping system of white mustard combined with buckwheat (Fig. 2).

In 2007, due to the rainy and warm period, the main crops developed intensively and thus utilized most available nutrients. Moreover, the post-harvest period was droughty. Therefore all catch crops accumulated little dry matter yield, with a mean of ~0.25 of that produced in 2006. In the soil low in humus, the largest dry matter yield was accumulated by the combined white mustard-buckwheat sustainable cropping system. Yield was 40.0% more than white mustard grown as a sole crop. Such results might have been influenced by the low mineral N

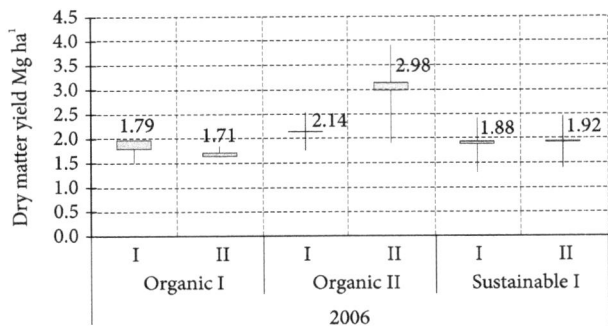

LSD$_{05}$: 0.478

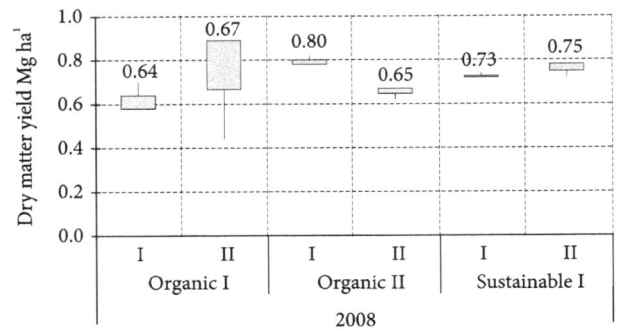

LSD$_{05}$: 0.079

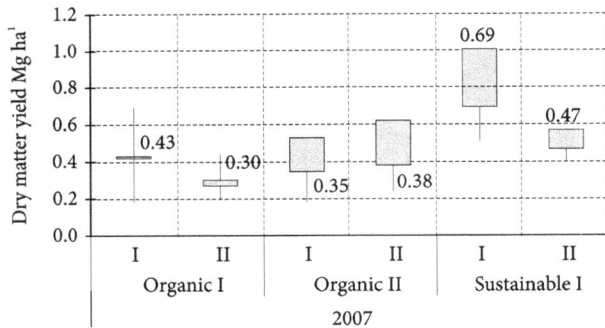

LSD$_{05}$: 0.122

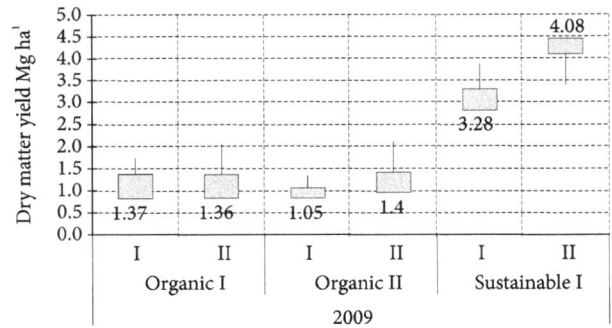

LSD$_{05}$: 0.387

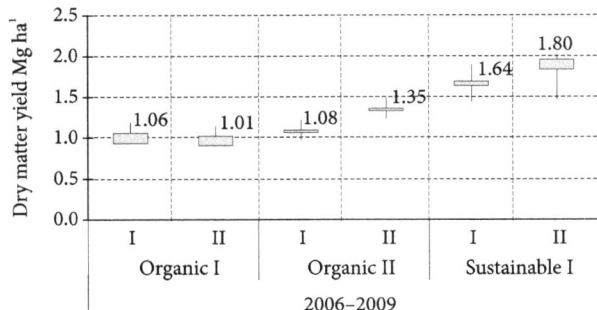

LSD$_{05}$: 0.188

Fig. 2. Dry matter yield of catch crops

Notes: I – Standard error;
I – low humus content in the soil;
II – moderate humus content in the soil.

fertilizer (ammonium nitrate) rate (N_{30}) applied to promote winter wheat straw decomposition and by the buckwheat specific root system due to which it can assimilate more varied nutrient forms.

The literature indicates that although buckwheat roots mass is much smaller than that of some cereals, its sucking power can be 2.7–14.0 times greater, due to its long root hair. Moreover, buckwheat roots exude formic, acetic and citric acids capable of dissolving stable compounds present in the soil, especially those of phosphorus (P), which is of special relevance in fine-textured soil deficient in P (Marcinkonis *et al.* 2007).

In the soil with moderate humus content the same trend of catch crop dry matter yield variation persisted. The largest dry matter yield was accumulated in the sustainable cropping system with the white mustard–buckwheat combination. This was 25.0% more than in the organic cropping system where only white mustard was grown. In 2006, in the soil low and moderate in humus content, much greater dry matter yield was accumulated by white mustard grown in combination with buckwheat compared with that of mustard grown as a sole crop. These results were determined in the sustainable cropping system by mineral N fertilizer (N_{30}) applied to promote straw decomposition and which created better establishment nutritional conditions for catch crops during the droughty post-harvest period than for those grown in the organic cropping system. Moreover, in the year less favourable for catch crop growth, higher crop productivity was achieved when growing them in binary agroecosystems (mixtures).

The literature indicates that plants possessing biologically different root nutrition assimilate nutrients at different rates and intensities, which enables plants to better exploit various soil and environmental conditions (Möller *et al.* 2008). Blue lupine grown in the organic cropping system combined with oilseed radish, as a crop with a longer growing season, develop more slowly and, due to the small leaf area, accumulate a low content of assimilates. Thus, although lupine fixed atmospheric N when grown in mixture, it accumulated less dry matter yield than white mustard-buckwheat mixture.

In the 2008, although the growing season of the main crop (winter wheat) was dry, it removed little nutrients from the soil. However, in rainy August leaching was pronounced, since nearly double the monthly mean amount of rain fell. Catch crops developed poorly and, due to the prolonged drought, dry matter yield was small. However, the mean yield was 1.6 times larger compared with catch crops grown in 2007. In the soil with low humus content, more dry matter yield accumulation was noted in the organic cropping system for white mustard grown as a sole crop or in the sustainable cropping system grown in combination with buckwheat. Their dry matter yields, compared with that of the blue lupine-oilseed radish mixture were 33.3 and

16.7%, respectively, higher. In the soil with moderate humus content, there were no significant differences between the dry matter yield accumulated by catch crops grown in the organic and sustainable cropping systems. The differences might have been alleviated by better nutritional conditions.

In the 2009 growing year, which was similar in weather respect to the 2007, analysis of catch crop dry matter yield showed that markedly higher biomass contents accumulated in the sustainable I cropping system when growing white mustard-buckwheat compared with the organic II system when only growing white mustard. The difference in low humus soil was 2.4-fold and in moderate humus soil was 2.9-fold.

According to the mean data of 2006–2009, analysis of catch crop dry matter yield during the winter wheat (field pea pre-crop) post-harvest period showed that markedly higher dry matter yield accumulated in the sustainable I cropping system when growing white mustard mixed with buckwheat compared with the organic II system when growing only white mustard. The difference in low humus content was 54.7% and in moderate humus content soil was 33.3%. Such results may have been influenced both by the biological properties of catch crops and the low N rate (N_{30}) applied in the sustainable I cropping system for straw mineralization, which promoted catch crop development. The least dry matter yield of catch crops in the soil with low and moderate humus content was accumulated when growing blue lupine mixed with oil radish.

Analysis of the data from separate years suggests the catch crop white mustard grown as a sole crop or mixed with buckwheat under the conditions of cereal post-harvest period produced greater dry matter yields compared with the blue lupine-oilseed radish system. The data showed that the complex effects of environmental factors in Lithuania not favourable for post-harvest catch crop growth and development every year. By late summer with days becoming shorter lighting of plant parts actively participating in the process of photosynthesis decreases along with above-ground mass and roots. Consequently, catch crop development and accumulation of nutrients in biomass is largely determined by weather conditions (temperature, precipitation amount and distribution, solar radiation intensity and cloudiness). Due to unfavourable conditions during the initial growth stages, catch crops accumulate less biomass, and later with days becoming shorter, they form generative organs (Mattson, Erwin 2005; Odhiambo, Bomke 2007).

Having applied catch crops after winter wheat, fertilizer efficacy differed between years. Efficacy was related to rainfall during the growing season, which influenced the mineralization of organic fertilizers. The strength of the correlation between catch crop dry matter yield and rainfall amount was related to the soil humus status and fertilizer applications.

In the soil with low humus content, in the organic I and II cropping systems, the relationship between catch crop dry matter yield and precipitation was stronger (r = 0.628** and 0.871**, respectively) than that in the soil with moderate humus content (r = 0.616** and 0.815**, respectively; all n = 16, ** = P < 0.01). Consequently, results were affected by the application of manures in the organic cropping systems. In turn, mineralization rate and nutrient release for plant nutrition largely depend on soil water content. With the prolonged dry growing season, manures release few nutrients and thus have little influence on catch crop yield.

In the soil with both low and moderate humus contents, in the sustainable I cropping systems, incorporated N_{30} had positive effects on catch crop growth and development. Therefore, there was no significant correlation between dry matter yield of catch crops and rainfall.

Table 2. Nitrogen concentration and nitrogen uptake in the above-ground biomass of selected catch crops

Cropping system (Factor B)	Catch crops and fertilization	Soil humus content (Factor A)				Mean factor B	
		low		moderate			
		concentration g kg⁻¹	uptake kg ha⁻¹	concent-ration g kg⁻¹	uptake kg ha⁻¹	concent-ration g kg⁻¹	uptake kg ha⁻¹
				2006			
Organic I	blue lupine + oilseed radish	34.3	61.6	35.9	61.3	35.1	61.4
Organic II	white mustard	37.6	80.6	34.5	102.7	36.1	91.7
Sustainable I	white mustard + buckwheat + N_{30}	36.2	67.6	37.3	71.0	36.7	69.3
Sustainable II	N_{30}	–	–	–	–	–	–
	Mean factor A	36.0	69.9	35.9	78.3	36.0	74.1
	LSD$_{05}$: A – 1.71; B – 2.09; AB – 2.96						
				2007			
Organic I	blue lupine + oilseed radish	20.0	9.8	19.4	4.1	19.7	6.9
Organic II	white mustard	21.9	7.7	20.1	7.2	21.0	7.5
Sustainable I	white mustard + buckwheat + N_{30}	20.2	13.3	22.2	10.3	21.2	11.7
Sustainable II	N_{30}	–	–	–	–	–	–
	Mean factor A	20.7	10.2	20.6	7.2	20.6	8.7
	LSD$_{05}$: A – 2.57; B – 3.15; AB – 4.45						
				2008			
Organic I	blue lupine + oilseed radish	23.0	14.8	21.2	10.9	22.1	12.8
Organic II	white mustard	17.5	13.9	19.3	12.4	18.4	13.2
Sustainable I	white mustard + buckwheat + N_{30}	20.6	14.9	25.4	18.4	23.0	16.7
Sustainable II	N_{30}	–	–	–	–	–	–
	Mean factor A	20.3	14.5	22.0	13.9	21.2	14.2
	LSD$_{05}$: A – 2.62; B – 3.21; AB – 4.54						
				2009			
Organic I	blue lupine + oilseed radish	22.5	31.4	21.3	29.3	21.9	30.4
Organic II	white mustard	17.8	18.6	16.7	23.9	17.3	21.2
Sustainable I	white mustard + buckwheat + N_{30}	20.5	69.1	15.6	72.1	18.1	70.6
Sustainable II	N_{30}	–	–	–	–	–	–
	Mean factor A	20.3	39.7	17.9	41.8	19.1	40.7
	LSD$_{05}$: A – 1.48; B – 2.10; AB – 3.31						
				Mean 2006–2009			
Organic I	blue lupine + oilseed radish	25.0	29.4	24.5	26.4	24.7	27.9
Organic II	white mustard	23.7	30.2	22.7	36.6	23.2	33.4
Sustainable I	white mustard + buckwheat + N_{30}	24.4	41.2	25.1	43.0	24.8	42.1
Sustainable II	N_{30}	–	–	–	–	–	–
	Mean factor A	24.3	33.6	24.1	35.3	24.2	34.5
	LSD$_{05}$: A – 8.639; B – 12.217; AB – 19.316						

Nitrogen uptake in the biomass of catch crops. Experiments conducted over four years showed that in different cropping systems N concentration in the above-ground biomass of catch crops during the post-harvest period was highest in 2006. This proved favourable for catch crop growth. In the soil with moderate humus content, the highest N concentration in the above-ground biomass was established in the sustainable cropping system. In the white mustard-buckwheat mixture, in the soil low in humus, the data were inconsistent (Table 2). In 2006, during the favourable post-harvest period, catch crops intensively used the remaining nutrients and produced much biomass and accumulated high N contents. In the soil low and moderate in humus content, the highest N content accumulated in the white mustard grown as a sole crop was 30.8 and 67.5%, respectively, more than the organic I cropping system, when growing blue lupine mixed with oilseed radish. In the sustainable I cropping system, a low mineral N fertilizer rate (N_{30}) did not have any significant effect on white mustard and buckwheat biomass N content.

In 2007, during the main crop post-harvest period, during the catch crop growing season, there was little rainfall in August and September; 11.1 and 6.2%, respectively, less than the long-term mean, both in the organic and sustainable cropping systems catch crops developed poorly and accumulated little biomass and N.

In soil with both low and moderate humus contents, N concentration in above-ground biomass of catch crops and their mixtures varied from 19.4–22.2 g kg^{-1}. There was a slightly higher N uptake in the sustainable cropping system in the white mustard-buckwheat combination, where mineral N fertilizer (N_{30}) had been applied. The largest amount of N was incorporated in the sustainable cropping system with the biomass of catch crops. In the white mustard and buckwheat mixture in soil with low and moderate humus N was 35.7% and 2.5 times more, respectively, than when growing blue lupine with oilseed radish.

In 2008, after sowing catch crops, at the seedling stage, heavy rainfall in August (almost double the mean monthly total at 116.5 mm) caused high N leaching. The dry September had a negative effect on catch crop development and their biomass accumulated little N. In soil with both low and moderate humus contents, white mustard grown as a catch crop in the organic cropping system had the lowest N concentration. This was reflected in the lower total N uptake in biomass. During the droughty, unfavourable period for catch crop growth, higher N uptake was noted for the white mustard-buckwheat mixture. This was reflected in significantly higher (68.8%) N content in blue lupine and oilseed radish in soil with moderate humus content. It is consistent that catch crop biomass and N content were markedly higher in the sustainable I cropping system, in which N fertilizer N_{30} was applied. September 2009 had a prolonged warm spell. Buckwheat developed intensively, thus the largest amount of N accumulated in the biomass of white mustard-buckwheat mixture. In the soil with low and moderate humus contents, N was by 2.2 and 2.5 times more, respectively, than blue lupine mixed with oilseed radish and 3.7 and 3.0 times more, respectively, than only white mustard.

During the warm and favourable growing season for catch crops in 2006 and 2009, when soil organic matter (SOM) mineralization was intensive, higher N uptake in biomass was noted in the organic cropping system. This was especially the case in soil with moderate humus content, for white mustard as short-season catch crop.

According to the mean data of 2006–2009, the biomass of the white mustard-buckwheat mixture grown in the soil with low and moderate humus contents had the highest N contents, compared with white mustard as a sole crop. The difference was 36.4 and 19.4% more, respectively. These data indicate that in the soil low in humus content, in poor nutrition conditions, the positive effects of catch crop biological characteristics manifested themselves more tangibly.

In the sustainable I cropping system, in soil with low and moderate humus contents, the biomass of the white mustard-buckwheat mixture contained 40.1 and 62.9%, respectively, more N than the biomass of blue lupine-oil radish mixture in the organic I cropping system. Although blue lupine fix N from the atmosphere and is superior to *Brassicaceae* family plants in organic agrosystems, according to its genetic origin it is a long-day plant. Therefore shortening days in autumn have marked negative impacts.

Higher soil humus status in most cases promoted more intensive N uptake in the biomass of catch crops. However, differences were not significant compared with the soil low in humus content. However, in dry periods in the growing season, more N was accumulated by binary catch crops. Usually, the white mustard-buckwheat mixture was superior to the blue lupine-oilseed radish mixture. During autumn, catch crops can reduce nitrate losses in soil by 10–20 kg ha^{-1}, by accumulating N in biomass (Xu et al. 2006; Zhao et al. 2012). The more N-rich the biomass incorporated into the soil is, the greater the risk of nitrate leaching (Vinther et al. 2004; Möller et al. 2008).

Mineral nitrogen content in soil before catch crop incorporation for green manure by ploughing in autumn. With climate warming and increasing risks of N leaching, after growing the main crops, it is vital to estimate N uptake in soil and plant residues and to apply appropriate crop and soil management practises (Ergstöm, Linden 2009). In autumn, cultivation of catch crops and incorporation of their biomass (green manure) makes it possible to include mineral N into biological nutrient cycling, retain it in the plough layer and prevent leaching. Hao et al. (2001) suggested leaving straw on the soil surface over winter and incorporating into soil only in spring, like mineral N fertilizer, otherwise nitrous oxide emission will increase.

In experiments (2006), catch crops produced much biomass. However, in warm and normally wet autumns, organic matter mineralization was intensive and before autumn ploughing, mineral N concentration in soil (0–40 cm depth) remained rather high and varied from 8.07–10.67 mg kg^{-1} soil (Table 3). These results were influenced by the dry growing season of the main crops and the sufficiently wet and warm growing season of catch crops in 2006. After harvest of the main crops, intensive SOM mineralization processes in soil caused comparatively high mineral N release.

Significantly more (mean 1.21 mg kg^{-1} soil or 13.8%) mineral N was present in soil with moderate humus content compared with soil with low humus content. When estimating only catch crops, it was found that residual mineral N from the soil was utilized less by white mustard sole crop. When growing white mustard only, mean mineral N concentration in soil with moderate and

Table 3. Effects of catch crops and applications of mineral nitrogen fertilizer on the mineral nitrogen content in soil before autumn ploughing, mg kg^{-1}

Cropping system (Factor B)	Catch crops and fertilization	Soil humus content (Factor A)		Mean factor B
		low	moderate	
		2006		
Organic I	blue lupine + oilseed radish	8.80	10.19	9.50
Organic II	white mustard	8.07	8.80	8.44
Sustainable I	white mustard + buckwheat + N$_{30}$	8.59	10.21	9.40
Sustainable II	N$_{30}$	9.59	10.67	10.13
	Mean factor A	8.76	9.97	9.37
	LSD$_{05}$: A – 1.044; B – 1.476; AB – 2.088			
		2007		
Organic I	blue lupine + oilseed radish	6.38	6.00	6.21
Organic II	white mustard	5.29	7.32	6.30
Sustainable I	white mustard + buckwheat + N$_{30}$	6.07	8.44	7.27
Sustainable II	N$_{30}$	7.42	8.64	8.03
	Mean factor A	6.30	7.61	7.00
	LSD$_{05}$: A – 0.318; B – 0.450; AB – 0.636			
		2008		
Organic I	blue lupine + oilseed radish	3.39	3.89	3.64
Organic II	white mustard	3.33	4.50	3.91
Sustainable I	white mustard + buckwheat + N$_{30}$	4.05	4.90	4.47
Sustainable II	N$_{30}$	6.89	7.57	7.23
	Mean factor A	4.41	5.21	4.81
	LSD$_{05}$: A – 0.685; B – 0.968; AB – 1.369			
		2009		
Organic I	blue lupine + oilseed radish	3.31	3.51	3.41
Organic II	white mustard	3.44	3.32	3.38
Sustainable I	white mustard + buckwheat + N$_{30}$	3.43	3.58	3.51
Sustainable II	N$_{30}$	4.14	4.15	4.15
	Mean factor A	3.58	3.64	3.61
	LSD$_{05}$: A – 0.103; B – 0.178; AB – 0.272			
		Mean 2006–2009		
Organic I	blue lupine + oilseed radish	5.47	5.90	5.69
Organic II	white mustard	5.03	5.99	5.51
Sustainable I	white mustard + buckwheat + N$_{30}$	5.54	6.78	6.16
Sustainable II	N$_{30}$	7.01	7.76	7.39
	Mean factor A	5.76	6.61	6.19
	LSD$_{05}$: A – 0.198; B – 0.343; AB – 0.524			

low humus content was significantly less (11.2 and 10.2%, respectively), compared with mixtures of blue lupine- oilseed radish and white mustard-buckwheat. The highest soil mineral N content was in plots where to promote straw decomposition N_{30} was applied in the form of ammonium nitrate and no catch crops were grown. In 2007, during the main crop post-harvest period, catch crops developed and grew poorly, accumulated little biomass and used little soil N. Before the incorporation of catch crop biomass in soil the mean mineral N content was 25.3% less than 2006, in soils with both low and moderate humus contents. However, the main patterns of mineral N variation remained the same as in the favourable year (2006). Significantly more mineral N (mean 20.8%) was in the soil with moderate humus content compared with the soil with low humus content. This trend might have been influenced by more intensive organic matter mineralization processes in soil with more humus. The highest content of mineral N before autumn ploughing was in the sustainable II cropping system. In this system, straw mineralization was promoted by applying N_{30} in the form of ammonium nitrate and no catch crops were grown. However, significant mineral N differences in the sustainable II compared with the organic I system were only identified in the soil with a low humus content. In the soil with moderate humus content, the difference was alleviated by higher contents of N compounds released during the decomposition of organic matter. These data disagree with Bučienė (2003), who suggested that N applied for straw decomposition is effectively utilized and included in organic compounds.

In 2008, the heavy rainfall at the catch crops emergence stage (116.5 mm) promoted N leaching from soil, and the dry September inhibited soil microbiological processes. Consequently, mineral N concentration in the soil was a mean 48.7 and 31.3% less, respectively, compared with 2006 and 2007. However, changes due to agricultural practices remained similar. The highest mineral N contents before autumn ploughing were in the sustainable cropping system having applied mineral N fertilizer (N_{30}) to promote straw decomposition, in both soils with low and moderate humus contents. In 2008, the positive effect of catch crops on decreased mineral N leaching during the winter period was evident, since significantly less mineral N remained having applied N_{30} and with catch crop cultivation in the sustainable I cropping system compared with the same N rate applied without catch crops in sustainable II cropping system. The differences in the soil with low and moderate humus contents were 41.2 and 35.3%, respectively.

In the autumn 2009, during the period of intensive organic matter mineralization, before autumn ploughing, mineral N concentration in the 0–40 cm soil layer was 3.3–4.2 mg kg⁻¹ soil. In the soil in the sustainable II cropping system with low and moderate humus contents,

mineral N concentration was 25.1 and 18.2%, respectively, higher than in the organic I system. This higher mineral N concentration was influenced by applications of ammonium nitrate (N_{30}), applied to promote straw decomposition in soil.

According to the mean data of 2006–2009, in the autumn, during intensive organic matter mineralization before autumn ploughing, mineral N concentration in the 0–40 cm soil layer remained rather high and totalled 5.0–7.8 mg kg⁻¹ soil. In the soil low in humus content, before the incorporation of catch crops into the soil, soil mineral N concentration was lower in the organic II cropping system compared with the organic I system. The highest mineral N concentration in the 0–40 cm soil layer (28.2% higher than in the organic I system) was in the sustainable II cropping system. In this system, the higher mineral N concentration was influenced by ammonium nitrate (N_{30}) applications. Before the incorporation of catch crops, during late autumn, significantly higher (14.8%) mineral N concentration was in soil with moderate humus content, compared with soil with low humus content. In the soil with moderate humus content, the trends of mineral N concentration were similar and significantly higher (by 31.5%) in the sustainable II system, compared with the organic I system.

Soil mineral nitrogen variation during the spring-autumn period. The data of changes in mineral N concentration in spring before sowing field pea compared with that in autumn before the incorporation of catch crops for green manure in the 0–40 cm soil layer are presented in Figure 3. In autumn 2006, after incorporation of N-rich catch crops, and with prolonged spells of mild weather in November and December (when mean monthly temperature was 4.4 and 4.2 °C, respectively), the soil did not freeze and conditions for mineral N migration in soil were favourable. Consequently, in spring 2007 before sowing field pea, a marked reduction in mineral N occurred compared with soil in autumn before the incorporation of catch crop for green manure by ploughing. N contents varied from 5.17–5.84 mg kg⁻¹ and there were no significant differences between the cropping systems. However, analysis of mineral N losses from autumn to spring showed that the greatest reduction in N content occurred in the sustainable I and II cropping systems, where mineral N fertilizer (N_{30}) had been applied. Mineral N decreased most in the sustainable cropping system, where mineral N fertilizer (N_{30}) had been applied to promote straw decomposition and no catch crops had been grown. The mean N content was 16.7% higher on both low and moderate soil humus contents compared the treatment with the same fertilizer rate and catch crop. The least mineral N differences, compared with its status in autumn, were identified in the organic cropping system where white mustard was grown as a catch crop. Blue lupine mixed with oilseed radish had insignificant effects on

N retention in soil, since their biomass accumulated least N. In 2007, during the droughty post-harvest period, catch crops developed poorly, especially in the organic cropping system, and accumulated little biomass and low N content. Since there was little rainfall in August and September (11.1 and 6.2 mm, respectively, less than the long-term mean), in both treatments (with and without catch crops) soil mineral N content before autumn ploughing was similar. Thus, mineral N content in spring 2008 remained fairly consistent. Changes in soil mineral N concentration at 0–40 cm depth during the period from autumn 2007 to spring 2008 showed a similar trend to that in the previous year and the differences were not significant. The greatest difference between mineral N concentration in autumn and that remaining in spring was in the sustainable II cropping system, where ammonium nitrate was applied. In both low and moderate humus content soils, the mean N content was 0.92 mg kg^{-1}. In the organic cropping systems where catch crops had been grown, especially white mustard, which produced more biomass, an increase in mineral N concentration was evident in spring, however, only in soil with low humus content.

In spring 2009, there were the lowest mineral N concentrations in both the organic and sustainable cropping systems compared with earlier years. These results might have been influenced by the heavy rainfall in August 2008. At the catch crop emergence stage in late August there was nearly double the mean rainfall. Analysis of N losses shows that before the incorporation of catch crops losses were greatest where mineral N fertilizer (N$_{30}$) had been applied for straw decomposition without growing catch

crop. Highest N losses were measured in spring and differences in N concentrations between treatments almost disappeared. This year (2009) the trend remained the same to that in the previous year, mineral N losses in the sustainable II cropping system, where mineral N fertilizer was applied for straw decomposition and with no catch crops amounted to a mean 2.29 mg kg^{-1} in soils with both low and moderate humus contents. In organic cropping systems, after catch crops, there was a slight increase in mineral N. Kramberger *et al.* (2009) discussed the high importance of weather conditions in influencing mineral N migration.

Although mineral N variations in separate years were highly affected by weather conditions, analysis of data in spring 2010 indicated positive trends of catch crops in reducing mineral N migration.

In the soil with low humus content, comparison of mineral N concentration in autumn before incorporating catch crops with that in spring before sowing field pea, changes were small in all cropping systems. In the soil with moderate humus content, in the organic II cropping system with white mustard grown as a sole crop, mineral N concentration in soil was higher in spring than autumn. In the organic cropping systems, the positive immobilization effect of catch crops on mineral N concentration in soil manifested itself during the late autumn-early spring period of intensive leaching. In the sustainable II cropping system, where mineral N fertilizer was applied for straw decomposition, in the next year's spring, before sowing field pea, mineral N concentration in soil markedly

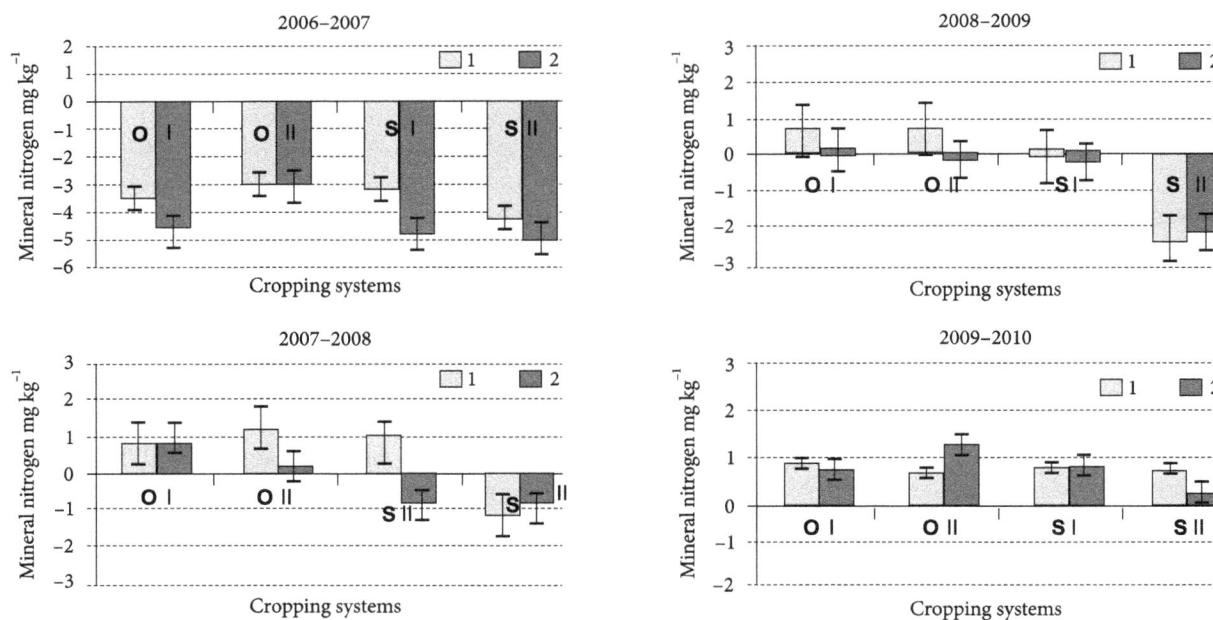

Fig. 3. Changes in mineral N concentration in spring before sowing field pea compared with autumn before the incorporation of catch crops for green manure in the 0–40 cm soil layer

Notes: 1 – Low humus content in the soil; 2 – Moderate humus content in the soil; Cropping systems: O I – organic I; O II – organic II; S I – sustainable I; S II – sustainable; $\perp$ – Standard error.

declined, in soil with low and moderate humus contents, compared with organic cropping systems. In the sustainable II cropping system, although before autumn ploughing N concentrations were high, in the next year's spring it was markedly lower than in the systems in which catch crops were grown during the winter wheat post-harvest period. This shows that application of mineral N fertilizer for straw decomposition increased mineral N concentration in the 0–40 cm layer before autumn ploughing. N concentrations markedly declined until spring due to its rapid leaching into deeper soil layers. These results show that some applied N for winter wheat straw decomposition remains unbound to organic compounds and remain in soil in the form of mineral N in cropping systems without catch crops. Torstensson *et al.* (2006) and Möller *et al.* (2008) suggest that the most effective N utilization occurs in cropping systems with catch crops.

Conclusions

Investigations were conducted on organic and sustainable cropping systems over the period 2006–2010 on a clay loam soil with low and moderate humus contents. Catch crops grown during the winter wheat post-harvest period markedly influenced nitrogen dynamics both in soils and crops:

1. Markedly higher dry matter yield of catch crop accumulated in the sustainable I cropping system (white mustard mixed with buckwheat) compared with the organic II system (white mustard only). The difference in low-humus content soil was 54.7%, in moderate-humus content soil it was 33.3%.

2. Higher dry matter yield of catch crops and their N content were accumulated when white mustard (short-vegetation plant) was grown in a mixture with buckwheat or as a sole crop, compared with oilseed radish grown with blue lupine (long-day plant).

3. Mineral N concentrations in the 0–40 cm soil layer late in autumn were higher in soil with moderate humus content, compared with soil with low humus content. The highest mineral N concentration in soil with both low and moderate humus contents, late in autumn and the risk of leaching into deeper layers were recorded having incorporated mineral N fertilizer (N_{30}) for winter wheat straw decomposition. The lowest mineral N concentration was measured in organic cropping systems with catch crops.

4. In the organic cropping system, the incorporation of catch crops into soil resulted in higher mineral N reserves in soil in spring than in the sustainable cropping system, having applied mineral N fertilizer for straw decomposition in autumn.

Applying organic cropping systems with catch crops on a clay loam soil with low and moderate humus contents is an efficient tool to promote environmental sustainability.

Acknowledgments

The paper presents research findings, which have been obtained through the long-term research programme "Productivity and Sustainability of Agricultural and Forest Soils" implemented by the Lithuanian Research Centre for Agriculture and Forestry. The authors thank Professor Dr Michael A. Fullen (The University of Wolverhampton, United Kingdom) for the English corrections and advices.

References

Arlauskienė, A.; Maikštėnienė, S.; Šlepetienė, A. 2010. Effect of cover crops and straw on the humic substances in clay loam, Cambisol // *Agronomy Research* 8(2): 397–402.

Arlauskienė, A.; Maikštėnienė, S.; Šlepetienė, A. 2011. Application of environmental protection measures for clay loam *Cambisol* used for agricultural purposes, *Journal of Environmental Engineering and Landscape Management* 19(1): 71–80. http://dx.doi.org/10.3846/16486897.2011.557266

Arlauskienė, A.; Maikštėnienė, S. 2010. The effect of cover crops and straw applied for manuring on spring barley yield and agrochemical soil properties, *Zemdirbyste – Agriculture* 97(2): 61–72.

Bučienė, A. 2003. *Žemdirbystės sistemų ekologiniai ryšiai* [Ecological relations of cropping systems]. Klaipėda. 176 p. (in Lithuanian).

Catt, J. A.; Howse, K. R.; Christian, D. G.; Lane, P. W.; Harris, G. L.; Goss, M. J. 1998. Strategies to decrease nitrate leaching in the Brimstone farm experiment, Oxfordshire, UK, 1988–1993: the effects of winter cover crops and unfertilised grass leys, *Plant and Soil* 203(1): 57–69. http://dx.doi.org/10.1023/A:1004389426718

Crandall, S. M.; Ruffo, M. L.; Bodlero, G. A. 2005. Cropping system and nitrogen dynamics under a cereal winter cover crop preceding corn, *Plant and Soil* 268(1): 209–219. http://dx.doi.org/10.1007/s11104-004-0272-x

Crawley, M. J. 2007. *The R book*. Chichester: John Wiley & Sons Ltd. 942 p. http://dx.doi.org/10.1002/9780470515075

Dawson, J. C.; Huggins, D. R.; Jones, S. S. 2008. Characterizing nitrogen use efficiency in natural and agricultural ecosystems to improve the performance of cereal crops in low-input and organic agricultural systems, *Field Crops Research* 107(2): 89–101. http://dx.doi.org/10.1016/j.fcr.2008.01.001

Di, H. J.; Kameron, K. C. 2002. Nitrate leaching in temperate agroecosystems: sources, factors and mitigating strategies, *Nutrient Cycling in Agroecosystems* 64(3): 237–256. http://dx.doi.org/10.1023/A:1021471531188

Diacono, M.; Montemurro, F. 2010. Long-term effects of organic amendments on soil fertility, *Agronomy for Sustainable Development* 30(2): 401–422. http://dx.doi.org/10.1051/agro/2009040

Ergstöm, L.; Linden, B. 2009. Importance of soil mineral N in early spring and subsequent net N mineralization for winter wheat following winter oilseed rape and peas in a milder climate, *Acta Agriculturae Scandinavica, Section B – Plant Soil Science* 59: 402–413.

Fullen, M. A.; Booth, C. A.; Panomtaranichagul, M.; Subedi, M.; Mei, L. Y. 2011. Agro-environmental lessons from the 'sustainable highland agriculture in South-East Asia' (Shasea)

project, *Journal of Environmental Engineering and Landscape Management* 19(1): 101–113. http://dx.doi.org/10.3846/16486897.2011.557476

Hao, X.; Chang, C.; Carefoot, J. M.; Janzen, H. H.; Ellert, B. H. 2001. Nitrous oxide emissions an irrigated soil as affected by fertilizer and straw management, *Nutrient Cycling in Agroecosystems* 60(1–3): 1–8. http://dx.doi.org/10.1023/A:1012603732435

Hasegawa, H.; Denison, F. 2005. Model predictions of winter rainfall effects on N dynamics of winter wheat rotation following legume cover crop or fallow, *Field Crop Research* 91(2–3): 251–261. http://dx.doi.org/10.1016/j.fcr.2004.07.019

Komatsuzaki, M.; Ohta, H. 2007. Soil management practices for sustainable agro-ecosystems, *Soil Management Practices for Sustainable Agro-Ecosystems,* 103–120.

Koutroubas, S. D.; Papakosta, D. K.; Doitsinis, A. 2008. Nitrogen utilization efficiency of safflower hybrids and open-pollinated varieties under Mediterranean conditions, *Field Crops Research* 107(1): 56–61. http://dx.doi.org/10.1016/j.fcr.2007.12.009

Kramberger, B.; Gselman, A.; Janzekovic, M.; Kaligaric, M.; Brauko, B. 2009. Effects of cover crops on soil mineral nitrogen and on the yield and nitrogen content of maize, *European Journal of Agronomy* 31(2): 103–109. http://dx.doi.org/10.1016/j.eja.2009.05.006

Lahti, T.; Kuikman, P. J. 2003. The effect of delaying autumn incorporation of green manure crop on N mineralization and spring wheat (*Triticum aestivum* L) performance, *Nutrient Cycling in Agroecosystems* 65(3): 265–280. http://dx.doi.org/10.1023/A:1022617104296

Larsson, M. H.; Kyllmar, K.; Jonasson, L.; Johnsson, H. 2005. Estimating reduction of nitrogen leaching from arable land and the related costs, *A Journal of the Human Environment* 34(7): 538–543.

Loges, R.; Kelm, M.; Taube, F. 2006. Nitrogen balances, nitrate leaching and energy efficiency of conventional and organic farming systems on fertile soils in Northern Germany, *Advances in GeoEcology* 38: 407–414.

Marcinkonis, S.; Pranaitis, K.; Lisova, R. 2007. Studies on various buckwheat biomasses, mineral versus organic fertilisation: conflict or synergism, in *16th International Symposium of the International Scientific Centre of Fertilisers (CIEC)*, 16–19 September, 2006, Ghent, Belgium, 336–341.

Mattson, N. S.; Erwin, J. E. 2005. The impact of photoperiod and irradiance on flowering of several herbaceous ornamentals, *Scientia Horticulturae* 104(3): 275–292. http://dx.doi.org/10.1016/j.scienta.2004.08.018

Möller, K.; Stinner, W.; Leithold, G. 2008. Growth, composition, biological N$_2$ fixation and nutrient uptake of a leguminous

cover crop mixture and the effect of their removal on field nitrogen balances and nitrate leaching risk, *Nutrient Cycling in Agroecosystems* 82(3): 233–249. http://dx.doi.org/10.1007/s10705-008-9182-2

Odhiambo, J. J. O.; Bomke, A. A. 2007. Cover crop effects on spring soil water content and the implications for cover crop management in south coastal British Columbia, *Agricultural Water Management* 88(1–3): 92–98. http://dx.doi.org/10.1016/j.agwat.2006.09.001

Shevtsova, L.; Romanenkov, V.; Sirotenko, O.; Smith, P.; Smith Jo, U.; Leech, P.; Kanzyvaa, S.; Rodionova, V. 2003. Effect of natural and agricultural factors on long-term soil organic matter dynamics in arable soddy-podzolic soils – modelling and observation, *Geoderma* 116(1–2): 165–189.

Stadler, C.; Heuwinkel, H.; von Tucher, S.; Gutser, R.; Scheu-Hekgert, M. 2005. Beeinflusst der Boden die N-Freisetzung aus pflanzlichen Dünger?, in *Ende der Nische* (Hrsg.: Heß J., Rahmann G.): Beiträge zur 8. Wissenschaftstagung zum Ökologischen Landbau Kassel, 217–219.

Tarakanovas, P.; Raudonius, S. 2003. *The program package "Selekcija" for processing statistical data.* Akademija, Kėdainiai. 56 p. (in Lithuanian).

Tonitto, C.; David, M. B.; Drinkwater, L. E. 2006. Replacing bare fallows with cover crops in fertilizer-intensive cropping systems: a meta-analysis of crop yield and N dynamics, *Agriculture, Ecosystems & Environment* 112(1): 58–72. http://dx.doi.org/10.1016/j.agee.2005.07.003

Torstensson, G.; Aronsson, H.; Bergström, L. 2006. Nutrient use efficiencies and leaching of organic and conventional cropping systems in Sweden, *American Society of Agronomy* 98(3): 603–615. http://dx.doi.org/10.2134/agronj2005.0224

Tripolskaja, L. 2005. *Organic fertilizers and their impact on the environment.* Akademija, Kėdainiai. 216 p. (in Lithuanian).

Velykis, A.; Satkus, A. 2008. Applicability of various amendments to improve clayey soil properties under reduced tillage management in Northern Lithuania, *Agronomijas Vestis – Latvian Journal of Agronomy* (10): 73–77.

Vinther, F. P.; Hansen, E. M.; Olesen, J. E. 2004. Effects of plant residues on crop performance, N mineralization and microbial activity including field CO$_2$ and N$_2$O fluxes in unfertilised crop rotation, *Nutrient Cycling in Agroecosystems* 70(2): 189–199. http://dx.doi.org/10.1023/B:FRES.0000048477.56417.46

Xu, Z. Z.; Yu, Z. W.; Wang, D. 2006. Nitrogen translocation in wheat plants under soil water deficit, *Plant and Soil* 280(1): 291–303. http://dx.doi.org/10.1007/s11104-005-3276-2

Zhao, W.-F.; Gao, Z.-Q.; Sun, M.; Deng, L.-F.; Li, Q. 2012. Effect of fertilization and mulching in fallow period on nitrogen accumulation of dryland wheat, *African Journal of Agricultural Research* 7(12): 1849–1854.

Laura MASILIONYTĖ. She works as a Scientist in Joniškėlis Experimental Station of the Lithuanian Research Centre for Agriculture and Forestry. She is a Doctor of Biomedical (Agronomy) Sciences, and is author and co-author of over 14 research papers, co-author of 2 monographs. She has participated in 15 scientific conferences. Her research interests include organic matter of soil, organic and sustainable agriculture.

Stanislava MAIKŠTĖNIENĖ. She works as the Chief Scientist and the Director in Joniškėlis Experimental Station of the Lithuanian Research Centre for Agriculture and Forestry. She is a doctor of Biomedical (Agronomy) Sciences. She is the author and co-author of over 80 research papers, as well as compiler and co-author of 3 monographs; has participated in 21 international conferences. Her research interests include crop fertilization, soil organic matter, organic and sustainable agriculture.

Aleksandras VELYKIS. He works as a Senior Scientist in Joniškėlis Experimental Station of the Lithuanian Research Centre for Agriculture and Forestry. He is a doctor of Biomedical (Agronomy) Sciences. He is the author and co-author of over 60 research papers, and co-author of 2 monographs. He has participated in 41 scientific conferences. His research interests include soil physical, chemical and biological properties; soil tillage systems; crop rotations; organic agriculture.

Antanas SATKUS. He works as a Senior Scientist in Joniškėlis Experimental Station of the Lithuanian Research Centre for Agriculture and Forestry. He is a Doctor of Biomedical (Agronomy) Sciences. He is the author and co-author of over 50 research papers, and co-author of 2 monographs. He has participated in 22 scientific conferences. His research interests include soil tillage, seedbed preparation, organic matter of soil, organic and sustainable agriculture.

7

EVALUATION OF MASS HOUSING SETTLEMENTS IN TERMS OF WIND AND NOISE CONTROL: ISTANBUL AND DIYARBAKIR AS A CASE

Gülay ZORER GEDİK [a], Neşe YÜĞRÜK AKDAĞ [b], Fatih KİRAZ [c], Bekir ŞENER [d], Raşide ÇAÇAN [e]

[a, b]Faculty of Architecture, Yildiz Technical University, Istanbul, 34349, Turkey
[c]Faculty of Fine Arts and Design, Nuh Naci Yazgan University, Kayseri, 38040, Turkey
[d]Naval Architecture and Maritime Faculty, Yildiz Technical University, Istanbul, 34349, Turkey
[e]Yildiz Technical University, Istanbul, 34349, Turkey

Abstract. In mass housing apartments, the comfort and quality of living conditions may be adversely affected by wind and noise especially on balconies, terraces, gardens and around swimming pools etc. The quantitative and empirical testing of building models according to physical conditions with regard to wind and noise parameters directly affects the formation of buildings in the design process. In this paper, two cities (Istanbul and Diyabakır), which are selected from two different climatic zones in Turkey are considered as examples to create maximum comfortable usage areas depending on wind and noise effects. For mass housing settlement scenarios, common comfortable areas in terms of wind and noise were determined by using Urbawind and soundPLAN softwares. The relevant data and acceptance criteria related to wind and noise and applied procedure are presented in the work. Performed studies show, if the settlements have dominant wind and noise directions, it is possible to find solutions using the geometric properties of the settlement in terms of wind and noise. In general, better comfort results appear in alternatives with L-C-U shaped design features rather than point-type and linear block layouts.

Keywords: mass housing, wind control, noise control, environmental processes modeling, environmental impact assessment.

Introduction

Wind and noise are both physical environment factors that cause serious problems in urban areas and should be considered important in design and planning processes. They also have many negative effects on human health and comfort. While the numbers of mass-housing developments in different climatic zones increase rapidly, the effect of wind speed in open areas especially due to building design has caused an increase in complaints. Additionally, because of locating mass housing close to roads, noise is also a parameter that increases complaints. Searching for solutions to the problems after the design process due to unacceptable wind and noise levels leads to additional costs and unnecessary expenditure of resources. Often, any changes bring additional costs to owners and cause conflicts.

Although the negative effects of wind and noise occur together in plenty of settlements, no study has been presented that provides guidance for combining these parameters in the design of mass housing. Consequently, studies for the optimization of wind and noise parameters in mass housing areas are needed. A great number of studies are available concerning wind and noise individually. When the relevant studies are analysed it can be seen that in urban areas, air flow distribution differs greatly depending on the dimensions of structures, relative positions of one to another and their distribution in a city (Blocken, Carmeliet 2004; Hong, Lin 2015; Shi et al. 2015). Wind has additive, reducing and directing effects on this distribution. Several researches were made studying the creation of comfortable spaces for pedestrians around buildings depending on wind speed and the placing of wind barriers to reduce uncomfortable situations (Bu et al. 2009; Aanen, Van Uffelen 2009; Willemsen, Wisse 2007; Stathopoulos 2009; Koss 2006). Wind is one of the main physical causes of erosion. Researchers concentrate on wind barriers designed to protect agricultural land against erosion in areas with strong and continuous wind (Cornelis, Gabriels 2005;

Corresponding author: Bekir Şener
E-mails: bsener@yildiz.edu.tr; brsener@gmail.com

Nordstrom, Hotta 2004; Campi *et al.* 2009). Wind tunnel experiments and computer simulation methods are used in studies related to wind (He, Song 1999; Yoshie *et al.* 2007; Hu, Wang 2004; Hagen *et al.* 1981; Kubota *et al.* 2008). Nowadays, it can be seen that simulations are preferred to wind tunnel experiments, since simulations, are faster and cheaper to modify than physical models. Studies of wind barriers, to reduce the negative effects of wind, have focused on engineering calculations rather than architectural design.

Studies on noise and noise control can be collected under many different headings such as the effects of noise on human health and efficiency, evaluation of noise as a design parameter in city and building planning, studies of noise mapping and action plans, and the design of noise barriers. Researches on the effects of noise show that noise starts to disturb when it exceeds 55 Leq dB(A), and once greater than 65 Leq dB(A) such disturbance increases significantly and causes serious problems over time (Future Noise Policy (European Commision 1996)). In noise control, starting to evaluate the problem from the city planning level and locating the noise-sensitive outdoor areas and buildings at the required distance from the source of noise is of vital importance. In most developed countries, noise-compatible land use planning decisions are taken at the urban planning stage (FHWA 2006; Chevallier *et al.* 2009; Murphy, King 2010; King *et al.* 2011). In making decisions, the noise emitted from industrial facilities is evaluated and definitions specified for noise-sensitive areas and structures as to how far away they are positioned from noise emitting facilities. According to current regulation in Turkey, "The Regulation of Environmental Noise Assessment and Management", it is recommended that exterior noise should not exceed 55 Leq dB(A) in settlements including sensitive structures (RENAM 2010). Arrangement of buildings according to noise sources is an important parameter in terms of noise level that can reach both outdoor areas and building structures. In the planning stage, positioning the narrow side of the buildings in the direction of noise sources; placing the high and long parallel structures at specific angles to minimize possible sound reflections from structure surfaces are among important design parameters. There are several studies related to determine the effect of the factors on noise environment such as placement and shape of structures, green fields, width and density of the roads and topography (Makarewicz 1991; Thorsson *et al.* 2004; Baltrenas *et al.* 2011; Guedes *et al.* 2011). Guedes *et al.*, have found that the physical characteristics of the urban shape such as construction density, the existence of open spaces, and the shape and physical position of buildings have a significant influence on environmental noise (Guedes *et al.* 2011). On the other hand, especially for mass housing settlements, there is limited number of studies that deal with design

configurations of outdoor activity areas to protect them from noise (FHWA 2006; Desanghere 2007; Montana 2008; Newman, Thornley 1996). A wide range of studies have been performed on the subject of traffic noise control. These studies are mainly related with the usage and importance of barriers in terms of city acoustics (Crombie, Hothersall 1994; Akdağ 2001; Bootby *et al.* 2001; Watson 2006), selection of suitable equipment/sections for barriers, the form of barriers (Cianfrini *et al.* 2007; Oldham, Egan 2011; Ishizuka, Fujiwara 2004; Naderzadeh *et al.* 2011; Venckus *et al.* 2012), relation between traffic noise, the barrier and the receiver (Hong Kong Environmental Protection Department 2003; Ekici, Bougdah 2003; Watts, Godfrey 1999), visual design criteria for barriers (Kotzen, English 2004; Bendtsen 1994; Maffei *et al.* 2013), and usage of plants as a noise barrier (Kotzen 2002).

In the architectural design process, solutions in order to prevent wind and noise induced problems can be considered in three groups:
  – Settlement
  – Barriers around structures
  – The building envelope of structures.

In Yıldız Technical University, a research project supported by TUBITAK (The Scientific and Technological Research Council of Turkey) has been completed with the aim of developing a methodology for determining the most suitable settlement formations in terms of wind and noise control for outdoor spaces (Gedik *et al.* 2014). This project considers five different climatic zones (cold, hot humid, hot dry, temperate humid and temperate dry). The scope of this project is limited to the consideration of human comfort in terms of wind and noise levels for outdoor activities in mass housing settlements. Other specific effects of wind (including turbulence and noise vibration) and noise within the structure are outside the scope of this study. In this paper produced from the TUBITAK project, design configurations of two cities, which are selected from two different climatic zones (temperate humid-Istanbul and hot dry-Diyarbakır) are considered as examples to create maximum comfortable usage areas depending on wind and noise effects taking account of the building bylaws for mass housing in Turkey. The comparative results of other cities will be presented in a later article.

1. Methodology, data and acceptances

The methodology of this study consists of three stages:
  1. Creating settlement scenarios for mass housing.
  2. Determining the relevant data and acceptance criteria related to wind and noise.
  3. Determining comfortable areas in terms of wind and noise individually and defining common comfortable areas.

1.1. Creating settlement scenarios for mass housing

Within this paper, the size of a unit house that will be used is taken as 150 sqm. This is the upper size limit of a residential zone determined by Turkey Housing Development Fund (Prime Ministry Housing... 1997). Based on this, the width, length and height of the house is specified as 10×15×3 m, respectively.

The joining styles of unit house are defined as point-type block and linear block, as shown in Figure 1. For point-type blocks, the quad-joining style is selected since it is the most common type used mainly for economic reasons. Analyses are made for four different number of floors (3, 5, 7 and 10) to assess the effects on the results obtained.

Figure 2 shows the working area (1 hectare/10,000 sqm). The width and length of the land is determined as 80×125 m, respectively. To limit the analysis stages, the position of the road relative to the land is chosen to be on one side and parallel to the long edge. "The Turkish Zoning Regulation of Planned Areas" is used to determine the front, side and back yards of the houses (The Zoning Regulation... 1985). The 25 different mass housing settlement configurations considered are shown in Figure 3.

1.2. Determining the relevant data and acceptance criteria related to wind and noise

Wind:

It is necessary to determine criteria to define comfortable areas for different climatic conditions, taking into consideration the effects of the wind on a human body. When reviewing the literature in this context the studies on wind speeds that provide comfortable conditions in open areas can be classified by the studies focused on the mechanical effects of wind on the human body (Willemsen, Wisse 2007; Koss 2006; Pendwarden, Wise 1975; Isyumov, Davenport 1975), on thermal effects (Arens et al. 1986; Koch 2002) or both (Stathopoulos 2009; Stathopoulos et al. 2004; Hoppe 2002; Szucs 2004; Szucs et al. 2007). There is no standard published on the mechanical effects of wind. In the quoted studies, it is mostly the wind effects on pedestrian comfort that are discussed. In studies as from Pendwarden (Pendwarden 1973), usually 5m/sec is accepted as a threshold value for uncomfortable situations in terms of the mechanical effects of wind.

Studies concerning the thermal effect of wind are works that consider wind with other microclimate parameters and formulations developed from laboratory and statistical experiments in different parts of the world (Arens et al. 1986; Koch 2002; Stathopoulos et al. 2004).

Among the studies that are focused on both the mechanical and thermal effects of wind, the studies realized by Szucs et al. concerning audience comfort in stadiums are guidelines with their methodological approach (Szucs 2004; Szucs et al. 2007). Required wind speeds are determined for relative humidity of 40%, 60% and 80% from

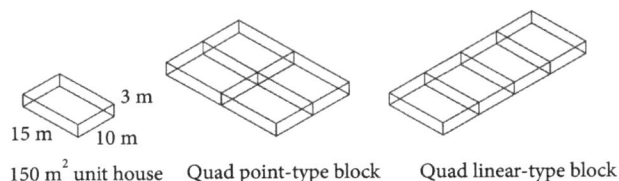

Fig. 1. Joining types of unit house

15 m　10 m　3 m
150 m² unit house　　Quad point-type block　　Quad linear-type block

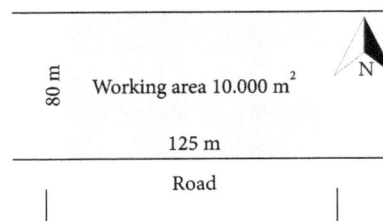

80 m　Working area 10.000 m²　N
125 m
Road

Fig. 2. Dimension and position of land

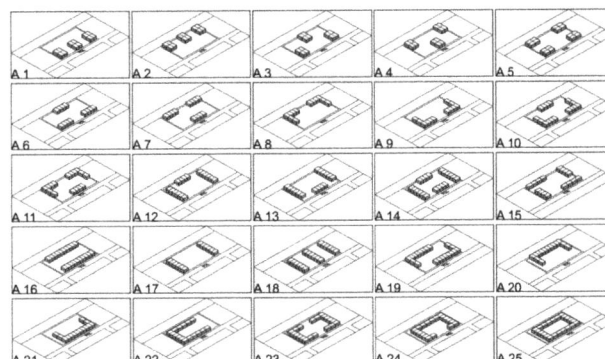

Fig. 3. Mass housing settlement configurations (A1-A5: Point type blocks, A6-A25: Linear type blocks)

rearranged psychometric diagram of Olgyay's bioclimatic comfort chart (Arens et al. 1986). It is considered that Willemsen and Wise used 5 m/sec wind speed as a threshold value in their study depending on activity levels (Willemsen, Wisse 2007). A threshold value of wind speed is determined according to climatic regions taking into account maximum acceptable values of wind speed for mechanical effects on the human body and minimum values based on relative humidity. Evaluating the climatic data, 5 m/sec wind (light breeze, Beaufort scale) for temperate humid climate region (Istanbul) and 3.6 m/sec wind (mechanical threshold) for hot dry climate region (Diyarbakır) are accepted as a threshold values.

In the next stage, to determine the annual acceptable percentage of hours that threshold value is exceeded (exceedance frequency) depending on activity level is important. To specify the annual acceptable exceedance frequency of threshold values indicated by F in Table 1, the studies in the literature have been examined. The study by Willemsen and Wisse is considered as significant with regards to both defining discomfort potential caused by the mechanical effects of wind on pedestrians and presenting the preparation of new wind comfort regulations in The Netherlands and notes about current studies on wind comfort in this regulation (Willemsen, Wisse 2007). Also by considering the annual acceptable wind speed exceedance frequency values (%) given in the study by Caniot (Caniot *et al.* 2011; acceptable exceedance frequencies (*F*) are created depending on activity level for Istanbul and Diyarbakır, as shown in Table 1.

In Table 1 wind speeds are expressed as threshold values (3.6 m/sec and 5 m/sec) and the acceptable exceedance frequencies indicated as *F* (5%; 10%) are given depending on activity type. In short, *F* is the annual acceptable percentage of hours that treshold value of wind speed is exceeded.

Table 1. Acceptable exceedance frequencies (*F*) of wind comfort threshold values (m/sec) dependent on activity levels (Caniot *et al.* 2011)

Activity type	Diyarbakir	Istanbul
Sitting	$F(V > 3.6 \text{ m/sec}) < 5\%$	$F(V > 5 \text{ m/sec}) < 5\%$
Walking	$F(V > 3.6 \text{ m/sec}) < 10\%$	$F(V > 5 \text{ m/sec}) < 10\%$

In simulations, the wind data received from the Turkish State Meteorological Service for Istanbul and Diyarbakır was used containing daily and hourly measurements over a 30 year period (Archive of Turkish State Meteorological Service 2013). The seasonal wind speeds and directions for Istanbul and Diyarbakır at 6:00, 9:00, 12:00, 15:00, 18:00, 21:00 and 24:00 hours are shown in Figure 4. UrbaWind software has been used for wind comfort simulations (UrbaWind 2013) Wind measurements are gathered at 10 m height in Turkey. Therefore, a reference point at 10 m height was defined and calculations were made according to this reference condition. In calculations, the height from ground was taken as 1.5 m considering the height range that affects the human body. The roughness ratio of the soil directly affects the wind velocity. Urbawind has four types of Roughness length for the inlet profile: 0.05 m open country, 0.001 m water, 0.25 m small density city and 0.7 m for high denstiy city or forest. 0.25 m small density city roughness profile has been selected for calculations. Climatologic data can be used as Tab File, Topowind File and TIM File formats in

Urbawind. Due to data taken from Turkish State Meteorological Service, Tab File format was used in calculations.

Noise:

SoundPLAN 7.3 has been used for simulations (Soundplan Manual 2012). In simulations, NMPB-Routes-96 method is used for noise propagation as suggested in EU Noise Directive (EU 2002) and RENAM (RENAM 2010).

In various national and international standards and regulations regarding noise, the noise levels that should not be exceeded are specified based on the region where the structure is located. According to current "Environmental Noise Assessment and Regulation" in Turkey, suggested maximum value of exposure from road traffic is 68 Leq dB(A) according to settlement region type (RENAM 2010). On the other hand, in many studies conducted in Turkey, the noise generated by roadside was determined to vary between 55 Leq dB(A) ile 85 Leq dB(A) by means of noise level measurements and prepared noise maps (Kumbay *et al.* 2006; Akdağ, Candemir 2009; Aknesil, Akdağ 2011; Dal, Akdağ 2011). In this study, in order to reveal the changes of comfort state of buildings located in different noise environment, the examinations were performed for four different situations that the road noise generate 55, 65, 75 and 85 Leq dB(A) noise at 1 m away from the road. Simulations and calculations were realized with the road sound power levels corresponding to these values; 70 dB(A), 79 dB(A), 113 dB(A) and 123 dB(A), respectively. For 55 and 65 Leq dB(A) the road width is specified as 14 m and for 75 and 85 Leq dB(A) the road width is set at 21 m. Individual lane width is taken to be 3.5 m, with refuge and pavement widths of 2 m.

In NMPB Routes method based ISO 9613-2 standard, in noise level calculations from source to receiver it is suggested to use G:0 (where G is the sound absorption features of ground) for low sound absorption surfaces like concrete, G:1 for land areas covered with grass, trees or another vegetation and between 0 and 1 for hard and porous mixed grounds. In this study, ground was assummed to be of mixed type and G was taken to be 0.6 (ISO 9613-2:1996). Road was chosen as asphalt surface. In the calculations, grid space was taken as 10 m, and the height from

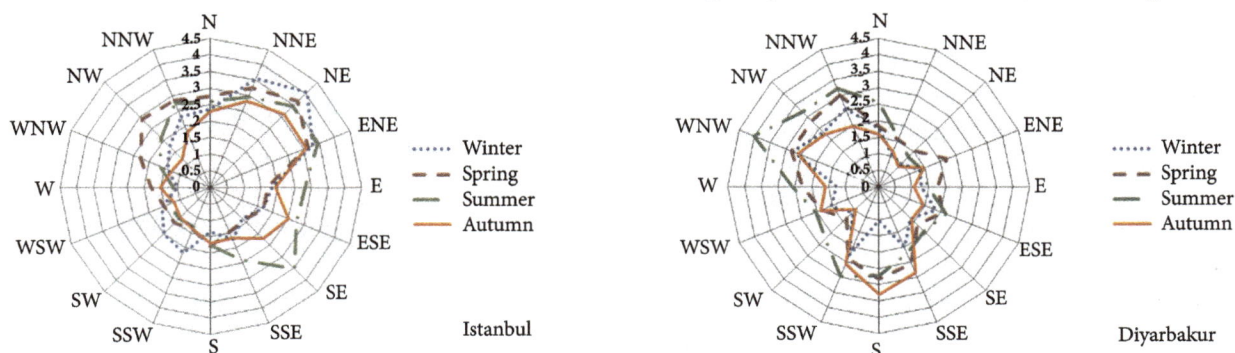

Fig. 4. The seasonal wind speeds (m/sec) for outdoor use hours for İstanbul and Diyarbakır

ground as 1.5 m. The effect of wind to noise propagation was taken into account by entering the annual wind frequency distribution to software in 18 directions with angle of 20 degrees and for 2 different time zone, day and night. The wind data was received from the Turkish State Meteorological Service and include last 30 years. In Turkey, wind measurements are performed for 16 directions in wind rose with angle of 22,5 degrees (Archive of Turkish State Meteorological Service 2013). Data for angle of 20 degrees, that required to enter to the noise software, was obtained by interpolation of existing data.

The values of the noise value for annoyance are different during the day, evening and night period. The open spaces in mass housung settlements are often used during daylight hours. So, the noise level that should not be exceeded for daylight hours was taken 55 Leq dB(A), considering the values given in related documents (WHO 1996; RENAM 2010).

Noise Measurements and Validation of the Models

SoundPLAN, used for noise calculations, is a highly accurate simulation software that both noise sources and environmental data can be defined in detail. It can be found various studies in literature that shows the sensitivity of the software (King, Rice 2009; Guedes *et al.* 2012; Dal, Akdağ 2011). In order to demonstrate the accuracy of the results obtained in this study, noise map of a part of mass housing located along a road with heavy traffic was prepared, noise level calculations was made at some points and compared with noise level measurements. 1/1000-scaled maps of the working area was provided in electronic form from Istanbul Metropolitan Municipality and transferred to SoundPLAN 7.3 software. These maps are with UTM coordinate system and include terrain elevations (x, y, z (height)) and building information system (intended use of building, number of floors etc.). There is not a remarkable change in traffic density of the road during the day, the number of vehicles, percentage of heavy vehicles and average speeds were determined based on observation for one hour period, from 10:00 to 11:00 and entered to the software. During the preparation of noise maps, NMPB Routes 96 (Guide de Bruit) standard for highways that was proposed by EU Noise Directive (EU 2002) and RENAM (RENAM 2010) was used. The temperature and relative humidity was measured as 25 °C and 70%, respectively. The average wind directions and speeds in summer for Istanbul was introduced to the software as wind rose (Archive of Turkish State Meteorological Service 2013). Grid spacing was selected as 10×10 m in calculations of maps for 1.5 m height. Noise map of the working area can be seen in Figure 5. Noise level calculations was also performed at points that shown in Figure 6.

All measurements were performed in accordance with the ISO 1996-1 standard (ISO 1996-1:2003) in the

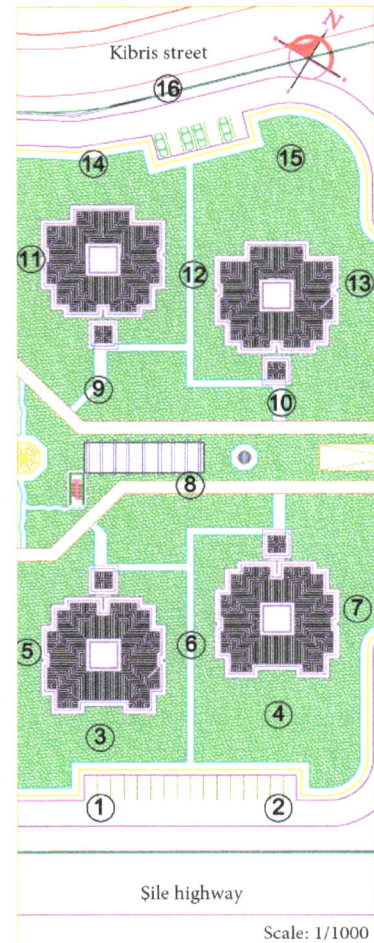

Fig. 5. Noise map of selected area

Fig. 6. Noise measurement points

Table 2. Comparison between measured and calculated noise levels

Evaluated points	1	2	3	4	5	6	7	8
Measured Leq dB(A)	67.5	68.0	66.5	66.0	62.0	65.0	65.5	55.0
Calculated Leq dB(A)	67.0	67.2	66.0	65.5	62.6	64.8	64.0	54.2
Difference Leq dB(A)	+0.7	+0.8	+0.5	+0.5	−0.6	+0.2	+1.5	+0.8
Evaluated points	9	10	11	12	13	14	15	16
Measured Leq dB(A)	56.0	56.7	53.0	53.0	52.2	49.0	51.5	51.5
Calculated Leq dB(A)	56.8	57.5	52.0	53.4	53.5	48.5	50.8	50.5
Difference Leq dB(A)	−0.8	−0.8	+1.0	−0.4	−1.2	+0.5	+0.7	+1.5

frequency range A, by use of Brüel&Kjær Type 2250 sound level meter. Measurements were also carried out 1.5 m above the ground with using a microphone wind-shield and at least 2 m away from buildings to prevent any surface reflection. The results of 15 minutes-long measurements and simulations realized for same points were given in Table 2.

According to the Good Practice Guide for Strategic Noise Mapping and the Production of Associated Data on Noise Exposure (WG-AEN 2006), the difference between actual measurement results and the calculated values must not exceed 1 dB(A) in a distance of 300 m from the source, 3 dB(A) in a distance of 600 m from the source and 10 dB in a distance of 2,000–3,000 m from the source. As shown in Table 2, the difference of the measurement and simulation results are below 1 dB(A) at 14 points and slightly above only at 2 points. Therefore it reveals the reliability of simulation results.

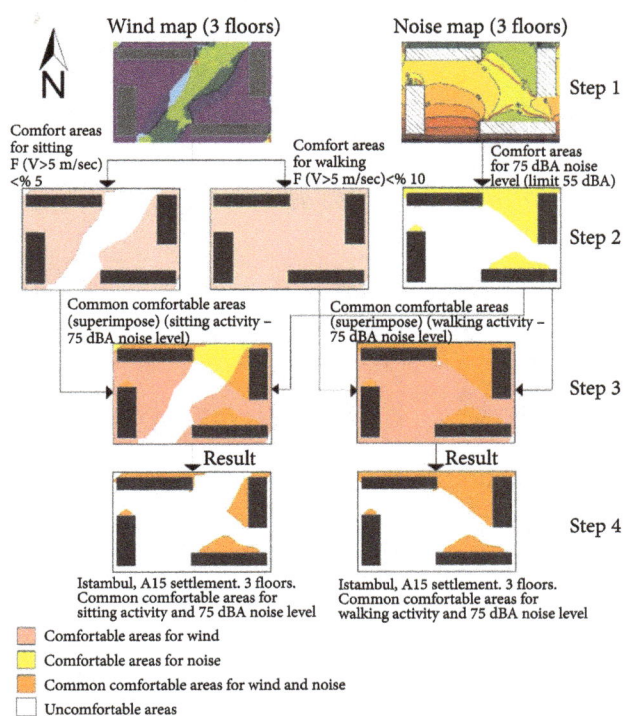

Fig. 7. The schematic procedure to determine the common comfortable areas in terms of wind and noise

1.3. Determining comfortable areas in terms of wind and noise individually and defining common comfortable areas

Comfort maps were created for 25 settlement alternatives in Istanbul and Diyarbakir according to different number of floors, noise levels and activity type. The results obtained were evaluated using a diagrammatic approach. Because of differences in calculation methods for the two physical parameters (acceptances, standards, equations, etc.), the comfort maps were initially prepared separately from the different simulations for both parameters and finally superimposed to specify the common comfortable areas. The ratio of common comfortable area to the total open area was calculated and given as a comfort percentage. The schematic procedure to determine the common comfortable areas in terms of wind and noise is given in Figure 7 an example for 3 floors, for an A15 type settlement.

The steps are as follows to determine the comfortable areas (Fig. 7):
- Step 1. to prepare the wind and noise maps separately by using Urbawind and SoundPLAN simulation programs.
- Step 2. to transfer the wind and noise maps to the AutoCAD software and to scan the comfortable areas on maps.
- Step 3. to superimpose the wind and noise comfort maps.
- Step 4. to create a new map by selecting common comfortable areas in terms of wind and noise and determine the ratio of common comfortable area to the total open area.

Following the procedure given above, comparison maps were prepared for 25 settlement alternatives in Istanbul and Diyarbakir. As an example of Step 2, maps showing the comfortable areas in terms of wind and noise are given in Figure 8 and Figure 9 for two alternatives in İstanbul. For the fourth step, sample maps showing the common comfortable areas are given in Figure 10.

As indicated in Section 1.2, in order to take account the effect of wind to noise propagation, the wind data of Istanbul and Diyarbakir was entered to SoundPLAN

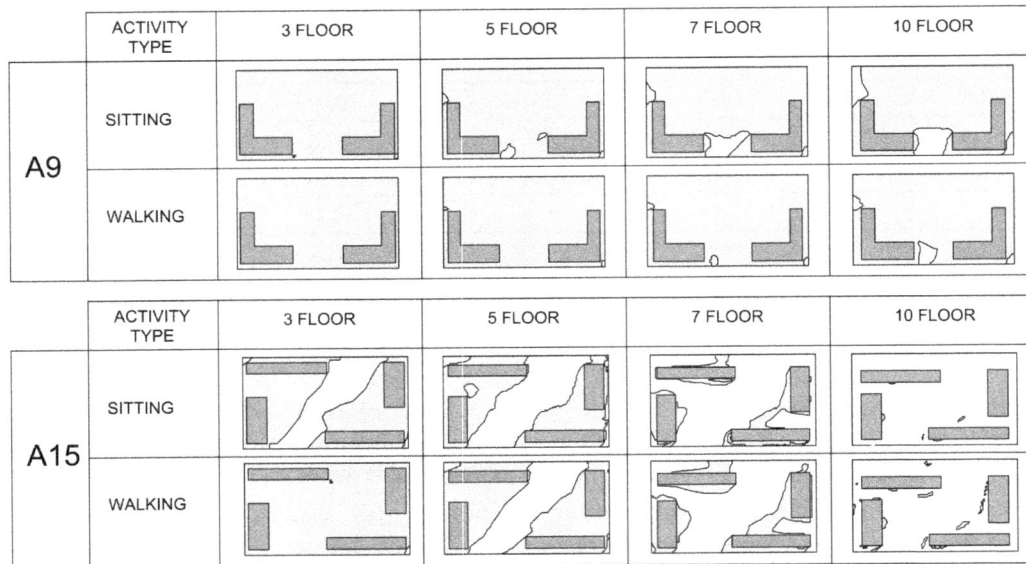

Fig. 8. The variation of comfortable outdoor areas according to mass housing alternatives in terms of wind (Dark grey: house blocks, light grey: comfortable areas, white: uncomfortable areas)

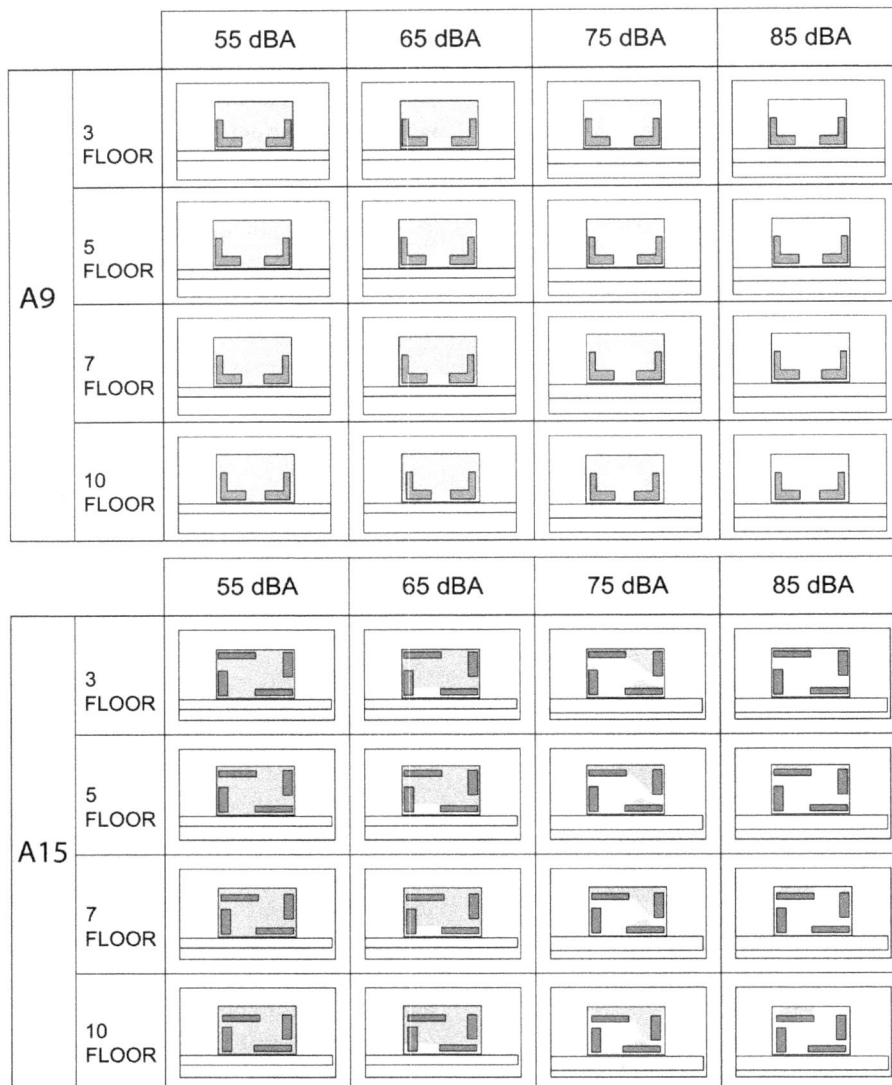

Fig. 9. The variation of comfortable outdoor areas according to mass housing alternatives in terms of noise (Dark grey: house blocks, light grey: comfortable areas, white: uncomfortable areas)

software. The noise maps generated according to results were compared and the results were found identical. It was presented in previous works that the effect of

the wind on noise level is important after 50 m (Harris 1994). In this study, the effect of wind on noise level is very low that cannot be noticed in results because of both the location and size of the parcel and direction and low annually-average speed of wind for Istanbul and Diyarbakir. Therefore, sample maps given in Figure 9 are valid for both Istanbul and Diyarbakir.

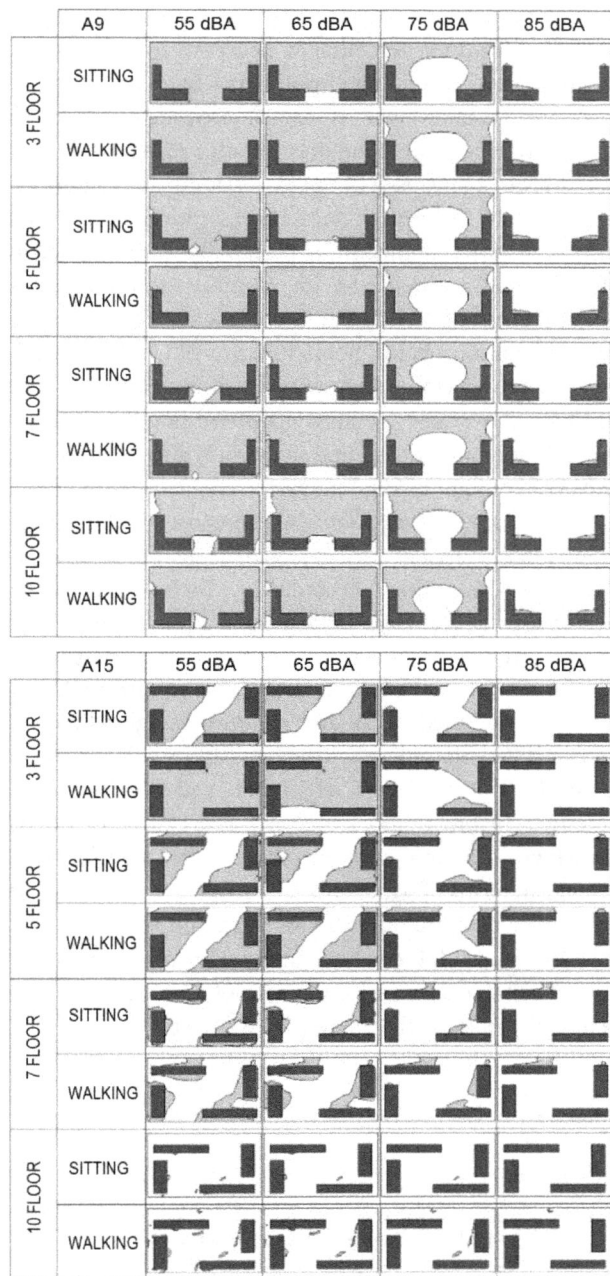

2. Evaluation of open comfortable areas in terms of both wind and noise

To carry out a general evaluation of common comfortable areas in terms of wind and noise, for each alternative settlement, the average results obtained are grouped according to the number of floors (3–5 floors for low buildings and for 7–10 floors high-rise buildings) and different noise levels. Grouping the percentages of common comfortable areas in terms of both factors in the diagrams for buildings with low and high numbers of floors provides a more robust assessment because it was determined that the comfort percentages in terms of wind show significant variations on high-rise buildings. Results are presented as graphics and evaluated as follows.

55 Leq dB(A) noise level is the comfort limit value, and 100% comfortable results were obtained for all settlement alternatives in terms of noise. Therefore, the percentages shown in Figure 11 are values obtained depend on wind.

For Istanbul, A9 and A17 configurations provide maximum comfort percentages both for low and high number of floors because any increase not occurs on outdoor wind speeds due to the structures. (Formation of the settlements do not influence the outdoor wind speed effects with height). A9 configuration with L-shaped structures that close the road side and do not use one-piece block on dominant wind directions (which can increase external wind speed and create channelling effects) gives good results in terms of wind and noise. In addition, configurations where the open area is located in the south and the structures constitute a barrier to the dominant wind direction, northeast and north (A2, A3, A5 from point-type blocks; A8, A12 and A20 from linear type blocks) are comfortable. The configuration with minimum comfortable area percentage is A13, where the north side is open, consisting of linear blocks vertical to the road and solid blocks that create channel effects. The findings are

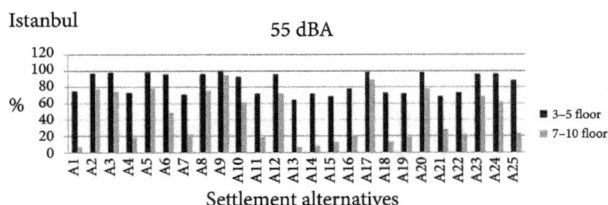

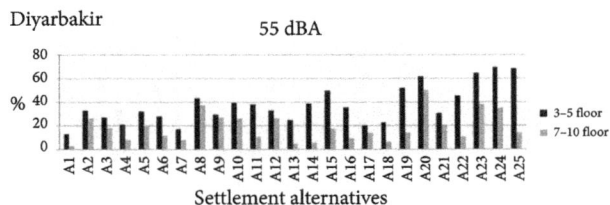

Fig. 10. The variation of comfortable outdoor areas according to mass housing alternatives in terms of wind and noise (Dark grey: house blocks, light grey: comfortable areas, white: uncomfortable areas)

Fig. 11. Averaged comfort percentages of the results of wind and 55 Leq dB(A) noise level

different for Diyarbakir as it can be clearly seen that comfort percentages decrease dramatically, especially for point-type blocks. The reason is that the average wind speed values in Diyarbakır are higher compared to Istanbul throughout the year. An increase at comfort percentages can be seen in Figure 11 going from point-type settlements to linear-type. A20 type configuration, where the north side is closed and the open areas positioned on the south side, provide higher comfort percentages for high number of floors. Consequently, A20, A23 and A24 configurations would seem to be good choices. It can also be seen, in configurations where the NE and NW directions are closed with L-shaped structures and in partially open or yard-type houses where the northern directions are completely closed with open areas arranged on the south side, better results are obtained.

In Figure 12, it can be seen that the comfort percentages decrease at 65 Leq dB(A) noise level and this result is mainly due to wind.

For Istanbul, again A9 and A17 configurations provide maximum comfort percentages both for low and high number of floors. It can be seen that the comfort percentages obtained in terms of noise start to affect common comfortable area percentages. Configurations like A23 with linear type blocks, closed off from the road are better performers than even configuration A17 when considering low numbers of floors. The A5 configuration has good performance no matter how many floors are considered whilst A11 and A13 configurations are the poorest performers. For Diyarbakır, comfort percentages decrease at 65 Leq dB(A) level but the Figure 12 is very similar to the 55 Leq dB(A) conditions. Comfort percentages of point-type block configurations are quite low both for low and high number of floors. L-shaped and yard-type linear, C-shaped, inverse closed U-shaped and fully closed

O-shaped configurations (A20, A23, A24, and A25) are good performers for prevailing northerly winds.

In Figure 13, it can be seen that the comfort percentages decrease considerably at 75 Leq dB(A) noise level and this is mainly due to noise. With increasing noise level, common comfort percentages decrease specifically for configurations consisting of point-type blocks.

For Istanbul, just as for 55 Leq dB(A) and 65 Leq dB(A) conditions, the performance of A9 configuration is remarkably good (51%) for all numbers of floors (3-5-7-10) at 75 Leq dB(A). A25 configuration, fully closed yard-type, has a high comfort percentage at 68% in terms of both wind and noise for low number of floors, but its percentage drops to 17% because of the unfavorable situation for high number of floors in terms of wind. For yard-type structures with a high number of floors, the air flow that passed over the structure channeled into the yard and created discomfort. The same situation is also valid for A21 and A22 configurations. The performance of the A17 and A20 configurations are poor when compared with configurations that consist of yard-type structures and completely closed road sides like A21 and A25 which provide better performance. For Diyarbakır, especially for point-type blocks and discrete linear-type settlements, the common comfortable area percentages drop to below 10% in 75 Leq dB(A)of compared to 20–30% for wind-weighted 55 and 65 Leq dB(A) noise levels. The level drops below 20% for most of the closed linear settlements and close to 30% is obtained for low number of floors. Only the fully closed yard-type A25 configuration provides 55% comfort percentage in terms of wind with low number of floors but the percentage drops to only 10% for high numbers of floors.

In Figure 14, it can be seen that the comfort percentages of suitable usage areas in terms of wind and noise are defined by noise at the 85 Leq dB(A) level.

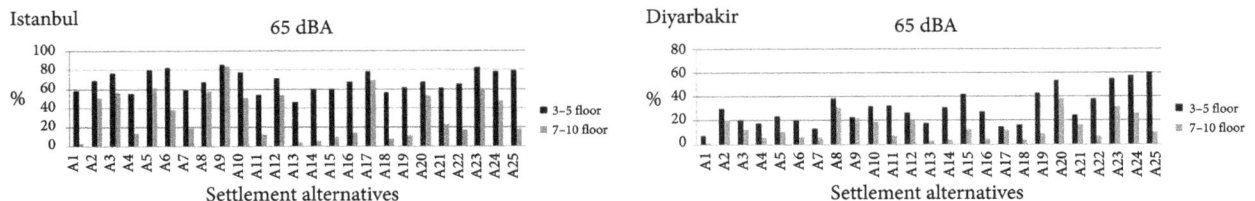

Fig. 12. Averaged comfort percentages of the results of wind and 65 Leq dB(A) noise level

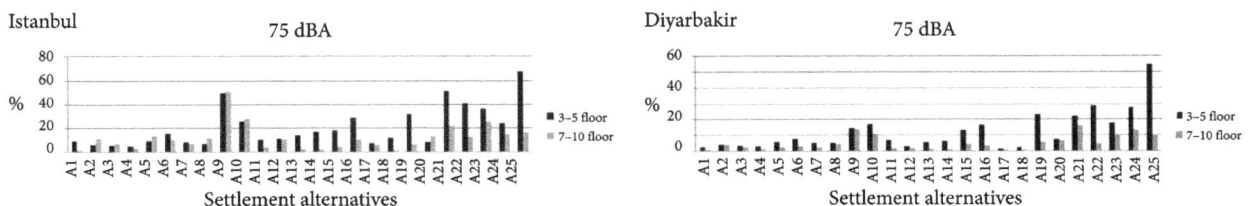

Fig. 13. Averaged comfort percentages of the results of wind and 75 Leq dB(A) noise level

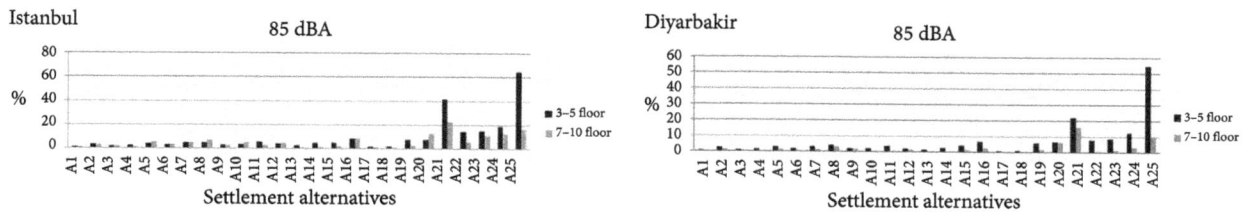

Fig. 14. Averaged comfort percentages of the results of wind and 85 Leq dB(A) noise level

As illustrated in Figure 14, for Istanbul, there are large decreases in comfort percentages for all settlement alternatives. Except for yard-type A21 and A25 type configurations with low numbers of floors and closed on the road side, comfort percentages are quite low, even dropping to just 1%. The results for comfort percentages are severely affected by the influence of noise. For Diyarbakır, except in linear U-shaped A21 configurations where the road side is completely closed and fully closed yard-type A25 configurations with low numbers of floors, all settlement alternatives have quite low comfort percentages, the majority close to zero. Only A25 configuration with a low number of floors provides around 55% comfort percentage.

A general review of the results obtained for Istanbul and Diyarbakir for common suitable usage areas in terms of wind and noise conditions is summarized below:

For Istanbul, at 55 Leq dB(A) and 65 Leq dB(A) where the wind is dominant in comfort percentages, results are similar. Including point-type blocks, most of settlements are considered to have good performance for both low and high numbers of floors. It can be seen that with increasing noise level, the performance of discrete point-type blocks reduces considerably; linear-type settlements come to the fore and the number of configurations with high comfort percentage decreases. For 3–5 floor settlements, linear configurations come to the fore, whilst for 7–10 floor settlements the comfort percentages are quite low. It can be seen that floor height adversely affects wind-related results and particularly in configurations with 7 floors or more, this effect seems to be very significant.

Since the average wind speed values in Diyarbakir are higher than those recorded at Istanbul throughout the year, comfort percentages and the number of comfortable configurations decrease dramatically. Similar results are obtained in both the 55 and 65 Leq dB(A) diagrams. For mainly northerly dominated wind directions closed L-shaped and yard-type linear, C-shaped, inverse closed U-shaped and fully closed O-shaped configurations (A20, A23, A24, and A25) perform well.

It is possible to determine configurations with both point-type and linear-type blocks that have high percentages of comfort for İstanbul (an exception being at the 85 Leq dB(A) noise level in Istanbul). On the other hand, there are significant differences between point-type

and linear-type blocks in all the noise level diagrams for Diyarbakır because of the high reduction of comfort level for point-type blocks. Diyarbakır is a difficult city to provide comfort using the geometric properties of a settlement. Point-type and discrete linear settlements perform poorly in Diyarbakır. Only a few settlements with yard-type linear configurations perform acceptably.

At 75 Leq dB(A) noise level and above, in the options where the blocks are arranged vertical to the road (in a north-south axes), the comfort ratio decreases dramatically. If the blocks are placed parallel to the road as a noise barrier, comfortable areas increase behind the structures. Therefore, improvements occur in comfort values for an alternative arrangement with blocks placed parallel to the road. Better comfort results appear in alternatives with L-C-U shaped design features rather than point-type and linear block layouts. This is related to the sheltered areas created by these layout alternatives from the effects of wind and noise.

Conclusions

Wind and noise are both physical environmental factors affecting user comfort directly. Because of them open and semi-open spaces cannot be used efficiently. This situation applies particularly in modern-day mass housing developments. Both wind and noise are important components of building physics in the field of determining the sizing for buildings, orienting structures in terms of climatic and noise factors and the design of structural shape and positioning with respect to other nearby structures. In this study, it has been shown that the configuration and positioning of structures in layout plan could be a major source of problems for open areas in terms of wind and noise.

Today, especially high rise identical apartment blocks are seen in the new mass housing settlements in all cities of Turkey like most of other countries. Therefore, this article has revealed the situation experienced in the present and developed solutions and recommendations in terms of wind and noise subjects. The general conclusions of this study are:

1. It has been demonstrated that linear-type settlements give better results compared with point-type block settlements. For block settlements, confi-

gurations parallel to the road have more comfortable areas than vertical alternatives. Increasing of floor numbers affects both wind and noise comfort adversely. From the perspective of noise control, the main reasons for this situation are the distance of buildings from the road, distance from one another and reflections from buildings surfaces. The configuration and positioning of structures should be determined by consideration of the dominant wind direction and annual variation.

2. Configurations that can create channelling effects along wind directions should be strictly avoided, particularly for high numbers of floors. High apartment blocks are totally unsuited to this kind of settlements as they are more exposed to solar radiation and do not provide sheltered outdoor spaces for the occupants as well.

3. If the settlements have dominant wind and noise directions, it is possible to find solutions using the geometric properties of the settlement in terms of wind and noise.

4. The results presented in this paper show it will be possible to select suitable configurations respecting wind and noise levels based on wind data for a particular city and the position of the structural design according to road position and noise level. However, the scope of this study is limited to providing a guide to human comfort in terms of wind and noise for outdoor activities in mass housing settlements. The effects of wind and noise within the structure could be considered as the subject of another study.

5. It was generally found that the performance of comfortable open spaces is not sufficiently provided for by components such as structural configuration, position and height, so additional considerations are needed. In order to improve the comfort performance of these spaces, recommendations are made to guide the design of wind and noise barriers. It is anticipated that related studies will be presented in a subsequent article.

Funding

This work was supported by the TUBITAK under Grant 111M560.

References

Aanen, L.; Van Uffelen, G. M. 2009. The evaluation of the application of CFD on pedestrian wind comfort in engineering practice, a validation study, in *EACWE 5*, 19–23 July 2009, Florence, Italy. 9 p.

Akdağ, N. Y. 2001. The use of noise barriers for urban noise control, *Journal of Architect and Engineer* 30: 80–82 (in Turkish).

Akdağ, N. Y.; Candemir, N. 2009. *Importance of noise mapping in urban information systems*. Report of state planning organization project. Project No. 26-DPT-03-01-01 (in Turkish).

Aknesil, A. E.; Akdağ, N. Y. 2011. Legal regulations governing noise control in Turkey and examples of environmental noise determination studies in Istanbul, in *6th Annual International Symposium on Environment*, 16–19 May 2011, Athens, Greece.

Archive of Turkish State Meteorological Service [online]. 2013 [cited 15 June 2013]. Available from Internet: http://www.mgm.gov.tr

Arens, E. A.; Gonzales, R.; Berglund, L. 1986. *Thermal comfort under an extended range of environmental conditions*. Center for the Built Environment, 01 January 1986, UC Berkeley.

Baltrėnas, P.; Petraitis, E.; Januševičius, T. 2011. Noise level study and assessment in the southern part of Panevėžys, *Journal of Environmental Engineering and Landscape Management* 18(4): 271–280. https://doi.org/10.3846/jeelm.2010.31

Bendtsen, H. 1994. Visual principles for the design of noise barriers, *Science of the Total Environment* 146–147: 67–71. https://doi.org/10.1016/0048-9697(94)90221-6

Blocken, B.; Carmeliet, J. 2004. Pedestrian wind environment around buildings: literature review and practical examples, *Journal of Thermal Envelope and Building Science* 28: 107–159. https://doi.org/10.1177/1097196304044396

Bootby, T. E.; Burroughs, C. B.; Bernecker, C. A.; Manbeck, H. B.; Ritter, M. A.; Grgurevich, S.; Cegelka, S.; Lee, P. H. 2001. *Design of wood highway sound barriers*. USDA Forest Service. Laboratory Research Paper FPL-RP-596, 1–66.

Bu, Z.; Kato, S.; Ishida, Y.; Huang, H. 2009. New criteria for assessing local wind environment at pedestrian level based on exceedance probability analysis, *Building and Environment* 44: 1501–1508. https://doi.org/10.1016/j.buildenv.2008.08.002

Campi, P.; Palumbo, A. D.; Mastrorilli, M. 2009. Effects of tree windbreak on microclimate and wheat productivity in a Mediterranean environment, *European Journal of Agronomy* 30: 220–227. https://doi.org/10.1016/j.eja.2008.10.004

Caniot, G.; Li, W.; Dupont, G. 2011. *Validations and applications of a CFD tool dedicated to wind assessment in urban areas* [online], [cited 10 August 2013]. Available from Internet: http://meteodyn.com/wp-content/uploads/2012/06/Validations-and-applications-of-a-CFD-tool-dedicated-to-wind-assessment-in-urban-areas.pdf

Chevallier, E.; Can, A.; Nadji, M.; Leclercq, L. 2009. Improving noise assessment at intersections by modeling traffic dynamics, Transportation research part D, *Transport and Environment* 14/2: 100–110. https://doi.org/10.1016/j.trd.2008.09.014

Cianfrini, C.; Corcione, M.; Fontana, L. 2007. Experimental verification of the acoustic performance of diffusive roadside noise barriers, *Applied Acoustics* 68(11–12): 1357–1372. https://doi.org/10.1016/j.apacoust.2006.07.018

Cornelis, W. M.; Gabriels, D. 2005. Optimal windbreak design for wind-erosion control, *Journal of Arid Environments* 61: 315–332. https://doi.org/10.1016/j.jaridenv.2004.10.005

Crombie, D. H.; Hothersall, D. C. 1994. The acoustic performance of multiple edge noise barriers, in *Inter-noise* 94, 29–31 August 1994, Yokohama, Japan.

Dal, Z.; Akdağ, N. Y. 2011. Noise disturbance caused by outdoor activities-a simulated environment study for Ali Sami Yen stadium, Istanbul, *Environmental Monitoring and Assessment* 174/1: 347–360. https://doi.org/10.1007/s10661-010-1462-z

Desanghere, G. 2007. QCITY: Providing cities a guide for noise action plans, in *Inter-noise 2007*, 28–31 August, Istanbul, Turkey.

Ekici, I.; Bougdah, H. A. 2003. Review of research on environmental noise barriers, *Building Acoustics* 10/4: 289–323. https://doi.org/10.1260/135101003772776712

EU. 2002. *EU Noise Directive, 2002/49/EC* [online], [cited 20 July 2014]. Available from Internet: http://ec.europa.eu/environment/noise/directive_en.htm

European Commision. 1996. *Future Noise Policy*. European Commission Green Paper. November, Brussels, Belgium.

FHWA. 2006. *Noise compatible land use curriculum* [online]. US Federal Highway Administration, Department of Transportation [cited 15 July 2014]. Available from Internet: http://www.fhwa.dot.gov/environment/noise/nnoise _compatible_planning/ workshops/ ncp_curr.pdf

Gedik, G. Z.; Akdağ, N. Y.; Sener, B.; Kiraz, F.; Çaçan, F. 2014. *Optimization of mass housing settlements in terms of wind and noise control*. Report of TUBITAK 1001 Project, Pr. No. 111M560 (in Turkish).

Guedes, I. C. M.; Bertoli, S. R.; Zannin, P. H. T. 2011. Influence of urban shapes on environmental noise: a case study in Aracaju – Brazil, *Science of the Total Environment* 412: 66–76. https://doi.org/10.1016/j.scitotenv.2011.10.018

Hagen, L. J.; Skidmore, E. L.; Miller, P. L.; Kipp, J. E. 1981. Simulation of effect of wind barriers on airflow, *ASAE* 24: 1002–1008.

Harris, C. M. 1994. *Noise control in buildings*. McGraw-Hill Inc., USA.

He, J.; Song, C. S. 1999. Evaluation of pedestrian winds in urban area by numerical approach, *Journal of Wind Engineering and Industrial Aerodynamics* 81: 295–309. https://doi.org/10.1016/S0167-6105(99)00025-2

Hong, B.; Lin, B. 2015. Numerical studies of the outdoor wind environment and thermal comfort at pedestrian level in housing blocks with different building layout patterns and trees arrangement, *Renewable Energy* 73: 18–27. https://doi.org/10.1016/j.renene.2014.05.060

Hong Kong Environmental Protection Department. 2003. *Guidelines on design of noise barriers* [online], [cited 10 December 2013]. Available from Internet: http://www.hyd.gov.hk/en/publications_and_publicity/publications/technical_document/guidelines_on_noise_barriers/index.html

Hoppe, P. 2002. Different aspects of assessing indoor and outdoor thermal comfort, *Energy and Buildings* 34: 661–665. https://doi.org/10.1016/S0378-7788(02)00017-8

Hu, C.; Wang, F. 2004. Using a CFD approach for the study of street-level winds in a built-up area, *Building and Environment* 40: 617–631. https://doi.org/10.1016/j.buildenv.2004.08.016

Ishizuka, T.; Fujiwara, K. 2004. Performance of noise barriers with various edge shapes and acoustical conditions, *Applied Acoustics* 65: 125–141. https://doi.org/10.1016/j.apacoust.2003.08.006

ISO 1996-1:2003. *Acoustics – description, measurement and assessment of environmental noise – Part 1: Basic Quantities and Assessment Procedures.*

ISO 9613-2:1996. *Acoustics, attenuation of sound during propagation outdoors – Part 2: General method of calculation.*

Isyumov, N.; Davenport, A. G. 1975. The ground level wind environment in built up areas, in *4th International Conference on Wind Effects on Buildings and Structures*, 1975, London, UK.

King, E. A.; Murphy, E.; Rice, H. J. 2011. Implementation of the EU environmental noise directive: lessons from the first phase of strategic noise mapping and action planning in Ireland, *Journal of Environmental Management* 92(3): 756–754. https://doi.org/10.1016/j.jenvman.2010.10.034

King, E. A.; Rice, H. J. 2009. The development of a practical framework for strategic noise mapping, *Applied Acoustics* 70(8): 1116–1127. https://doi.org/10.1016/j.apacoust.2009.01.005

Koch-Nielsen, H. 2002. *Stay cool*. A Design Guide for the Built Environment in Hot Climates. Earthscan Publications Ltd.

Koss, H. H. 2006. On differences and similarities of applied wind comfort criteria, *Journal of Wind Engineering and Industrial Aerodynamics* 94: 781–797. https://doi.org/10.1016/j.jweia.2006.06.005

Kotzen, B. 2002. Plants and environmental noise barriers, in *International Conference on Urban Horticulture*, 2–6 September 2002, Waedenswil, Switzerland.

Kotzen, B.; English, C. 2004. *Environmental noise barriers, a guide to their acoustic and visual design*. 2nd ed. USA: Taylor&Francis.

Kubota, T.; Miura, M.; Tominaga, Y.; Mochida, A. 2008. Wind tunnel test on the relationship between building density and pedestrian-level wind velocity: development of guidelines for realizing acceptable wind environment in residential neighborhoods, *Building and Environment* 43: 1699–1708. https://doi.org/10.1016/j.buildenv.2007.10.015

Kumbay, A.; Yüksel, Z.; Akdağ, N. Y.; Can, C. 2006. Evaluation of urban noise problems: historical peninsula case study, in *Euronoise 2006*, 30 May – 1 June, Tempere, Finland.

Maffei, L.; Masullo, M.; Aletta, F.; Aletta, F.; Di Gabriele, M. 2013. The influence of visual characteristics of barriers on railway noise perception, *Science of the Total Environment* 445–446: 41–47. https://doi.org/10.1016/j.scitotenv.2012.12.025

Makarewicz, R. 1991. Traffic noise in a built-up area, *Applied Acoustics* 31(3): 37–50. https://doi.org/10.1016/0003-682X(91)90045-G

Montana Department of Transportation. 2008. *Growing neighborhoods in growing corridors: land use planning for highway noise*. Report no. FHWA/MT-08-002/8117-36, US.

Murphy, E.; King, E. A. 2010. Strategic environmental noise mapping: methodological issues concerning the implementation of the EU noise directive and their policy implications, *Environment International* 36–3: 290–98.

Naderzadeh, M.; Monazzam, M. R.; Nassiri, P.; Bellah Fard, S. M. 2011. Application of perforated sheets to improve the efficiency of reactive profiled noise barriers, *Applied Acoustics* 72(6): 393–398. https://doi.org/10.1016/j.apacoust.2011.01.002

Newman, P.; Thornley, A. 1996. *Urban planning in Europe*. 1st ed. London: Routledge, 60–61. https://doi.org/10.4324/9780203427941

Nordstrom, K. F.; Hotta, S. 2004. Wind erosion from cropland in the USA: a review of problems, solutions and prospects, *Geoderma* 121: 157–167. https://doi.org/10.1016/j.geoderma.2003.11.012

Oldham, D. J.; Egan, C. A. A. 2011. Parametric investigation of the performance of T-profiled highway noise barriers and the identification of a potential predictive approach, *Applied Acoustics* 72: 803–813. https://doi.org/10.1016/j.apacoust.2011.04.012

Pendwarden, A. D. 1973. Acceptable wind speeds in towns, *Building Science* 8: 259–267. https://doi.org/10.1016/0007-3628(73)90008-X

Pendwarden, A. D.; Wise, A. F. E. 1975. *Wind environment around buildings*. Building Research Establishment Digest.

RENAM. 2010. The Regulation of Environmental Noise Assessment and Management. *The Official Gazette*, No. 25862, Turkey.

Shi, X.; Zhu, Y.; Duan, J.; Shao, R.; Wang, J. 2015. Assessment of pedestrian wind environment in urban planning design, *Landscape and Urban Planing* 140: 17–28. https://doi.org/10.1016/j.landurbplan.2015.03.013

SoundPLAN manual V 7.3. 2012. Backnang: Braustain+Berndt GMBH.

Stathopoulos, T. 2009. Wind and Comfort, in *Eacwe 5*, 19–23 July 2009, Florence, Italy.

Stathopoulos, T.; Wu, H.; Zacharias, J. 2004. Outdoor human comfort in an urban climate, *Building and Environment* 39: 297–305. https://doi.org/10.1016/j.buildenv.2003.09.001

Szucs, A. 2004. Stadia in the environment – environment in stadia, in *Plea2004 – The 21ᵗʰ Conference on Passive and Low Energy Architecture*, 19–22 September, Eindhoven, The Netherlands.

Szucs, A.; Moreau, S.; Allard, F. 2007. Spectators' aerothermal comfort assessment method in stadia, *Building and Environment* 42: 2227–2240. https://doi.org/10.1016/j.buildenv.2006.03.009

The zoning regulation of planned areas, *The Official Gazette* No. 18916, 1985, Turkey.

Prime Ministry housing development administration regulations for the implementation of the public housing loans, *The Official Gazette* No. 23019, 14 June, 1997, Turkey.

Thorsson, P. J.; Ogren, M.; Kropp, W. 2004. Noise levels on the shielded side in cities using a flat city model, *Applied Acoustics* 65/4: 313–323. https://doi.org/10.1016/j.apacoust.2003.11.005

UrbaWind [online]. 2013 [cited July 2016]. Available from Internet: http://meteodyn.com/en/logiciels/cfd-wind-pedestrian-comfort-safety-urbawind-software/#.WB8wGvmLTDc

Venckus, Z.; Grubliauskas, R.; Venslovas, A. 2012. Research on the effectiveness of the inclined top type of a noise barrier, *Journal of Environmental Engineering and Landscape Management* 20(2): 155–162. https://doi.org/10.3846/16486897.2011.634068

Watson, D. 2006. *Evaluation of benefits and opportunities for innovative noise barrier designs*. Report for Arizona Department of Transportation, Report no. 572, November, USA.

Watts, G. R.; Godfrey, N. S. 1999. Effects on roadside noise levels of sound absorption materials in noise barriers, *Applied Acoustics* 58: 385–402. https://doi.org/10.1016/S0003-682X(99)00007-9

WG-AEN. 2006. *European Commission Working Group. Good practice guide for strategic noise mapping and the production of associated data on noise exposure* [online], [cited July 2016]. Available from Internet: http://ec.europa.eu/environment/noise/pdf/wg_aen.pdf

WHO. 1996. *Guidelines for community Noise-5 noise management*. World Health Organization.

Willemsen, E.; Wisse, J. A. 2007. Design for wind comfort in the Netherlands: procedures, criteria and open research issues, *Journal of Wind Engineering and Ind. Aerodynamics* 95: 1541–1550. https://doi.org/10.1016/j.jweia.2007.02.006

Yoshie, R.; Mochida, A.; Tominaga, Y.; Kataoka, H. 2007. Co-operative project for CFD prediction of pedestrian wind environment in the architectural institute of Japan, *Journal of Wind Engineering and Industrial Aerodynamics* 95: 1551–1578. https://doi.org/10.1016/j.jweia.2007.02.023

Gülay ZORER GEDİK is a Professor at the Faculty of Architecture of Yildiz Technical University, Turkey. She received her PhD in Building Physics from Yildiz Technical University, Istanbul in 1995. She is especially interested in the whole process of Climatic Building Design and Energy Efficient Building Principles and Wind and Solar Architecture. She has completed eight research projects. Her last (project coordinator) research project's title is "Optimization of mass-housing settlements in terms of wind and noise control". She has worked for the National Building Energy Calculation Methodology Improvement Project (as a coordinator of Net Energy part). She has published many peer-reviewed scientific articles in high impact journals and presented many papers at international and national conferences.

Neşe YÜĞRÜK AKDAĞ. She studied at Yildiz Technical University and obtained a degree in Architecture in 1984, an MSc degree in "Building Physics" in 1987, and received her PhD in "Architectural Acoustics" from Yildiz Technical University in 1995. She is full professor since 2011 in Building Physics Unit of the Faculty of Architecture in Yildiz Technical University. She has range of articles, papers, books and applications on room and building acoustics. She has also several researches on noise mapping. She prepared and organized some national projects and contributed to several national and international projects. She is a member of the Turkish Acoustical Society.

Fatih KİRAZ. He is an Assistant Professor at the Faculty of Fine Arts and Design of Nuh Naci Yazgan University, Turkey. He received his PhD in Building Physics from Yildiz Technical University, Istanbul in 2015. He is especially interested in the whole process of Climatic Building Design and Noise Control and Wind and Solar Architecture. He has participated as a scholar to the research Project "Optimization of mass-housing settlements in terms of wind and noise control".

Bekir ŞENER. He is an Assistant Professor at the Naval Architecture and Maritime Faculty of Yildiz Technical University, Turkey and currently the Vice Head of Department of Naval Architecture and Marine Engineering. He received his PhD in Naval Architecture from Yildiz Technical University, in 2012. His research interest include design of ship and yachts, hydrodynamics, 3D modeling and CFD. He has participated as a researcher to the "Optimization of mass-housing settlements in terms of wind and noise control" project. He also prepared and contributed several national projects.

Raşide ÇAÇAN. She studied at Yildiz Technical University and received Bachelor degree in Architecture in 2008, and MSc degree in "Building Physics" in 2014. She has participated as a scholar to the "Optimization of mass-housing settlements in terms of wind and noise control" project.

8

CHANGES IN PHYSICO-CHEMICAL PARAMETERS DURING FORCED-AERATION STATIC-PILE CO-COMPOSTING OF CATTLE MANURE WITH CALCIUM CYANAMIDE

Huasai Simujide[a], Chen Aorigele[a], Chun-Jie Wang[b], Jun-E Yu[c], Bai Manda[a], Ma Lina[d]

[a] College of Animal Science, Inner Mongolia Agricultural University, 010018 Hohhot, P. R. China
[b] College of Veterinary Medicine, Inner Mongolia Agricultural University, 010018 Hohhot, P. R. China
[c] Ulanqab Vocational College, 012000 Ulanqab, P. R. China
[d] College of Life Sciences, Inner Mongolia Agricultural University, 010018 Hohhot, P. R. China

Abstract. The goal of this research was to determine the effect of $CaCN_2$ addition into manure mixed with sawdust on the composting process under forced-aeration static condition, especially on nitrogen (N). The changes in the physical and chemical parameters over the entire composting period were evaluated. The profile of temperature, pH, and NO_3^--N was improved in the piles mixed with 2%, 3% and 4% $CaCN_2$ (the test piles). The NH_4^+-N met the limit value accepted for mature compost. Finally, the additive contents of not less than 2% but not more than 3% in $CaCN_2$ were recommended for this system based upon a comprehensive evaluation of the measured parameters.

Keywords: nitrogen, composting, manure, waste management technologies.

Reference to this paper should be made as follows: Simujide, H.; Aorigele, C.; Wang, C.-J.; Yu, J.-E.; Manda, B.; Lina, M. 2014. Changes in physico-chemical parameters during forced-aeration static-pile co-composting of cattle manure with calcium cyanamide, *Journal of Environmental Engineering and Landscape Management* 22(02): 125–131. http://dx.doi.org/10.3846/16486897.2013.852557

Introduction

Composting is a low-cost, effective and natural way of recycling organic materials, and it is included in sustainable agriculture and recommended for organic agriculture (Peigné, Girardin 2004). Composting has been shown to have some advantages, including pathogen suppression, weed seed killing and improvement of pesticide degradation (Dorahy *et al.* 2009; Karanasios *et al.* 2010), and its end products (composts) have been used as soil amendments due to their high concentration in organic matter (Ohsowski *et al.* 2012). Moreover, compost application in soil has been reported to reduce losses of N (Kelln *et al.* 2012). Despite these benefits, composting also can have several disadvantages. Several studies demonstrated that nutrients would be lost during composting (Venglovsky *et al.* 2011; Webber *et al.* 2009) and possible odors associated with composting would be generated (Li *et al.* 2008; Hanajima *et al.* 2010). Additionally, nutrient losses are an agronomic problem for

organic farmers because they attempt to compensate for N scarce organic farms by their compost (Peigné, Girardin 2004). One of the nutrient losses during composting is N which is mainly lost as ammonia (NH_3) but may also as N_2, N_2O and NO_x (Bueno *et al.* 2009; Velasco-Velasco *et al.* 2011). Fukumoto and Inubushi (2009) noted that most of the N losses resulted from NH_3 emission which was 9.5% of the initial total N (T–N) during active composting of swine manure, when total N_2O emissions were 9.3% of T–N, and total N losses were 27.8% of the initial T–N which was markedly greater than the sum total of NH_3 and N_2O. The N losses during composting can be influenced by several factors including temperature, pH, C/N ratio, and turning (Parkinson *et al.* 2004; Bueno *et al.* 2009).

Calcium cyanamide ($CaCN_2$) is an environmentally friendly N fertilizer with pesticide effects (Shi *et al.* 2009). And our previous studies showed that manure composting would quickly reach the sanitary standard and the quality of the composting products would be improved with the

Corresponding author: Chen Aorigele
E-mail: aori6009@163.com

addition of $CaCN_2$ (Simujide *et al.* 2012ab). Overall, these studies indicated the possibility of use of $CaCN_2$ as an amendment during manure composting to improve composting efficiency. However, N fate during composting of this kind of compostable mixtures had not been evaluated. Therefore, the aim of this study was to evaluate the effect of $CaCN_2$ addition on the changes in the physico-chemical parameters during manure composting, paying special attention to the evolution of N, which will provide further information about the feasibility of selection of $CaCN_2$ for composting.

1. Materials and methods

1.1. Experimental design

The experiment was carried out for 63 d from 21 April to 23 October 2012. Fresh manure of apparently healthy dairy cows from a 900-cow dairy farm was collected and mixed with sawdust (bulking agent) at a ratio of 4:1. Composting was conducted on a concrete apron, subdivided into four separate compartments, each with floor dimensions 50×50 cm. Compartments were isolated from each other by 40 cm high walls, and were all roofed. An iron screen mesh was installed about 10 cm above the reactor bottom to segregate the compost pile from the aeration channel. Sawdust was laid 2 cm thick over the iron screen mesh to distribute air equally, and air was blown to the piles using air pumps (one pump for one pile). Forced-aeration was conducted intermittently every day from the beginning of the composting process to 35 d, and then the piles were turned once a week. In addition, the air supply was enriched by turning of piles after each sampling. Compost piles contained about 25 kg compostable mixtures per compartment at the beginning of the experiment, and were maintained in a roughly conical shape during composting. The piles were classified into three test piles (test I, test II and test III) and control pile. Solid $CaCN_2$ was respectively mixed with test I, test II and test III at a rate of 2%, 3% and 4% by weight, while was not added into the control pile. Compost samples in duplicate were collected from each pile by using five-spot analyses at days 0, 4, 7, 14, 21, 28, 35, 42, 56, and 63 for the analysis of different parameters. The characteristics of the composting materials are shown in Table 1.

Table 1. Characteristics of composting materials

	Fresh manure	Sawdust	$CaCN_2$
Moisture content (%)[a]	81.42±0.52[b]	7.95±0.44	–
C (%)[a]	33.80±1.60	42.43±1.78	1.71±0.97
N (%)[a]	1.71±0.05	0.25±0.01	21.24±0.08
C/N ratio	19.78	168.43	0.08

[a] On a wet weight basis; [b] Standard deviation.

1.2. Physico-chemical analysis

Ambient temperature around the compost bins and the temperature within each pile were measured daily at 9:00 AM and 16:00 PM. Daily temperature of the pile was the average temperature of the top, middle and bottom layer in the two measurements. The moisture content of the samples was determined after oven drying at 105 °C to a constant weight (Li 1983). The pH was determined by a Mettler-Toledo EL20 pH-meter (Mettler-Toledo international trading (Shanghai) Co., Ltd.). Total nitrogen (T–N) and total carbon (T–C) was respectively measured by kjeldahl method and $K_2Cr_2O_7$ volumetric method (Li 1983). The concentration of NH_4^+–N was determined by extracting the sample with 10% NaCl and the extracts were distillated (Nanjing Agricultural College 1980). Phenol-disulfonic acid colorimetric method was used to obtain the concentration of NO_3^-–N (Nanjing Agricultural College 1980).

1.3. Statistical analysis

Three replicates were used for each analysis. Data were presented as the mean values of triplicates.

2. Results and discussion

2.1. Temperature

The goal of the temperature control during a composting process is to achieve to the greatest extent the harmlessness and stabilization of the compost materials after composting. Temperature changes reflect microbial activities of the compost pile and the state of the composting process. As shown in Fig. 1, all piles presented a change of temperature rise, temperature drop and maturation. The test piles reached their highest temperature 44.80 °C at 13 d in test I, 55.00 °C at 11 d in test II and 52.00 °C at 14 d in test III, respectively, while the control pile recorded the maximum temperature 43.80 °C at 1 d. The result indicated that the time to reach the high temperatures during composting was delayed with the addition of $CaCN_2$ resulting from the reduced porosity of the compost piles in the beginning. However, the maximum temperatures in the test piles were higher than the control pile, and the duration of the high temperatures was also longer in the former than in the latter. But generally, the duration time in all piles was not long enough by comparison with the results of others. The temperatures of 55 to 60 °C for 7 to 14 d were regarded as the ideal temperatures for effective composting (Tang *et al.* 2004; Johannessen *et al.* 2005; Sylla *et al.* 2006). As for lower temperatures, 33.5 to 41.5 °C for 7 d was reported to be enough for efficient inactivation of bacterial populations such as *E. coli* during composting (Larney *et al.* 2003). In other aspect, a number of factors always have an effect on the change of composting temperature, including the composition of the composting materials, composting method, and environmental condition, etc. (Changa *et al.* 2003; de Guardia *et al.* 2010; Tirado, Michel 2010). The short duration of the high temperatures in the present study were mainly caused by the long ventilation time at the first stage of composting and the poor insulating qualities of small masses of the piles.

2.2. Moisture content

Desirable moisture contents of mixtures at the beginning of composting fall within the range of from 40 to 65% with a preferred range of 50–60% (Agnew, Leonard 2003; Trémier *et al.* 2009). Nevertheless, in fed-batch composting of household biowaste, the optimum moisture content was in the range 30–40%, at which the microorganisms showed the highest protease activity (Narihiro *et al.* 2004a, b). And a successful composting with high initial moisture contents was also reported (Hanajima *et al.* 2006). So, the optimum moisture conditions during composting depend on the nature of the compostable materials. The initial moisture contents in this study were adjusted around 65%. The moisture content in all piles, on the whole, exhibited a declining trend (Fig. 2). The mean moisture content of the test piles decreased from an initial value of 64.69% to a final value of 60.78% in test I, from 64.13% to 58.44% in test II and from 63.77% to 61.94% in test III, respectively. In the control pile, it was from 67.08% to 61.50%. Among them, the evaporation of moisture was highest in test II, which was related to the longest duration of the high temperatures in it.

2.3. pH

The pH in the test piles showed the same change that followed a first sharp decline and then stabilization trend, when it increased greatly at the first 4 days and then went to stabilize in the control pile (Fig. 3). As the additive content of $CaCN_2$ increased, the pH increased accordingly. The mean pH went respectively from an initial value of 7.40, 9.38, 10.78 and 11.61 to 7.41, 7.71, 8.47 and 8.91 in the control pile, test I, test II and test III at the end of composting. And during the process, the pH was below 9 from 4d in test I and from 14d both in test II and test III, and then almost fluctuated between 8 and 9 which is considered to be the preferred range for a successful composting (Zeng *et al.* 2011). Comparatively, the pH in the control pile was in 8–9 from 4 d to 21 d and then decreased slowly and stabilized around 7.5. This is an acceptable condition but not the ideal one.

2.4. N transformation

Trends in percent T–N were shown in Fig. 4. At the first stage of composting, the T–N changed relatively little in

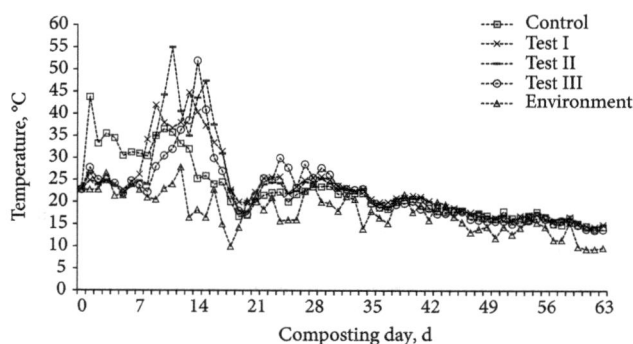

Fig. 1. The temperature profile during composting

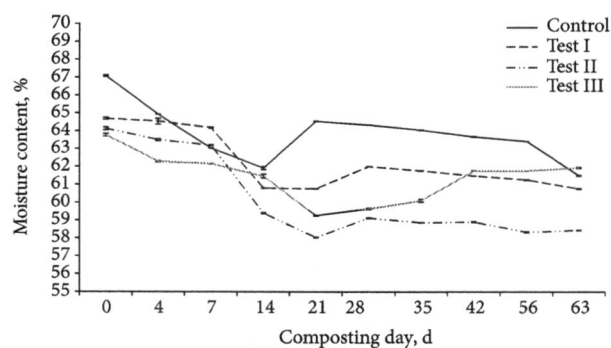

Fig. 2. The moisture content profile during composting

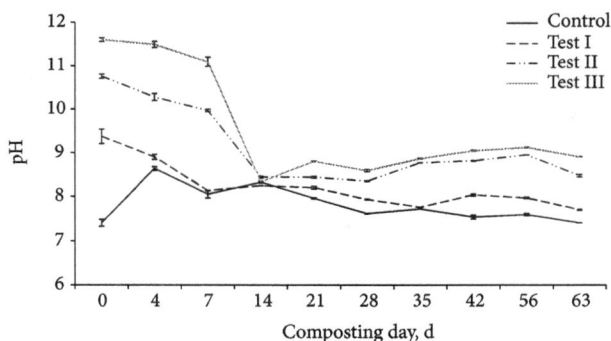

Fig. 3. The pH profile during composting

the control pile when decreased in the test piles, and then all began to increase. The decrease was due to the release of N in the form of NH_3. With the composting time increasing, organic matters constantly decompose into CO_2 and H_2O and have a continuous loss, and the volume of the composting mixtures decreases accordingly. So, the T–N is condensed and its content always has a slight increase at the end of composting (Rihani *et al.* 2010). The percent T–N increased from an initial value of 0.89%, 2.02%, 2.48% and 2.98% to 1.20%, 2.64%, 3.05% and 3.02% in the control pile, test I, test II and test III after 63d composting that the increasing rate was 34.83%, 30.39%, 22.98% and 1.34%, respectively. This showed that the

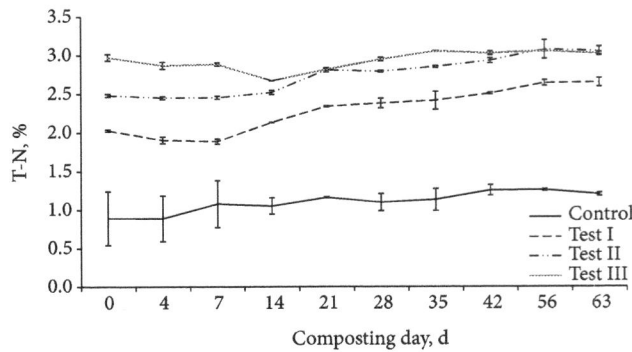

Fig. 4. The T-N profile during composting

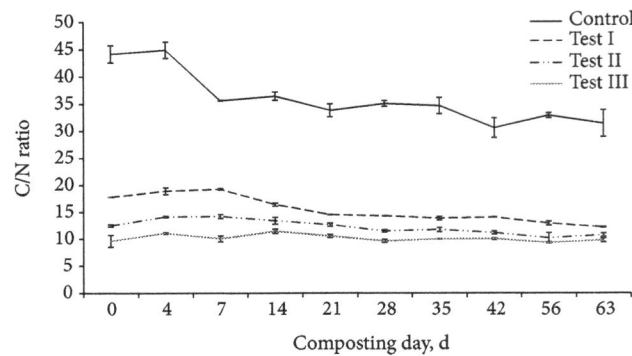

Fig. 5. The C/N ratio profile during composting

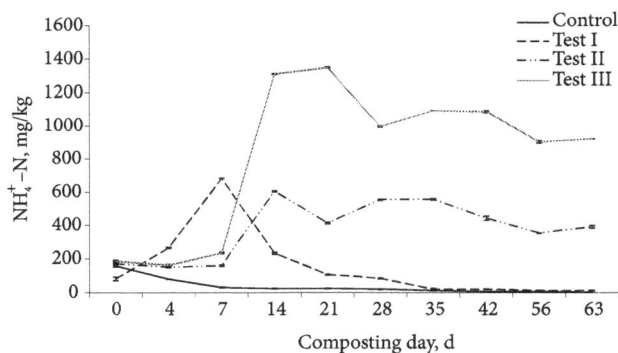

Fig. 6. The NH_4^+-N profile during composting

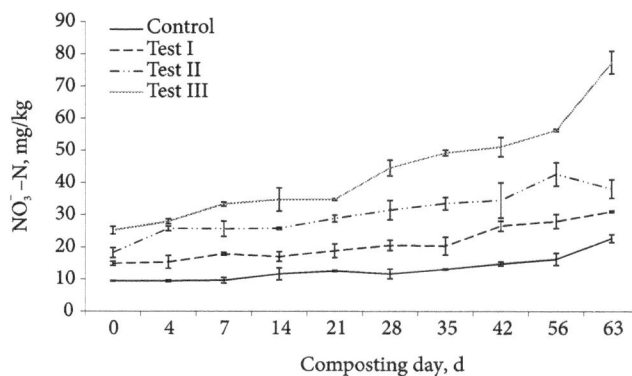

Fig. 7. The NO_3^--N profile during composting

increasing rate of T–N dropped with the increase of $CaCN_2$ additive content, it was because the increase of additive content resulted in the decrease of C/N ratio and when it went low, there was less C available for consumption, whereas N nutrients were relatively excess, causing a part of N to transform into ammoniacal N and have their volatilization losses. Paredes *et al.* (2000) found that during co-composting of olive mill wastewater with solid organic wastes, the high N losses through NH_3 volatilization occurred in the mixtures which had low initial C/N ratios (15.0–21.5), but it was reduced in the pile with the higher initial C/N ratio (31.0). In the current study, the initial C/N ratio was 44.13 in the control pile and 9.63–17.79 in the test piles. Then, it fell sharply in the control pile followed by test I and test II with a relatively little change in test III, before stabilizing and reaching values of 31.40 in the control pile and 9.75–12.17 in the test piles after the maturation phase (Fig. 5). The C/N ratio was so low at the start in these piles that it would not be evaluated by the normal suggested levels.

The concentration of NH_4^+-N always falls during the composting process (Paredes *et al.* 2000). However, different results have been found in some studies, for instance, Bueno *et al.* (2009) reported that the NH_4^+–N concentrations increased significantly at the initial stage and then decreased during composting of trimming residues at different levels of moisture and particle size. As shown in Fig. 6, the NH_4^+–N concentrations in all piles, although occurring at different times, increased significantly at the initial stage and reached their maximum level because of ammonification which is always associated with temperature increase and mineralization of organic N compounds (Bueno *et al.* 2009). Then, the NH_4^+–N concentrations decreased due to NH_3 volatilization and immobilization by microorganisms (Bueno *et al.* 2009), indicating that the inhibitor was gradually eliminated as composting process continued. Except in test III, the NH_4^+–N contents in other piles were <400 mg/kg at the end of composting, which is the maximum limit recommended for a mature compost. The NO_3^-–N contents showed a rising trend. During the thermophilic phase, lower increments were detected with respect to the mesophilic phase because of the inhibitory effect of NH_3 and high temperature on the growth of nitrifying bacteria (Huang *et al.* 2004; Bueno *et al.* 2009). The addition of N-rich $CaCN_2$ produced the composts with higher NO_3^-–N concentrations than the control pile (Fig. 7).

Conclusions

1. The forced-aeration static-pile composting process of cattle manure mixed with $CaCN_2$ is technically feasible,

and it could be considered as an effective way to transform N and inactivate *E. coli* in manure.

2. The addition of $CaCN_2$ delayed the time to reach the high temperatures during composting, but increased the high temperature levels and duration.

3. With addition of $CaCN_2$, the pH evolution was improved; the final values of the NH_4^+-N contents basically met the limit accepted for a mature compost; the final concentrations of NO_3^--N were enhanced.

4. With a comprehensive assessment of the physico-chemical parameter values, the addition of not less than 2% but not more than 3% $CaCN_2$ was recommended.

Acknowledgements

We thank the National Natural Science Foundation of China (Nos. 31060318 and 31260570).

References

Agnew, J. M.; Leonard, J. J. 2003. The physical properties of compost, *Compost Science & Utilization* 11(3): 238–264. http://dx.doi.org/10.1080/1065657X.2003.10702132

Bueno, P.; Yañez, R.; Caparrós, S.; Díaz, M. J. 2009. Evaluating environmental parameters for minimum ammonium losses during composting of trimming residues, *Journal of the Air & Waste Management Association* 59(7): 790–800.

Changa, C. M.; Wang, P.; Watson, M. E.; Michel, F. C.; Hoitink, H. A. J. 2003. Assessment of the reliability of a commercial maturity test kit for composted manures, *Compost Science & Utilization* 11(2): 127–145. http://dx.doi.org/10.1080/1065657X.2003.10702119

de Guardia, A.; Mallard, P.; Teglia, C.; Marin, A.; Le Pape, C.; Launay, M.; Benoist, J. C.; Petiot, C. 2010. Comparison of five organic wastes regarding their behaviour during composting: part 1, biodegradability, stabiliazation kinetics and temperature rise, *Waste Management* 30(3): 402–414. http://dx.doi.org/10.1016/j.wasman.2009.10.019

Dorahy, C. G.; Pirie, A. D.; McMaster, I.; Muirhead, L.; Pengelly, P.; Chan, K. Y.; Jackson, M.; Barchia, I. M. 2009. Environmental risk assessment of compost prepared from salvinia, *Egeria densa*, and alligator weed, *Journal of Environmental Quality* 38(4): 1483–1492. http://dx.doi.org/10.2134/jeq2007.0555

Fukumoto, Y.; Inubushi, K. 2009. Effect of nitrite accumulation on nitrous oxide emission and total nitrogen loss during swine manure composting, *Soil Science and Plant Nutrition* 55(3): 428–434. http://dx.doi.org/10.1111/j.1747-0765.2009.00376.x

Hanajima, D.; Kuroda, K.; Fukumoto, Y.; Haga, K. 2006. Effect of addition of organic waste on reduction of *Escherichia coli* during cattle feces composting under high-moisture condition, *Bioresource Technology* 97(14): 1626–1630. http://dx.doi.org/10.1016/j.biortech.2005.07.034

Hanajima, D.; Kuroda, K.; Morishita, K.; Fujita, J.; Maeda, K.; Morioka, R. 2010. Key odor components responsible for the impact on olfactory sense during swine feces composting, *Bioresource Technology* 101(7): 2306–2310. http://dx.doi.org/10.1016/j.biortech.2009.11.026

Huang, G. F.; Wong, J. W.; Wu, Q. T.; Nagar, B. B. 2004. Effect of C/N on composting of pig manure with sawdust, *Waste Management* 24(8): 805–813. http://dx.doi.org/10.1016/j.wasman.2004.03.011

Johannessen, G. S.; James, C. E.; Allison, H. E.; Smith, D. L.; Suander, J. R.; McCarthy, A. J. 2005. Survival of a Shiga toxin-encoding bacteriophage in a compost model, *FEMS Microbiology Letters* 245(2): 369–375. http://dx.doi.org/10.1016/j.femsle.2005.03.031

Karanasios, E.; Tsiropoulos, N. G.; Karpouzas, D. G.; Ehaliotis, C. 2010. Degradation and adsorption of pesticides in compost-based biomixtures as potential substrates for biobeds in Southern Europe, *Journal of Agricultural and Food Chemistry* 58(16): 9147–9156. http://dx.doi.org/10.1021/jf1011853

Kelln, B.; Lardner, H.; Schoenau, J.; King, T. 2012. Effects of beef cow winter feeding systems, pen manure and compost on soil nitrogen and phosphorous amounts and distribution, soil density, and crop biomass, *Nutrient Cycling in Agroecosystems* 92(2): 183–194. http://dx.doi.org/10.1007/s10705-011-9480-y

Larney, F. J.; Yanke, L. J.; Miller, J. J.; McAllister, T. A. 2003. Fate of coliform bacteria in composted beef cattle feedlot manure, *Journal of Environmental Quality* 32(4): 1508–1515. http://dx.doi.org/10.2134/jeq2003.1508

Li, Q. K. 1983. *Conventional methods of soil and agricultural chemistry analysis*. Beijing: Science Press. 457 p. (in Chinese).

Li, X.; Zhang, R.; Pang, Y. 2008. Characteristics of dairy manure composting with rice straw, *Bioresource Technology* 99(2): 359–367. http://dx.doi.org/10.1016/j.biortech.2006.12.009

Nagao, N.; Watanabe, K.; Osa, S.; Matsuyama, T.; Kurosawa, N.; Toda, T. 2008. Bacterial community and decomposition rate in long term fed-batch composting using woodchip and Polyethylene terephthalate (PET) as bulking agents, *World Journal of Microbiology & Biotechnology* 24(8): 1417–1424. http://dx.doi.org/10.1007/s11274-007-9625-y

Nanjing Agricultural College. 1980. *Soil and agricultural chemistry analysis*. Beijing: Agricultural Press. 395 p. (in Chinese).

Narihiro, T.; Abe, T.; Yamanaka, Y.; Hiraishi, A. 2004a. Microbial population dynamics during fed-batch operation of commercially available garbage composters, *Applied Microbiology and Biotechnology* 65(4): 488–495. http://dx.doi.org/10.1007/s00253-004-1629-z

Narihiro, T.; Takebayashi, S.; Hiraishi, A. 2004b. Activity and phylogenetic composition of proteolytic bacteria in mesophilic fed-batch garbage composters, *Microbes and Environment* 19(4): 292–300. http://dx.doi.org/10.1264/jsme2.19.292

Ohsowski, B. M.; Klironomos, J. N.; Dunfield, K. E.; Hart, M. M. 2012. The potential of soil amendments for restoring severely disturbed grasslands, *Applied Soil Ecology* 60: 77–83. http://dx.doi.org/10.1016/j.apsoil.2012.02.006

Paredes, C.; Roig, A.; Bernd, M. P. 2000. Evolution of organic matter and nitrogen during co-composting of olive mill wastewater with solid organic wastes, *Biology and Fertility of Soils* 32(3): 222–227. http://dx.doi.org/10.1007/s003740000239

Parkinson, R.; Gibbs, P.; Burchett, S.; Misselbrook, T. 2004. Effect of turning regime and seasonal weather conditions on nitrogen and phosphorus losses during aerobic composting of cattle manure, *Bioresource Technology* 91(2): 171–178. http://dx.doi.org/10.1016/S0960-8524(03)00174-3

Peigné, J.; Girardin, P. 2004. Environmental impacts of farm-scale composting practices, *Water Air and Soil Pollution* 153(1–4): 45–68.
http://dx.doi.org/10.1023/B:WATE.0000019932.04020.b6

Rihani, M.; Malamis, D.; Bihaoui, B.; Etahiri, S.; Loizidou, M.; Assobhei, O. 2010. In-vessel treatment of urban primary sludge by aerobic composting, *Bioresource technology* 101(15): 5988–5995. http://dx.doi.org/10.1016/j.biortech.2010.03.007

Shi, K.; Wang, L.; Zhou, Y. H.; Yu, Y. L.; Yu, J. Q. 2009. Effects of calcium cyanamide on soil microbial communities and *Fusarium oxysporum* f. sp. *Cucumberinum, Chemosphere* 75(7): 872–877.
http://dx.doi.org/10.1016/j.chemosphere.2009.01.054

Simujide, H.; Aorigele, C.; Wang, C. J.; Manda, B.; Ma, L. N. 2012a. Effect of calcium cyanamide on pathogenic *Escherichia coli* during mesophilic composting and impact on composting process, *Global Nest Journal* 14(4): 460–467.

Simujide, H.; Aorigele, C.; Wang, C. J.; Manda, B.; Ma, L. N.; Wu, M. Y.; Li, Y.; Bai, T. R. G. 2012b. Reduction of foodborne pathogens during cattle manure composting with addition of calcium cyanamide, *Journal of Environmental Engineering and Landscape Management* 21(2): 77–84.
http://dx.doi.org/10.3846/16486897.2012.721373

Sylla, Y. B.; Kuroda, M.; Yamada, M.; Matsumoto, N. 2006. Feasibility study of a passive aeration reactor equipped with vertical pipes for compost stabilization of cow manure, *Waste Management & Research* 24(5): 456–464.
http://dx.doi.org/10.1177/0734242X06066429

Tang, J. C.; Kanamori, T.; Inoue, Y.; Yasuta, T.; Yoshida, S.; Katayama, A. 2004. Changes in the microbial community structure during thermophilic composting of manure as detected by the quinone profile method, *Process Biochemistry* 39(12): 1999–2006. http://dx.doi.org/10.1016/j.procbio.2003.09.029

Tirado, S. M.; Michel, F. C. 2010. Effects of turning frequency, windrow size and season on the production of dairy manure/sawdust composts, *Compost Science & Utilization* 18(2): 70–80. http://dx.doi.org/10.1080/1065657X.2010.10736938

Trémier, A.; Teglia, C.; Barrington, S. 2009. Effect of initial physical characteristics on sludge compost performance, *Bioresource Technology* 100(15): 3751–3758.
http://dx.doi.org/10.1016/j.biortech.2009.01.009

Velasco-Velasco, J.; Parkinson, R.; Kuri, V. 2011. Ammonia emissions during vermicomposting of sheep manure, *Bioresource Technology* 102(23): 10959–10964.
http://dx.doi.org/10.1016/j.biortech.2011.09.047

Venglovsky, J.; Sasakova, N.; Gregová, G.; Lakticova, K.; Ondrasovicova, O.; Ondrasovic, M.; Köfer, J.; Schobesberger, H. 2011. Nitrogen loss during composting of poultry litter, in *Animal Hygiene and Sustainable Livestock Production, Proceedings of the XV International Congress of the International Society for Animal Hygiene,* 3–7 July, 2011, Vienna, Austria, Tribun EU, 1127–1129.

Webber, D. F.; Mickelson, S. K.; Richard, T. L.; Ahn, H. K. 2009. Effects of a livestock manure windrow composting site with a fly ash pad surface and vegetative filter strip buffers on sediment, nitrate, and phosphorus losses with runoff, *Journal of Soil and Water Conservation* 64(2): 163–171.
http://dx.doi.org/10.2489/jswc.64.2.163

Yao, H. C. 2002. *Veterinary Microbiology experiment guidance.* Beijing: Chinese Agricultural Press. 128 p. (in Chinese).

Zeng, G.; Yu, Z.; Chen, Y.; Zhang, J.; Li, H.; Yu, M.; Zhao, M. 2011. Response of compost maturity and microbial community composition to pentachlorophenol (PCP)-contaminated soil during composting, *Bioresource Technology* 102(10): 5905–5911. http://dx.doi.org/10.1016/j.biortech.2011.02.088

Huasai SIMUJIDE. Works in College of Animal Science, Inner Mongolia Agricultural University. Number of publications: 16; participated in two conferences in China. Research interests include waste management technologies, air pollution, landscape management, veterinary microbiology.

Chen AORIGELE. Works in College of Animal Science, Inner Mongolia Agricultural University. Number of publications: 54; participated in 22 conferences in China. He has been awarded the following awards: Inner Mongolia Autonomous Region Science and Technology Progress Award twice, Inner Mongolia Autonomous Region Youth Science and Technology Award once. He has been the Executive Director of Straw Resources Utilization Branch of Chinese Association of Agricultural Science Societies; the Executive Director of Inner Mongolia Dairy Association; the Vice General Secretary of Inner Mongolia Association of Animal Science; the Deputy Director of Inner Mongolia Breeding Professional Committee; the member of Japanese Society of Animal Science, Japanese Society of Grassland Science, Japanese Society of Application of Animal Behavior. Research interests include waste management technologies, air pollution, landscape management, environmental management.

Chun-Jie WANG. Works in College of Veterinary Medicine, Inner Mongolia Agricultural University. Number of publications: 59; has participated in 18 conferences in China. He has been awarded the following awards: Inner Mongolia Autonomous Region "321 talents project" the second level people, Inner Mongolia Autonomous Region Department of Education "111 talents project" the third level people. He has been the Council Member of Chinese Association of Animal Physiology, Northern China Association of Veterinary Pathology; the member of Chinese Association of Veterinary Pathology; the premium member of Chinese Association of Animal Science and Veterinary Medicine. Research interests include waste management technologies, landscape management, environmental management, medical microbiology.

Jun-E YU. Works in Ulanqab Vocational College. Number of publications: 3; has participated in one conference in China. Research interests include waste management technologies, air pollution, landscape management.

Bai MANDA. Works in College of Animal Science, Inner Mongolia Agricultural University. Number of publications: 5; has participated in two conferences in China. Research interests include waste management technologies, air pollution, landscape management.

Ma LINA. Works in College of Life Sciences, Inner Mongolia Agricultural University. Number of publications: 5; has participated in two conferences in China. Research interests include waste management technologies, air pollution, landscape management, food and fermentation engineering.

9

ASSESSMENT OF THE ADSORPTION KINETICS, EQUILIBRIUM AND THERMODYNAMICS FOR THE POTENTIAL REMOVAL OF NI²⁺ FROM AQUEOUS SOLUTION USING WASTE EGGSHELL

Sukru ASLAN, Ayben POLAT, Ugur Savas TOPCU

Department of Environmental Engineering, Cumhuriyet University, 58140, Sivas, Turkey

Abstract. In this study, Ni^{2+} sorption onto the waste eggshell was investigated under different operational conditions. Results indicated that the eggshell could be successfully used to remove Ni^{2+} ions from the water. Quick sorption process reached to equilibrium in about 2 hours with maximum sorption at pH 7.0. Based on the experimental data, Langmuir isotherm model with the q_m value of 1.845 mg Ni^{2+}/g eggshell was observed. The pseudo-second-order model provided the best correlation coefficient in comparison with other models. The calculated q_e values derived from the pseudo-second-order for sorption of Ni^{2+} ions were very close to the experimental (q_{exp}) values. Such thermodynamic parameters as $\Delta G°$, $\Delta H°$, and $\Delta S°$ were determined in order to predict the nature of adsorption. Results indicated that the adsorption of Ni^{2+} onto the eggshell was endothermically supported by the increasing adsorption of Ni^{2+} ions with temperature.

Keywords: adsorption isotherms, eggshell, nickel, thermodynamics.

Introduction

Heavy metal pollution of water sources is one of the most important problems. Various industries – such as mining and smelting of metalliferrous, electroplating, battery manufacture, textile production, refineries, and petrochemical factories etc. – produce wastewater that contains metals (Celekli, Bozkurt 2011) which is discharged to the water bodies after treatment. As heavy metals are non-biodegradable water pollutants, they tend to accumulate in living organisms (Gupta *et al.* 2010; Celekli, Bozkurt 2011).

In this experimental study, nickel was selected as an adsorbate, because nickel compounds have widespread application in many industrial processes, such as metal plating, silver refineries, zinc base casting and storage battery industries, and its concentration in industrial wastewaters range from 3.40 to 900 mg/L (Erdogan *et al.* 2005). Although low concentrations of nickel may be beneficial to organisms as a component in a number of enzymes and stimulate the activation of microorganisms (Aslan, Gurbuz 2011), an exceeded permissible exposure level of nickel causes various health problems such as unintentional weight loss, heart and liver damage, renal edema, lung and pulmonary fibrosis, skin dermatitis, and gastrointestinal discomfort (Bar-Sela *et al.* 1992; Celekli *et al.* 2010; Gupta *et al.* 2010; Kumar *et al.* 2011; Pahlavanzadeh *et al.* 2010).

Nickel (II) is one of the toxic pollutants in the industrial effluent wastewaters. Maximum contaminant limits of Ni (II) set by the EU and WHO as 0.02 mg/L and 0.07 mg/L, respectively (EU 2011; WHO 2005). Recommended limits for Ni (II) in reclaimed water for irrigation are 0.2 mg/L and 2.0 mg/L for the short and long-term usage (USEPA 2012).

Due to the accumulation of heavy metals in the food chain and persistence in the ecosystem, it causes the toxicity for living organisms. The wastewaters containing heavy metals are required to be properly treated prior to discharge into the receiving waters (Bhatnagar, Minocha 2010). Such conventional methods as the chemical precipitation, filtration, ion exchange, evaporation, reverse osmosis, solvent extraction, electrochemical treatment, membrane technologies, and adsorption could be applied in order to remove heavy metals from wastewater. Among these methods, adsorption is considered as an efficient and inexpensive method when the low concentration of heavy metal exists in the wastewater (Bermúdez *et al.* 2011; Ghazy *et al.* 2011; Kumar *et al.* 2011).

Corresponding author: Sukru Aslan
E-mail: saslan@cumhuriyet.edu.tr

Although the usage of activated carbon for sorption is an expensive method, commercially available activated carbon for the heavy metal removal from wastewater has been studied extensively (Erdogan *et al.* 2005). Consequently it is important to find new materials for removing by sorption of heavy metals from wastewaters.

In order to decrease the treatment cost of wastewater, researches have been focused on finding cheapest and effective sorbents. Among the numerous natural materials for removing nickel ions, biosorbents such as *Spirulina platensis* (Celekli, Bozkurt 2011), *Gracilaria caudata* and *Sargassum muticum* (Bermúdez *et al.* 2011), *Punica granatum* peel waste (Bhatnagar, Minocha 2010), cashew nut shell (Kumar *et al.* 2011), waste tea (Malkoc, Nuhoglu 2009; Shah *et al.* 2012) (*Camella cinencis)* (Aikpokpodion *et al.* 2010), acid–washed barley straw (Thevannan *et al.* 2011), Thespesia Populnea bark (Prabakaran, Arivoli 2012), activated locust bean husk (*Parkia biglobosa*) (Oladunni *et al.* 2012), *Sargassum filipendula* (Kleinubing *et al.* 2012), bael tree leaf powder (Kumar, Kirthika 2009), chitosan encapsulated *Sargassum sp.* (Yang *et al.* 2011), sugarcane bagasse pith (Krishnan *et al.* 2011), activated sludge (Liu *et al.* 2012), treated alga (*Oedogonium hatei*) (Gupta *et al.* 2010), *chlorella vulgaris* (Aksu, Donmez 2006), eggshell (Ghazy *et al.* 2011) have attracted attention as a low-cost sorbents from wastewater.

The aim of this study was to evaluate the adsorption capacity of the waste eggshell to remove Ni^{2+} ions in the synthetic waters. Effects of the initial pH, Ni^{2+} concentrations, temperature, contact time, and adsorbent dosage were determined in the batch experiments. The equilibrium isotherms; Langmuir, Freundlich, Temkin, and Dubinin-Radushkevich (D – R) were determined by applying various Ni^{2+} concentrations. Adsorption kinetic models were used to analyze the kinetic and mechanisms of Ni^{2+} adsorption.

1. Materials and methods

1.1. Preparation of adsorbent

After washing of the waste eggshell by tap and distilled waters, it was dried at 60 °C for 24 hours in an oven. The dry clean eggshells were crushed and screened through a set of sieves to get the size of 75–106 μm.

1.2. Sorption experimental studies

The synthetic waters were prepared by dissolving known masses of $NiCl_2 \cdot 6H_2O$ in the distilled water. A known amount of eggshell was used throughout the experiments and final volumes of the solution were 100 mL. Experimental studies were carried in 250 mL glass-stoppered Erlenmeyer flasks.

Experiments were performed at various initial concentrations of Ni^{2+} (5.0–50 mg/L) and waste eggshell amounts (0.1–1.0 g/L). Additionally, temperature and pH were tested for various levels to determine the optimal operational conditions. Kinetic constants were determined at the initial concentrations of 15–25–35 mg Ni^{2+}/L at the constant initial pH value and adsorbent dosage of $\cong 7.0$ and 0.5 g/L, respectively. Kinetic experiments were also carried out at the temperature of 298–308–318 K. Batch experimental studies were performed in an orbital incubator shaker (Gerhardt) at a constant speed of 150 rpm.

1.3. Analytical methods

In order to get supernatant liquids, the samples were centrifuged at 4000 rpm for 10 min (NF800, NUVE). Concentrations of Ni^{2+} in the solutions were determined with a Merck photometer (PHARO100). A spectraquant analytical kit (Merck, 14785) was used to measure Ni^{2+} concentrations in the initial and final solutions.

1.4. Calculations

The following equation was used to determine the amount of Ni^{2+} adsorbed onto eggshell:

$$q_e (mg/g) = (C_0 - C_e)(mg/L) \times \frac{V}{M} (mL/g). \qquad (1)$$

Adsorption process was quantified by calculating the sorption percentage (E %) as defined by the Eq. 2:

$$Sorption(E)(\%) = \frac{C_0 - C_e}{C_0} \times 100, \qquad (2)$$

where q_e (mg/g) is the maximum amount of Ni^{2+} adsorbed at equilibrium; the initial and equilibrium concentrations of Ni^{2+} in the solutions were shown as C_o and C_e (mg/L), respectively. M is the amount of eggshell (g), and V (mL) is the total solution volume in the Erlenmeyer flasks.

2. Results and discussion

Sorption experiments were performed in duplicate and the average values of samples were presented in the study. Blank samples (without Ni^{2+}) were used also to compare the results through all batch procedures. Data presented in the figures are mean values of standard deviation ($\leq 7\%$) from the experiments.

2.1. Effect of contact time

Figure 1 shows the variation of Ni^{2+} uptake with mixing time at pH 7.0 using 0.5 g eggshell. As can be seen from the figure that the equilibrium time for the sorption of Ni^{2+} was about 120 min. At the equilibrium point, the highest Ni^{2+} sorption efficiency of about 50% and the adsorption value of 1.95 mg/g were obtained.

The active sites of sorbents availability and the highest driving force for mass transfer at the beginning of the experiments (zero to 20 min) caused rapid uptake of Ni^{2+} ions from the solution. Due to the occupancy of eggshell active sites and the lower concentrations of Ni^{2+} in the solutions, Ni^{2+} sorption was slower after passing 20 min of agitation times.

2.2. Effects of pH

Since the initial pH of solution not only affects the reactive groups present on the surface of adsorbents (protonation/deprotonation effects), but also influences solubility of metals and the competition ability of hydrogen ions with metal ions, the initial pH of solution is an important parameter for the evaluation of sorption performances (Bermúdez et al. 2011; Celekli, Bozkurt 2011; Chojnacka, 2005). The sorption of Ni^{2+} was investigated as the function of pHs in the range of 3.0 to 7.0 with an increment of 0.5 pH units. Experiments were not extended to pH value of higher than 7.0 because the precipitation of Ni^{2+} ions forming hydroxides. Precipitation of Ni^{2+} may be applied for recovery when the wastewater is not including other pollutants.

Polat and Aslan (2014) determined the temperature and pH effects on the release of Ca^{2+} and HCO_3^- from the eggshell. Because the eggshell was composed mainly of calcium carbonate, pH were increased during the experimental study. However, the level of solution pHs were lower than 8.0 at the end of the batch experimental studies.

Sorption of Ni^{2+} on the eggshell sorbents at various pHs are presented in Figure 2. As can be seen in figure that uptake of Ni^{2+} was a function of initial solution pH. The adsorption of Ni^{2+} ions increases with increase in the solution pHs. Significant sorption was not observed at the pH values of 3.0 and 3.5. The lowest Ni^{2+} adsorption efficiency of about 11% was observed at the initial pH value of 3.0. It could be attributed to the higher concentration of hydrogen ions in the solution competing with Ni^{2+} for binding sites on the eggshell. Similar observations were reported in the literature for the sorption of Ni^{2+} ions on the various adsorbents (Bermúdez et al. 2011; Celekli, Bozkurt 2011; Kumar et al. 2011). At low pH values, H^+ ions occupy most of the adsorption sites of eggshell surface and Ni^{2+} sorption could be limited due to the electric repulsion with H^+ ions on the eggshell surface (Kumar et al. 2011). Increasing the pHs values from 3.5 to 6.5, sorption capacities (q_e) and the removal efficiencies of Ni^{2+} were increased significantly from 0.67 mg/g to 2.02 mg/g and 17.5% to 50.4% respectively. It was assumed that because the adsorbent surface was more negatively charged at high pHs, the sorption of heavy metal ions by eggshell increased. The q_e value and removal efficiency decreases slightly when the initial pH of solution was increased to 7.0.

The sorption capacities and removal efficiencies of Ni^{2+} were increased significantly from 0.43 to 2.02 mg

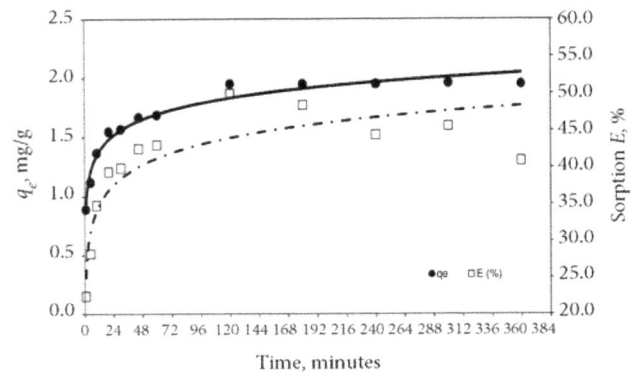

Fig. 1. Contact time effects on the sorption (C_0 = 20 mg Ni^{2+}/L)

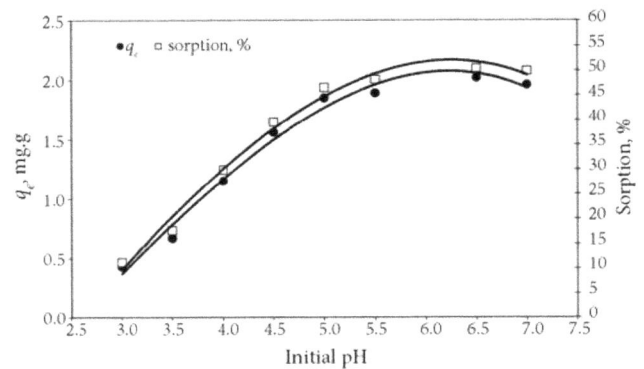

Fig. 2. Initial pH effects on the sorption of Ni^{2+}

Ni^{2+}/g.eggshell and 11% to 50.5%, respectively by increasing the pH from 3.0 to 6.5.

2.3. Effect of temperature

The effect of temperature on Ni^{2+} uptake capacity of eggshell was studied and results are presented in Figure 3. As can be seen, Ni^{2+} ions uptake capacity of eggshell increased with increasing temperature up to 318 K. An increase in temperature from 298 to 318 K, increases the q_e values from 1.96 to 2.2 mg/g. Sorption efficiency achieved about 56% at the temperature of 318 K. The removal efficiencies of Ni^{2+} ions at equilibrium were about 50% and 56% at the temperature of 298 K and 318K, respectively. It was assumed that it was probably related with the increase of Ca^{2+} release from the eggshells at higher temperatures. Polat and Aslan (2014) reported that elevating the temperature from 25 to 50 °C, release of Ca^{2+} ions into the aqueous solution increased about two times. Rising temperature might create new active sites by releasing Ca^{2+} ions from the eggshell (Polat, Aslan 2014) and enlarge the pore size of adsorbent (Demirbas et al. 2009). Additionally, the collision frequency between adsorbent and Ni^{2+} ions is elevated at high temperatures.

Experimental results indicating that the adsorption of Ni^{2+} ions was favored at higher temperature and the sorption of Ni^{2+} is endothermic in nature. As a results the adsorption capacity of eggshell is improved at high temperature.

2.4. Effect of adsorbent amount

The adsorbent dosage is an important parameter in the sorption process. At a given equilibrium concentration of pollutants, the adsorbent takes up more pollutants at lower adsorbent amount than at higher amounts (Al-Homaidan et al. 2014). Effect of eggshell doses on the removal efficiency of Ni^{2+} and q_e values are shown in Figure 4. It was observed that the Ni^{2+} removal efficiency of the eggshell was a function of eggshell amounts in the solution. The percent removal of Ni^{2+} declined along with the decrease in eggshell amount.

It can be seen from the figure that initially the removal efficiency increases gradually with the increase in eggshell amount in the aqueous solution while the q_e values decreases. The maximum adsorption efficiency of Ni^{2+} ion onto the eggshell was found to be 75.1% at the dose of 10 g/L eggshell. The increase in sorption efficiency of heavy metal could be attributed to the increased number and exchangeable sites of adsorbent available for the adsorption (Kumar et al. 2011).

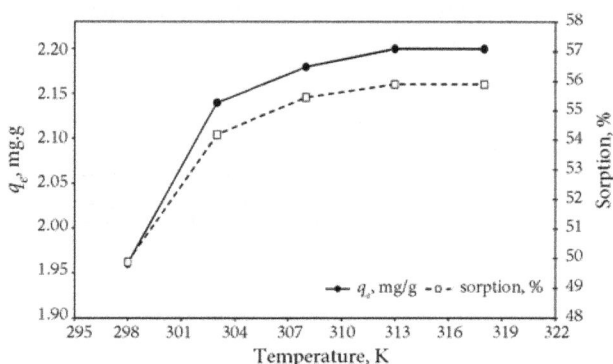

Fig. 3. Effects of temperature on the sorption of process

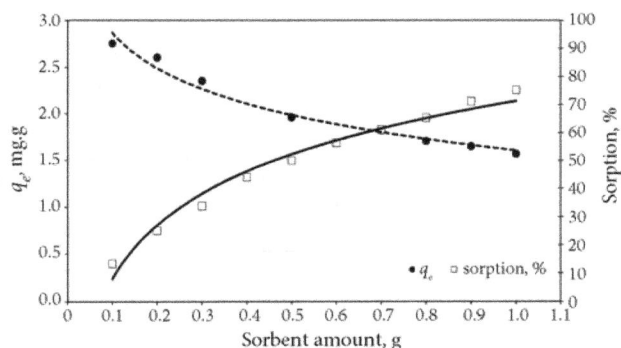

Fig. 4. Effects of eggshell amount on the Ni^{2+} ions sorption

2.5. Modeling of sorption equilibrium depending on Ni^{2+} concentrations

In order to understand the interaction between a sorbate and an adsorbent, it is important to establish the most appropriate correlation for the equilibrium curves. The experimental data were analyzed by applying the most commonly used equilibrium models namely Langmuir, Freundlich, Temkin, and D-R. The mathematical expressions are given in Table 1. Where q_m indicates the monolayer sorption capacity of adsorbate (mg/g). The constants b and E are the mean free energy and sorption per molecule of the sorbate, respectively. Sorption parameters for the isotherms are as follows; K_L (L/mg) Langmuir constant related to the energy of sorption, K_{Fi} (L/mg) Freundlich constant related to the sorption capacity of adsorbent, q_{max} (mg/g) is the maximum biosorption capacity of D–R. b_T and A_T (L/mg) Temkin isotherm parameters, R is the gas constant (8.314 joule.mol/K); T is the absolute temperature (K). The value of R_L indicates that the shape of the sorption process is; unfavorable ($R_L > 1$), linear ($R_L = 1$), favorable ($0 < R_L < 1$) or irreversible ($R_L = 0$) (Kilic et al. 2011; Sljivic et al. 2009).

The constants of isotherms equation are presented in Table 2. The best fit was obtained by Langmuir model as compared with the other isotherm models due to determine the highest correlation coefficient value of 0.993. Langmuir model is suggesting that the Ni^{2+} ions were adsorbed onto the eggshell in a monolayer. The maximum monolayer adsorption capacity was found to be 1.845 mg Ni^{2+}/g for the eggshell. The essential characteristic of the Langmuir isotherm can be used to predict the affinity between the sorbent and sorbate using separation factor, "R_L". The R_L was determined to be 0.0119–0.107 for the concentrations of 5.0–50 mg Ni^{2+}/L which indicated that the sorption of Ni^{2+} by waste eggshell sample was favorable.

Previous experimental studies indicated that the pretreatment procedure increase the uptake capacity of sorbents (Ahmad et al. 2010; Gupta et al. 2010; Ewecharoen et al. 2009) and most of the organic materials (Gupta et al. 2006; Bhatnagar, Minocka 2010; Celekli, Bozkurt 2011) have higher q_e value than an inorganic materials (Otun et al. 2006; Bayat 2002; Rao et al. 2002). In order to justify the validity of eggshell as an adsorbent for Ni^{2+} adsorption, adsorption potential is compared with other adsorbents and summarized in Table 3.

2.6. Kinetics of sorption

Sorption kinetics of Ni^{2+} ions on the eggshell were analyzed using four kinetic models for fitting sorption kinetic data: pseudo-first-order, pseudo-second-order, intraparticle diffusion, and Elovich models. Equations for the kinetic models are presented in Table 1. Experiments were

Table 1. Equations of isotherm and kinetic models

	Equations	Plot	Parameters	References
Equilibrium models				
Langmuir	$q_e (mg/g) = q_m \dfrac{K_L C_e}{1 + K_L C_e}$ $R_L = 1/(1 + K_L \times C_0)$	C_e/q_e vs $\cdot C_e$	q_m = 1/slope k_L = slope/intercept	Kilic *et al.* (2011); Sljivic *et al.* (2009)
Freundlich	$q_e (mg/g) = K_{Fi} C_e^{1/n}$	$\log q_e$ vs $\cdot \log C_e$	k_F = exp (intercept) n = 1/(slope)	Tsai *et al.* (2008)
Temkin	$q_e (mg/g) = B \ln A_T + B \ln C_e$	Q_e vs $\ln C_e$	q_e = slope A_t = exp(intercept)/ (slope)	Kilic *et al.* (2011)
Dubinin-Radushkevich	$\ln q_e = \ln q_{max} - \beta \varepsilon^2$ $\varepsilon = RT \ln\left(1 + \dfrac{1}{C_\varepsilon}\right)$	$\ln q_e$ vs ε^2	q_0 = exp(intercept) b = −(slope)	Baig *et al.* (2010)
Kinetic models				
Pseudo first-order	$\log(q_e - q_t) = \log q_e - \dfrac{k_1}{2.303} t$	$\log (q_e - q_t)$ vs t	q_e = exp(intercept) k_1 = −(slope)	Chiou, Li (2002); Chairat *et al.* (2005)
Pseudo second-order	$\dfrac{t}{q_t} = \dfrac{1}{k_2 q_e^2} + \dfrac{t}{q_e}$ $h = k_2 \times q_e^2$	t/q_t vs t	q_e = 1/(slope) k_2 = (slope)^2/(intercept)	Rao *et al.* (2002)
Intra particle diffusion	$q_t = k_{id} t^{1/2} + C$	q_t vs $t^{1/2}$	k_i = slope	Ghasemi *et al.* (2012)
Elovich	$q_t = \dfrac{1}{\beta} \ln \alpha\beta + \dfrac{1}{\beta} \ln t$	q_t vs $\ln t$	b = slope a = 1/(slope) exp(intercept/slope)	Demirbas *et al.* (2009)

repeated for different initial eggshell amounts (15–25–35 mg/L) and temperatures (298–308–318 K).

Sorption capacities (q_e) and the calculated values (q_e, k_1, k_2, R^2, and h) from the models are presented in Table 4. Comparison the results of kinetic data, it can be concluded that the pseudo-second-order model provided the best correlation coefficient. In addition, the calculated q_e values derived from the pseudo-second-order were very close to the experimental (q_{exp}) values.

As can be seen from the Table 4 that the equilibrium adsorption capacity, q_e, increases as the initial Ni^{2+} concentration, C_i, increased from 15 to 35 mg/L. However, it was found that the rate constant of pseudo-second-order (k_2) seem to have a decreasing trend with increasing the initial Ni^{2+} concentrations. Similar trends were also observed by applying pseudo-second-order model at various temperatures. For example, the values of q_e increased from 1.96 mg/g to 2.22 mg/g at the temperature of 298 K and 318 K, respectively. The reason for this situation might be attributed to the less competition for the sorption surfaces sites at lower concentration. At higher concentrations, the competition for the surface active sites is high and consequently lower sorption rates are achieved (Kumar *et al.*

Table 2. Correlation coefficient and adsorption parameters for the models

Model	Sorption Parameters	
Freundlich	R^2	0.688
	n	4.63
	K_F	1.061
Langmuir	R^2	0.993
	R_L	0.033
	q_m	1.845
	K_L	1.665
Temkin	R^2	0.690
	B_T	0.282
	$A_T (L/g)$	29.96
R-D	R^2	0.946
	q_0 (mg/g)	1.83
	β (mol^2/j^2)	0.519
	E (kj/mol)	0.98

2011). These results confirmed that the chemisorption mechanisms may play an important role for the sorption of Ni^{2+} on the eggshell.

2.7. Adsorption thermodynamics

In order to determine the thermodynamic parameters – such as free energy ($\Delta G°$, Kj/mol), entalphy ($\Delta H°$, Kj/mol), and entropy ($\Delta S°$, j/mol/K) – change of Ni^{2+} adsorption onto the eggshell, the batch experiments were performed at different temperatures of 298, 303, 308, 313, and 318 K. The thermodynamic parameters are calculated using the following equations (Mezenner, Bensmailli 2009):

$$\Delta G° = - RT \cdot \ln K_d; \qquad (3)$$
$$\ln K_d = - (\Delta H°/RT) + (\Delta S°/R), \qquad (4)$$

where K_d is the distribution coefficient for the adsorption.

Table 3. Comparison of maximum monolayer adsorption on Ni (II) ions onto various adsorbents

Adsorbents	q_m (mg/g)	K_L (ml/g)	References
Pomegranate peel	69.4	24.10^3	Bhatnagar, Minocka 2010
Spirulina platensis	69.04	0.0019	Celekli, Bozkurt 2011
Irradiation-grafted activated carbon	55.7	0.009	Ewecharoen et al. 2009
Acid-treated alga	44.247	0.063	Gupta et al. 2010
Activated carbon	44.1	0.005	Ewecharoen et al. 2009
Untreated alga	40.983	0.060	Gupta et al. 2010
Modified activated carbon II	37.175	0.091	Hasar 2003
Modified activated carbon I	30.769	0.025	Hasar 2003
Cashew nut shell	18.868	0.071	Kumar et al. 2011
Calcined phosphate	15.53	0.299	Hannachi et al. 2010
Clarified sludge	14.3	0.222	Hannachi et al. 2010
Red mud	13.63	0.102	Hannachi et al. 2010
Anode dust	8.64	$6.50 \cdot 10^{-3}$	Strkalj et al. 2010
Powdered eggshell	7.0	0.281	Otun et al. 2006
Smectite clay	6.68	0.586	Mbadcam et al. 2012
Sawdust	4.6	38.14	Bozic et al. 2009
Sepiolite	3.44	0.24	Sanchez et al. 1999
eggshell	2.36	0.478	Ghazy et al. 2011
eggshell	1.84	1.665	This study
Bael tree leaf powder	1.527	0.0622	Kumar, Kirthika 2009
Fly ash (Seyitomer)	1.160	1.839	Bayat 2002
Fly ash (Afşin-Elbistan)	0.787	2.092	Bayat 2002
Fly ash	0.249	0.0684	Agarwal et al. 2013
Fly ash	0.03	0.08	Rao et al. 2002
Bagasse	0.001	0.48	Rao et al. 2002

Table 4. Parameters of adsorption kinetics of Ni2+ by eggshell

Ni^{2+} (mg/L)	q_e (exp) (mg/g)	Pseudo-first-order			Pseudo-second-order			Intraparticle diffusion			Elovich		
		q_e	k_1	R^2	q_e	k_2	R^2	h	k_p	R^2	α	β	R^2
15	1.6	0.035	0.453	0.823	1.62	0.49	0.996	1.29	0.055	0.629	775	7.813	0.723
25	1.94	0.028	0.782	0.795	1.97	0.112	0.997	0.434	0.096	0.904	15.38	4.74	0.910
35	2.28	0.021	1.183	0.919	2.32	0.0613	0.993	0.33	0.141	0.918	2.76	3.14	0.986
Temperature (K)													
298	1.96	0.0138	0.723	0.685	1.71	0.304	0.999	0.89	0.09	0.797	20.07	5.21	0.955
308	2.39	0.0253	0.991	0.848	2.44	0.082	0.993	0.488	0.126	0.678	6.32	3.24	0.848
318	2.22	0.035	1.167	0.947	2.29	0.080	0.997	0.419	0.143	0.825	2.45	2.96	0.967

The calculated values of thermodynamic parameters are presented in Table 5. The value of $\Delta G°$ is small at 298 K and negative with increase in temperature. It indicates that the adsorption process leads to an increase in Gibbs energy. The negative $\Delta G°$ value means the Ni^{2+} sorption onto the eggshell is feasible and spontaneous in the nature. The value of $\Delta H°$ (13.66 kj/mol) and $\Delta S°$ (0.136 Kj/mol/K) were determined from the data. The positive values of $\Delta H°$ and $\Delta S°$ suggests the endothermic nature of process and randomness at the eggshell – solution interface during the sorption (Katal *et al.* 2012).

Table 5. Thermodynamic parameters for the adsorption of Ni^{2+} onto eggshell

$\Delta H°$ (kj/mol)	$\Delta S°$ (kj/mol/K)	$\Delta G°$ (kj/K.mol)				
		298 K	303 K	308 K	313 K	318 K
Ni^{2+} 13.66	0.136	0.013	−0.422	−0.561	−0.610	−0.629

Conclusions

As can be seen from the experimental results, the waste eggshells might be used for nickel removal from aqueous solution. Following conclusions could be drawn from the study:

1. The maximum sorption capacities were determined at the pH value of 6.5. Adsorption of Ni^{2+} was highly temperature dependent.

2. The Ni^{2+} ions were adsorbed onto the eggshell in a monolayer due to the highest correlation coefficient (0.9995) was determined using the Langmuir model comparing with the other models.

3. The sorption perfectly complies with pseudo-second order reaction than the others (pseudo first order kinetics, intraparticle diffusion, and Elovich models) and the sorption of Ni^{2+} onto the eggshell appeared to be controlled by the chemisorption process.

4. Of the thermodynamic point of view, the sorption mechanisms of Ni^{2+} ions onto the eggshell was endothermic.

Acknowledgements

This study was supported by The Research Fund of Cumhuriyet University (CUBAP) under Grant No. M-459, Sivas, Turkey.

References

Ahmad, R.; Kumar, R.; Haseeb, S. 2010. Adsorption of Cu^{2+} from aqueous solution onto iron oxide coated eggshell powder: evaluation of equilibrium, isotherms, kinetics, and regeneration capacity, *Arabian Journal of Chemistry* 5(3): 353–359. http://dx.doi.org/10.1016/j.arabjc.2010.09.003

Aikpokpodion, P. E.; Ipinmoroti, R. R.; Omotoso, S. M. 2010. Biosorption of nickel (II) from aqueous solution using waste tea (Camella cinencis) materials, *American – Eurasian Journal of Toxicological Sciences* 2(2): 72–82 [online], [cited July 2014]. Available from Internet: http://www.idosi.org/aejts/aejts2(2)10.htm

Agarwal, A. K.; Kadu, M. S.; Pandhurnekar, C. P.; Muthreja, I. L. 2013. Removal of nickel (II) ions from aqueous solution using thermal power plant fly ash as a low cost adsorbent, *International Journal of Environmental Protection* 3(3): 33–43 [online], [cited July 2014]. Available from Internet: http://www.ij-ep.org/paperInfo.aspx?ID=200

Aksu, Z.; Donmez, G. 2006. Binary biosorption of cadmium(II) and nickel(II) onto dried Chlorella vulgaris: co-ion effect on mono-component isotherm parameters, *Process Biochemistry* 41: 860–868. http://dx.doi.org/10.1016/j.procbio.2005.10.025

Al-Homaidan, A. A.; Al-Houri, H. J.; Al-Hazzani, A. A.; Elgaaly, G.; Moubayed, N. M. S. 2014. Biosorption of copper ions from aqueous solutions by Spirulina platensis biomass, *Arabian Journal of Chemistry* 7: 57–62. http://dx.doi.org/10.1016/j.arabjc.2013.05.022

Aslan, S.; Gurbuz, B. 2011. Influence of operational parameters and low nickel concentrations on partial nitrification in a submerged biofilter, *Applied Biochemistry and Biotechnology* 165: 1543–1555. http://dx.doi.org./10.1007/s12010-011-9374-0

Baig, J. A.; Kazi, T. G.; Shah, A. Q.; Kandhro, G. A.; Afridi, H. I.; Khan, S.; Kolachi, N. F. 2010. Biosorption studies on powder of stem of Acacia nilotica: removal of arsenic from surface water, *Journal of Hazardous Materials* 178: 941–948. http://dx.doi.org/10.1016/j.jhazmat.2010.02.028

Bar-Sela, S.; Levy, M.; Westin, J. B.; Laster, R.; Richter, E. D. 1992. Medical findings in nickel-cadmium battery workers, *Israel Journal of Medical Sciences* 28: 51–53 [online], [cited July 2014]. Available from Internet: http://www.ncbi.nlm.nih.gov/pubmed/1428813

Bayat, B. 2002. Comparative study of adsorption properties of Turkish fly ashes I. The case of nickel(II), copper(II) and zinc(II), *Journal of Hazardous Materials* B95: 251–273. http://dx.doi.org/10.1016/S0304-3894(02)00140-1

Bermúdez, Y. G.; Rico, I. L. R.; Bermúdez, O. G.; Guibal, E. 2011. nickel biosorption using Gracilaria caudata and Sargassum muticum, *Chemical Engineering Journal* 166: 122–131. http://dx.doi.org/10.1016/j.cej.2010.10.038

Bhatnagar, A.; Minocha, A. K. 2010. Biosorption optimization of nickel removal from water using Punica granatum peel waste, *Colloids and Surfaces B: Biointerfaces* 76: 544–548. http://dx.doi.org/10.1016/j.colsurfb.2009.12.016

Bozic, D.; Stankovic, V.; Gorgievski, M.; Bogdanovic, G.; Kovacevic, R. 2009. Adsorption of heavy metal ions by sawdust of deciduous trees, *Journal of Hazardous Materials* 171: 684–692. http://dx.doi.org/10.1016/j.jhazmat.2009.06.055

Celekli, A.; Bozkurt, H. 2011. Bio-sorption of cadmium and nickel ions using Spirulina platensis: kinetic and equilibrium studies, *Desalination* 275: 141–147. http://dx.doi.org/10.1016/j.desal.2011.02.043

Celekli, A.; Atmaca, M.; Bozkurt, H. 2010. An ecofriendly process: predictive modelling of copper adsorption from aqueous solution on Spirulina platensis, *Journal of Hazardous Materials* 173: 123–129. http://dx.doi.org/10.1016/j.jhazmat.2009.08.057

Chairat, M.; Rattanaphani, S.; Bremner, J. B.; Rattanaphani, V. 2005. An adsorption and kinetic study of lac dyeing on silk, *Dyes and Pigments* 64: 231–241. http://dx.doi.org/10.1016/j.dyepig.2004.06.009

Chiou, M. S.; Li, H. Y. 2002. Equilibrium and kinetic modeling of adsorption of reactive dye on cross-linked chitosan beads, *Journal of Hazardous Materials* B93: 233–248. http://dx.doi.org/10.1016/S0304-3894(02)00030-4

Chojnacka, K. 2005. Biosorption of Cr (III) ions by eggshells, *Journal of Hazardous Materials* B121: 167–173. http://dx.doi.org/10.1016/j.jhazmat.2005.02.004

Demirbas, E.; Dizge, E.; Sulak, M. T.; Kobya, M. 2009. Adsorption kinetics and equilibrium of copper from aqueous solutions using hazelnut shell activated carbon, *Chemical Engineering Journal* 148: 480–487. http://dx.doi.org/10.1016/j.cej.2008.09.027

Erdogan, S.; Onal, Y.; Basar, C. A.; Erdemoglu, S. B.; Ozdemir, C. S.; Koseoglu, E.; Icduygu, G. 2005. Optimization of nickel adsorption from aqueous solution by using activated carbon prepared from waste apricot by chemical activation, *Applied Surface Science* 252: 1324–1331. http://dx.doi.org/10.1016/j.apsusc.2005.02.089

EU. 2011. *European and National Drinking Water Quality Standards*. Northern Ireland Environment Agency, Department for Regional Development, October 2011.

Ewecharoen, A.; Thiravetyan, P.; Wendel, E.; Bertagnolli, H. 2009. Nickel adsorption by sodium polyacrylate-grafted activated carbon, *Journal of Hazardous Materials* 171: 335–339. http://dx.doi.org/10.1016/j.jhazmat.2009.06.008

Ghasemi, Z.; Seif, A.; Ahmadi, T. S.; Zargar, B.; Rashidi, F.; Rouzbahani, G. M. 2012. Thermodynamic and kinetic studies for the adsorption of Hg(II) by nano-TiO_2 from aqueous solution, *Advanced Powder Technology* 23: 148–156. http://dx.doi.org/10.1016/j.apt.2011.01.004.

Ghazy, S. E.; El-Asmy, A. A.-H.; El-Nokrashy, A. M. 2011. Batch removal of nickel by eggshell as a low cost sorbent, *International Journal of Industrial Chemistry* 2(4): 242–252 [online], [cited July 2014]. Available from Internet: http://www.ijichem.org/FCKUploaded/file/Archive%20Issue/2011/2/4/article_7.pdf.

Gupta, V. K.; Rastogi, A.; Saini, V. K.; Jain, N. 2006. Biosorption of copper(II) from aqueous solutions by Spirogyra species, *Journal of Colloid and Interface Science* 296: 59–63. http://dx.doi.org/10.1016/j.jcis.2005.08.033

Gupta, V. K.; Rastogi, A.; Nayak, A. 2010. Biosorption of nickel onto treated alga (Oedogonium hatei): application of isotherm and kinetic models, *Journal of Colloid and Interface Science* 342: 533–539. http://dx.doi.org/10.1016/j.jcis.2009.10.074

Hannachi, Y.; Shapovalov, N. A.; Hannachi, A. 2010. Adsorption of nickel from aqueous solution by the use of low-cost adsorbents, *Korean Journal of Chemical Engineer* 27(1): 152–158. http://dx.doi.org/10.1016/j.jcis.2009.10.074

Hasar, H. 2003. Adsorption of nickel(II) from aqueous solution onto activated carbon prepared from almond husk, *Journal of Hazardous Materials* B97: 49–57. http://dx.doi.org/10.1007/s11814-009-0303-7

Katal R.; Baei, M. S.; Rahmati, H. T.; Esfendian, H. 2012. Kinetic, isotherm, and thermodynamic study of nitrate adsorption from aqueous solution using modified rice husk, *Journal of Industrial Chemistry* 18: 295–302. http://dx.doi.org/10.1016/j.jiec.2011.11.035

Kilic, M.; Varol, E. A.; Putun, A. E. 2011. Adsorptive removal of phenol from aqueous solutions on activated carbon prepared from tobacco residues: equilibrium, kinetics and thermodynamics, *Journal of Hazardous Materials* 189: 397–403. http://dx.doi.org/10.1016/j.jhazmat.2011.02.051

Kleinubing, S. J.; Guibal, E.; Silva, E. A.; Silva, M. G. C. 2012. Copper and nickel competitive biosorption simulation from single and binary systems by Sargassum filipendula, *Chemical Engineering Journal* 184: 16–22. http://dx.doi.org/10.1016/j.cej.2011.11.023

Krishnan, K. A.; Sreejalekshmi K. G.; Baiju, R. S. 2011. Nickel(II) adsorption onto biomass based activated carbon obtained from sugarcane bagasse pith, *Bioresource Technology* 102: 10239–10247. http://dx.doi.org/10.1016/j.biortech.2011.08.069

Kumar, P. S.; Ramalingam, S.; Kirupha, S. D.; Murugesan, A.; Vidhyadevi, T.; Sivanesan, S. 2011. Adsorption behavior of nickel(II) onto cashew nut shell: equilibrium, thermodynamics, kinetics, mechanism and process design, *Chemical Engineering Journal* 167: 122–131. http://dx.doi.org/10.1016/j.cej.2010.12.010

Kumar, P. S.; Kirthika, K. 2009. Equilibrium and kinetic study of adsorption on nickel from aquous solution onto Bael Tree leaf powder, *Journal of Engineering Science and Technology* 4(4): 351–363 [online], [cited July 2014]. Available from Internet: http://jestec.taylors.edu.my/Vol%204%20Issue%204%20December%2009/Vol_4_4_351_363_P.%20Senthil%20Kumar.pdf

Liu, D.; Tao, Y.; Li, K.; Yu, J. 2012. Influence of the presence of three typical surfactants on the adsorption of nickel (II) to aerobic activated sludge, *Bioresource Technology* 126: 56–63. http://dx.doi.org/10.1016/j.biortech.2012.09.025

Malkoc, E.; Nuhoglu, Y. 2009. Removal of Ni(II) ions from aqueous solutions using waste of tea factory: adsorption on a fixed-bed column, *Journal of Hazardous Materials* B135: 328–336. http://dx.doi.org/10.1016/j.jhazmat.2005.11.070

Mbadcam, K.; Dongmo, J. S.; Ndaghu, D. D. 2012. Kinetic and thermodynamic studies of the adsorption of nickel (II) ions from aqueous solution by smectite clay from Sabga-Cameroon, *International Journal of Current Research* 4(5): 162–167 [online], [cited July 2014]. Available from Internet: http://www.journalcra.com/article/kinetic-and-thermodynamic-studies-adsorption-nickel-ii-ions-aqueous-solutions-smectite-clay-

Mezenner, N. Y.; Bensmaili, A. 2009. Kinetics and thermodynamic study of phosphate adsorption on iron hydroxide-eggshell waste, *Chemical Engineering Journal* 147: 87–96. http://dx.doi.org/10.1016/j.cej.2008.06.024

Oladunni, N.; Agbaji, E. B.; Idris, S. O. 2012. Removal of Pb^{2+} and Ni^{2+} ions from aqueous solutions by adsorption onto activated locust bean (Parkia biglobosa) husk, *Scholars Research Library Archives of Applied Science Research* 4(6): 2308–2321. http://dx.doi.org/10.3923/jas.2006.2376

Otun, J. A.; Oke, I. A.; Olarinoye, N. O.; Adie, D. B.; Okuofu, C. A. 2006. Adsorption isotherms of Pb (II), Ni (II) and Cd (II) ions onto PES, *Journal of Applied Sciences* 6(11): 2368–2376. http://dx.doi.org/10.3923/jas.2006.2368.2376

Pahlavanzadeh, H.; Keshtkar, A. R.; Safdari, J.; Abadi, Z. 2010. Biosorption of nickel(II) from aqueous solution by brown algae: equilibrium, dynamic and thermodynamic studies, *Journal of Hazardous Materials* 175: 304–310. http://dx.doi.org/10.1016/j.jhazmat.2009.10.004

Polat, A.; Aslan, S. 2014. Kinetic and isotherm study of cupper adsorption from aqueous solution using waste eggshell, *Journal of Environmental Engineering and Landscape Management* 22(2): 132–140.
http://dx.doi.org/10.3846/16486897.2013.865631

Prabakaran, R.; Arivoli, S. 2012. Adsorption kinetics, equilibrium and thermodynamic studies of Nickel adsorption onto Thespesia Populnea bark as biosorbent from aqueous solutions, *Scholars Research Library European Journal of Applied Engineering and Scientific Research* 1(4): 134–142 [online], [cited July 2014]. Available from Internet: http://scholarsresearchlibrary.com/archive.html

Rao, M.; Parwate, A. V.; Bhole, A. G. 2002. Removal of Cr^{6+} and Ni^{2+} from aqueous solution using bagasse and fly ash, *Waste Management* 22: 821–830.
http://dx.doi.org/10.1016/S0956-053X(02)00011-9

Sanchez, A. G.; Ayuso, E. A.; Blas, J. 1999. Sorption of heavy metals from industrial wastewater by low cost mineral silicates, *Clay Minerals* 34: 469–477.
http://dx.doi.org/101180/000985599546370

Shah, J.; Jan, M. R.; Haq, A.; Zeeshan, M. 2012. Equilibrium, kinetic and thermodynamic studies for sorption of Ni (II) from aqueous solution using formaldehyde treated waste tea leaves, *Journal of Saudi Chemical Society* (in press).
http://dx.doi.org/10.101016/j.jics.2012.04.004

Sljivic, M.; Smiciklas, I.; Plecas, I.; Mitric, M. 2009. The influence of equilibration conditions and hydroxyapatite physico-chemical properties onto retention of Cu^{2+} ion, *Chemical Engineering Journal* 148: 80–88.
http://dx.doi.org/10.1016/j.cej.2008.08.003

Strkalj, A.; Rađenović, A.; Malina, J. 2010. Adsorption of Ni (II) ions from aqueous solution on anode dust: effect of pH value, *RMZ – Materials and Geoenvironment* 57(2): 165–172 [online], [cited July 2014]. Available from Internet: http://www.rmz-mg.com/letniki/rmz57/RMZ57_0165-0172.pdf

Thevannan, A.; Hill, G.; Niu, C. H. 2011. Kinetics of nickel biosorption by acid-washed barley straw, *The Canadian Journal of Chemical Engineering*, 89(1): 176–182.
http://dx.doi.org/10.1016/j.biortech.2007.04.010

Tsai, W.-T.; Hsien, K.-J.; Hsu, H.-C. ; Lin, C.-M.; Lin, K.-Y.; Chiu, C.-H. 2008. Utilization of ground eggshell waste as an adsorbent for the removal of dyes from aqueous solution, *Bioresource Technology* 99: 1623–1629.
http://dx.doi.org/10.1016/j.biortech.2007.04.010

USEPA. 2012. *Guidelines for Water Reuse* EPA/600/R-12/618, September. United States Environmental Protecction Agency [online], [cited July 2014]. Available from Internet: http://water.epa.gov/drink/contaminants/index.cfm

WHO. 2005. *Nickel in drinking-water*. Background document for development of WHO Guidelines for Drinking-water Quality, World Health Organization, Geneva, Switzerland [online], [cited July 2014]. Available from Internet: http://www.who.int/water_sanitation_health/gdwqrevision/nickel2005.pdf

Yang, F.; Liu, H.; Qu, J.; Chen, J. P. 2011. Preparation and characterization of chitosan encapsulated *Sargassum sp.* biosorbent for nickel ions sorption, *Bioresource Technology* 102: 2821–2828. http://dx.doi.org/10.1016/j.biortech.2010.10.038

Sukru ASLAN, Dr Lecturer, Department of Environmental Engineering, Cumhuriyet University, Sivas, Turkey. Research interests: adsorption, biological nutrient removal, reuse.

Ayben POLAT, Environmental Engineer, Department of Environmental Engineering, Cumhuriyet University, Sivas, Turkey. Research interests: adsorption, biological nutrient removal.

Ugur Savas TOPCU, Environmental Engineer, Department of Environmental Engineering, Cumhuriyet University, Sivas, Turkey. Research interests: adsorption, biological nutrient removal.

10

EXPERIMENTAL RESEARCH INTO AERODYNAMIC PARAMETERS OF A CYLINDRICAL ONE-LEVEL 8-CHANNEL CYCLONE

Žilvinas VENCKUS, Albertas VENSLOVAS, Mantas PRANSKEVIČIUS

Institute of Environmental Protection, Vilnius Gediminas Technical University,
Saulėtekio al. 11, LT-10223 Vilnius, Lithuania

Abstract. The Environmental Protection Laboratory of Vilnius Gediminas Technical University has developed and installed a one-level 8-channel cyclone with different internal structure, which is used for separation of particulate matter from air streams. Airflow velocity was measured in five points of each channel: at the end and the beginning of the channel, in the middle of the channel and in points arranged at 45° angle from the channel's end and beginning. The highest airflow velocity determined in 8-channel device by regulating volumes of peripheral and transit airflows with curvilinear semi-rings was in cyclone's channel 1, while the lowest – in channel 2. Contrary to the above, the highest airflow velocity in 8-channel cyclone by using quarter-rings without holes and with 5° opening angle of plates in holes was recorded in channel 2. Tests on aerodynamic resistance were carried out in airflow inlet and outlet ducts. In all cases analysed the highest aerodynamic resistance in 8-channel cyclone was determined when airflow distribution was regulated at 75/25 ratio with semi-rings. When quarter-rings without holes and with 5° opening angle of plates in holes were used, higher aerodynamic resistance in the system was created by using quarter-rings without holes, i.e. 0° opening of plates.

Keywords: cyclone, semi-ring, quarter-ring, airflow, particulate matter, air cleaning.

Introduction

The database of patents on devices for dust removal from an air (gas) mixture is expanding. A new generation of structural devices can remove increasingly smaller particulate matter from an ambient air. All inventions have a common goal – to achieve a higher degree of separating particulate matter of 10 μm and smaller dispersion (Baltrėnas *et al.* 2012; Pushnov, Berengarten 2011).

Cyclones are widely applied due to their simple structure, reliable operation as well as low capital and operating costs. Cyclone type airflow cleaners are widely used to dry-clean polluted gas emitted during the certain technological processes (dehumidification, burning, agglomeration, fuel combustion, etc). There is a wide variety of cyclones. By air stream direction cyclones are divided into reverse stream and straight stream systems, by form they are classified as conical and cylindrical, and by air flow direction (clockwise and anticlockwise) – right and left cyclones (Altmeyer *et al.* 2004; Bernardo *et al.* 2006; Elsayed, Lacor 2010; Vaitiekūnas, Jakštonienė 2010).

The cyclone structure has four main elements: inlet, shell, hopper and outlet. Cyclones have no moving parts, and particulates are separated from gas streams by the inertial force which is created by the gas flow rotating inside the device (this force is often referred to as the centrifugal force). Since during the cleaning process particulates are affected by gravitational and frictional forces, an increasing number of tests is carried out on the influence of geometrical and operational parameters on the optimum performance of cyclones (Elsayed, Lacor 2011; Hu *et al.* 2005; Gujun *et al.* 2008).

Scientific papers emphasise that the cyclone's cleaning efficiency depends on a number of factors such as properties of particulates (size, density), the flow rate of gas in the cyclone, geometrical parameters of the cyclone and also the time of gas residence in the cyclone. For example, centrifugal forces increase when a cyclone diameter is smaller and therefore narrower cyclones achieve higher efficiency; gas residence time is longer in cyclones whose cylindrical and conical parts are longer than the diameter and therefore the cleaning efficiency of such cyclones is higher or increases when the gas rate increases in the cyclone. To achieve the gas rotation rate, different inlets are used: tangential, continual, helical or axial (of

Corresponding author: Žilvinas Venckus
E-mail: zilvinas.venckus@vgtu.lt

them, the most frequently used are tangential and continual inlet) (Burov *et al.* 2005, 2007; Gimbun *et al.* 2005; Kaya, Karagoz 2008; Veriankaitė *et al.* 2011).

Most scientific research concerns improvements to the structural elements of a cyclone inlet and outlet and the formation of swirling flows. Many experiments were carried out by reducing parameters of the cyclone shell, and also devices with a closed-loop system are created. The basis of such devices – curvilinear channels formed with semi-rings inside the device which create closed-loop systems. Loops of cylindrical shape have different diameters and are located at the angle φ = π, i.e. they are made of two walls of rings having different diameters. Each pair of adjacent channels forms one closed loop. Dust-polluted gaseous stream is filtered through a few layers, as particles circulate in closed loops due to the combination of dust-polluted stream filtering and centrifugal cleaning (Burov *et al.* 2011; Jakštonienė *et al.* 2011; Serebrensky 2011).

Scientific research has shown that theoretical relationships between particulate matter separation and the number of semi-rings in the cyclone can be approximately determined using formula 1:

$$\eta = 1 - \frac{1}{1+2^{n-1}}, \qquad (1)$$

where *n* – the number of semi-rings in multichannel cyclone, η – particulate matter separation coefficient (Burov *et al.* 2005; Serebrensky 2004).

One of the main technical requirements is to achieve the highest possible air cleaning efficiency by reducing aerodynamic resistance in the device in order to minimise the amount of energy necessary for cleaning larger amounts of passing airflows. Aerodynamic resistance of the device is influenced by the flow rate and pressure loss. Pressure loss (Δp) is directly dependent on losses generating at the time of gas entry into cyclone (Δp_in), losses generating inside the device (Δp_c) and losses generating at the time gas leaves the cyclone (Δp_out). The aforementioned losses in a multichannel cyclone can be theoretically described by the following relationship:

$$\Delta p = \underbrace{\left(1-\frac{S_1}{S_2}\right)^2 \frac{v_{in}^2 \rho}{2}}_{\Delta p_{in}} + \underbrace{\frac{4\,fl v^2}{2D_h}}_{\Delta p_c} + \underbrace{0,5\left(1-\frac{S_3}{S_4}\right)^2 \frac{v_{out}^2 \rho}{2}}_{\Delta p_{out}}, \qquad (2)$$

where v_{in} – incoming flow velocity, S_1 – cross-section area of gas inflow, S_2 – cross-section area of multichannel cyclone's channel 1 at outflow, v_{out} – outgoing flow velocity; S_3 – cross-section area of gas outflow, S_4 – cross-section area of the last channel at outlet, D_h – device's hydraulic diameter, *l* – flow distance, $f = \frac{0,341}{\sqrt[4]{Re}}$, $Re = \frac{vh\rho}{\mu}$, *h* – height of cyclone's cylindrical part.

The aim of such improvements to cyclone structures is to find the optimum relationship between aerodynamic

resistance, which depends on the airflow rate, and air cleaning efficiency.

The aim of this paper is to perform experimental tests on how different arrangement of semi-rings and quarter-rings in one-level cyclone's internal structure affects the distribution of airflow velocities and aerodynamic resistance in the device.

1. Methodology

The experimental laboratory stand was developed at the Environmental Protection Technology Laboratory of Vilnius Gediminas Technical University. The stand system comprises a centrifugal ventilator which creates airflow, a flow inlet duct (D = 200 mm), a flow cleaner, i.e. one-level 8-channel cyclone, and a cleaned airflow outlet duct (D = 160 mm) (Fig. 1).

Experimental tests on airflow rates were carried out in the air duct system in which test points were selected in a straight segment, a steady flow and inside the cyclone structure, in the special holes of 8 mm made in each channel in the cyclone's cover.

Polluted air travels to the cyclone's diffuser from the inlet duct, while the separated particulate matter enters the double hopper via segment circular slits, and cleaned air is released through the outlet duct. The values of airflow resistance in holes made in the cyclone's cover measured with a dynamic Pitot tube and multifunctional meter Testo-400 were recalculated into airflow rates.

Instruments used during the experiment:

1. Dynamic Pitot tube;
2. Multifunctional meter Testo-400;
3. Accessory Testo 0638.1445;
4. Rubber hoses;
5. Vane anemometer;
6. Aspirator, air suction rate error ±6%;

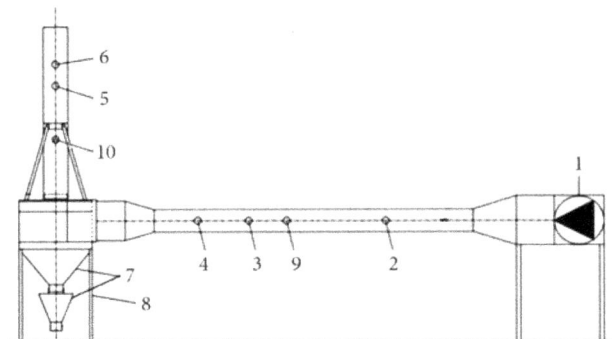

Fig. 1. Experimental stand for one-level 8-channel cyclone: 1 – centrifugal ventilator; 2 – particulates inlet; 3 – place for particulate matter concentration suction before cleaning; 4 – airflow rate measuring place in the inlet duct; 5 – place for particulate matter concentration suction after cleaning; 6 – airflow rate measuring place in the outlet duct; 7 – double hopper; 8 – bearing structure; 9, 10 – nozzles for measuring resistance created by the system

7. Electronic laboratory scale AG–204; graduation value 0.1 mg;

8. Filter holders;

9. Filters AFA–VP–20;

10. Pollutant feed sprayer;

11. Differential pressure meter DSM-1.

Airflow velocities in the device are evaluated taking into account different air flow rates created by the ventilator by changing air amounts in the ventilator control unit.

Devices of three different types were investigated by changing the type of the internal structure of one-level cyclone (see Fig. 2):

– eight-channel cyclone by making channels of different diameter and volume using quarter-rings without holes;

– eight-channel cyclone by making channels of different diameter and volume using quarter-rings having 5° opening angle of plates in holes;

– eight-channel cyclone by making channels in it using semi-rings.

Airflow rate was measured in five points of each channel: at the end and beginning of the channel, in the middle of the channel and at points arranged at 45° angle from the channel's end and beginning.

The airflow that enters and leaves each channel can be regulated by changing distances between semi-rings or quarter-rings of different diameter. As Figure 3 shows, three different positions of regulation were selected: a) peripheral (reciprocating) airflow volume equal to transit flow volume (moving to the next channel) (position 50/50); b) peripheral airflow volume by 50% larger than transit airflow volume (position 75/25); c) peripheral airflow volume by 50% smaller than transit airflow volume (position 25/75).

In order to avoid systematic mistakes and minimise the average error, three measurements were made in each measuring point.

Velocities in air ducts upstream and downstream the cyclone were measured by a vane anemometer and meter Testo-400. Due to improvements to the device structure the airflow velocity was measured with an anemometer in inlet and outlet ducts only. According to the measured velocities, airflow continuity and the exact amount of air entering the cyclone are determined. Measurements in the air duct were carried out in accordance with Lithuanian standard LAND 27-98/M-07 Measurement of gas flow velocity and volume rate in the air duct (1998).

To measure system resistance, special branch pipes, 7 mm in inner diameter and 10 mm in outer diameter,

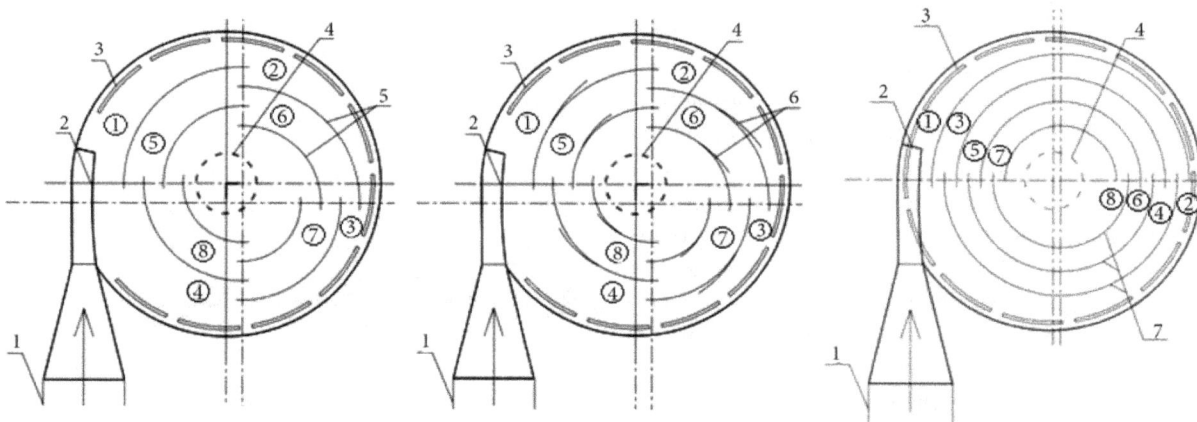

Fig. 2. Diagram of the internal structure of one-level 8-channel cyclone: 1 – polluted airflow supply duct, 2 – airflow inlet, 3 – segment circular slits, 4 – cleaned airflow outlet, 5 – curvilinear quarter-rings of different radius, 6 – opened plates, 7 – curvilinear semi-rings; 1–8 – sequential number of the channel

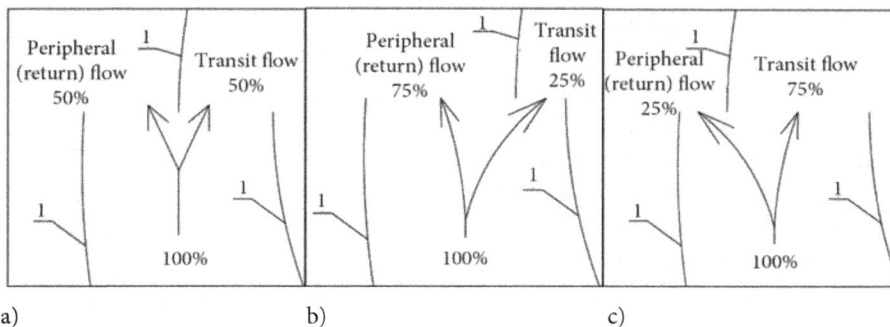

Fig. 3. Diagrams of flow distribution ratios in a one-level multichannel cyclone: a) 50/50%; b) 75/25%; c) 25/75%

were mounted in the air ducts of the experimental stand and hoses were connected to them for pressure measurement with the differential pressure meter DSM-1 (measuring range 0–20 000 kPa; error ±5 Pa).

Aerodynamic pressure shows the value of the sum of dynamic and static pressure. During the tests one hose was connected to the branch pipe upstream the air purifier and the other – downstream the purifier.

2. Results and their analysis

During the tests on flow rates in 8-channel device incoming and outgoing flows were regulated by curvilinear semi-rings, and also by using solid quarter-rings and quarter-rings with 5° opening angle of plates in holes. Tests were performed at an average flow rate in the cyclone of 16 m/s.

The results of tests on flow rates in 8-channel device obtained by regulating incoming and outgoing airflows with cylindrical curvilinear semi-rings are presented in Figure 4.

When transit and peripheral flows were regulated with semi-rings, a significant change in flow velocity was recorded in channel 1 of the cyclone. The highest value of airflow velocity was determined in the last measuring point of channel 1 in all semi-ring arrangement positions. When semi-rings are positioned so that the transit airflow is by 25% bigger than the peripheral flow (25/75 air volume distribution ratio), airflow rate reaches 20.0 m/s. When the volume of peripheral (reciprocating) airflow is equal to that of the transit airflow (moving to the next channels) (50/50 ratio), the measured flow rate reaches 23.6 ms/s. When semi-rings are arranged so that the peripheral air flow volume is 50% larger than the transit (75/25 ratio), the determined flow rate is 25.2 m/s. Velocity increases as a result of changing channel geometry, i.e. channel 1 is wider at the beginning of flow entry into cyclone than at the end of the cyclone and the flow rate increases when a uniform volume of passing airflow is maintained.

The lowest flow rate was determined in channel 2. Cylindrical curvilinear semi-rings in a cylindrically-shaped cyclone are arranged in the descending order of radii and due to such arrangement the highest volume is in channel 2 resulting in velocity change of up to 18.9–21.2%, compared to velocity recorded in channel 1. From channel 3, velocity gradually decreases by 3–5% in every channel from the beginning of the channel towards its end. Since each channel width remains stable, the passing airflow is uniform and is affected by aerodynamic resistance. This means that velocity decreases under the impact of aerodynamic resistance and frictional force, as the airflow evens up while moving and is in constant contact with the channel walls.

The results of airflow velocity distribution determined in one-level 8-channel cyclone using solid quarter-rings are presented in Figure 5.

In all cases analysed the highest airflow velocity was determined in the last measuring point of channel 2. Velocity determined in the aforementioned test point reaches 24.8–30.9 m/s. Velocity increases due to the fact that cylindrical curvilinear quarter-rings mounted in the cylindrically-shaped cyclone are arranged in the descending order of radii and are regulated considering the inlet width. Due to such arrangement quarter-ring 2 is moved up closer to the channel wall and the channel's volume suddenly drops at the end of channel 2.

Velocity in channels 1 and 3 increases in a similar way. Due to the structural arrangement the channels narrow and airflow velocity increases at the end of the channels respectively.

Velocity decreases in each channel starting from channel 4. A change in the average velocity in channel 4 and channel 5 reaches 2.8–3.8 m/s, and in channels 5 and 6 – 0.5–0.7 m/s. The tendency of decrease in

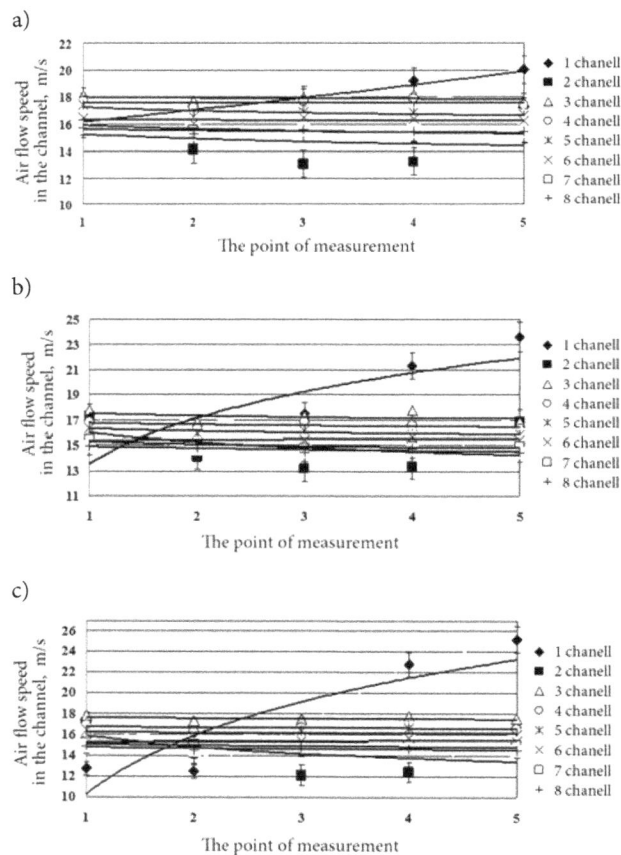

Fig. 4. Distribution of airflow velocities in 8-channel cyclone when air flows are regulated by curvilinear semi-rings:
a) peripheral airflow volume is by 50% smaller than transit airflow volume (25/75); b) peripheral airflow volume is equal to transit airflow volume (50/50 position); c) peripheral airflow volume is by 50% larger than transit air volume (75/25 position)

subsequent channels remains similar. From channel 4, the channel width does not change in any channel and therefore the moving airflow is affected only by aerodynamic resistance, which results in decreasing velocity.

The relationship between airflow velocity distribution and airflow ratios in one-level 8-channel cyclone by using quarter-rings with 5° opening of plates in holes is presented in Figure 6.

In all cases analysed the highest airflow velocity was determined in the last measuring point of channel 2 by regulating the internal position of quarter-rings. When flows are affected by 25/75 position, the determined velocity reaches 25.2 m/s, in case of 50/50 position – 28.1 m/s and 75/25–31.9 m/s.

Velocity continually falls in each channel starting from channel 4. The determined change in the average velocity in channel 4 and channel 5 reaches 3.4–3.8 m/s, and in channels 5 and 6 – 0.7–0.8 m/s. The tendency of decrease in subsequent channels remains similar. From channel 4, the channel width remains steady in each channel and therefore the passing airflow is uniform and is affected

only by aerodynamic resistance, i.e. velocity decreases under the impact of aerodynamic resistance.

In channels 5 to 8 velocity decreases from the beginning towards the end of the channel, the decrease reaches 1–6%. The biggest decrease in velocity, 3–7%, was in channel 5. The longer the airflow travels within a channel the higher the impact of frictional force on the flow and therefore a bigger drop in velocity was recorded in channel 5 whose width remains even.

A comparison of data has shown that after making holes with 5° opening angle of their plates in quarter-rings, airflow velocity in channels 1 to 4 rose by up to 1 m/s, while in internal channels (channels 5 to 8) respectively fell (up to 7%). Due to the holes with plates opened at 5° angle, part of the airflow is additionally returned and therefore velocity in external channels increases.

Figure 7 shows the aerodynamic resistance values of 8-channel cyclone when the peripheral to transit airflow ratio is regulated with curvilinear semi-rings.

Aerodynamic resistance in 8-channel cyclone was investigated at an average airflow velocity of 16 m/s. The

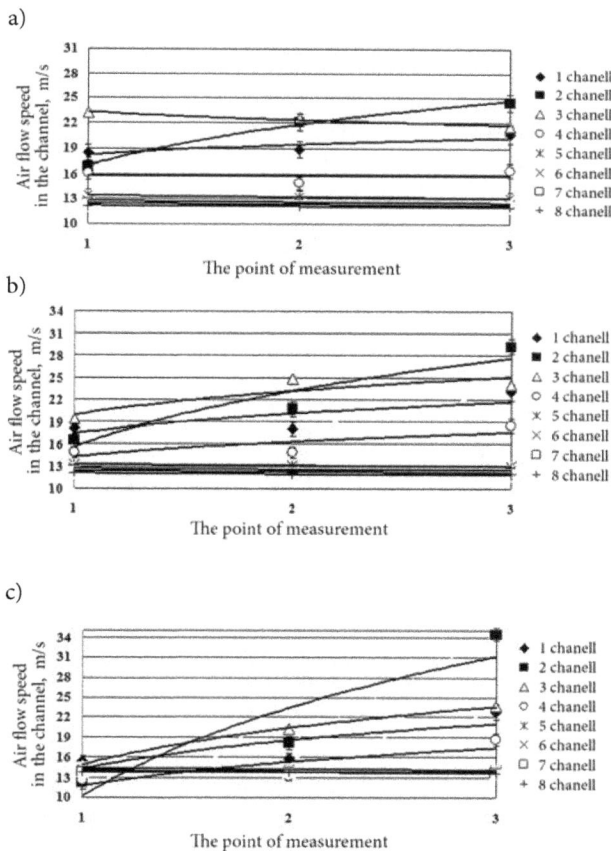

Fig. 5. Distribution of airflow velocities in 8-channel cyclone's channels when solid quarter-rings are used: a) peripheral airflow volume is by 50% smaller than transit airflow volume (25/75); b) peripheral airflow volume is equal to transit airflow volume (50/50 position); b) peripheral airflow volume is by 50% larger than transit air volume (75/25 position)

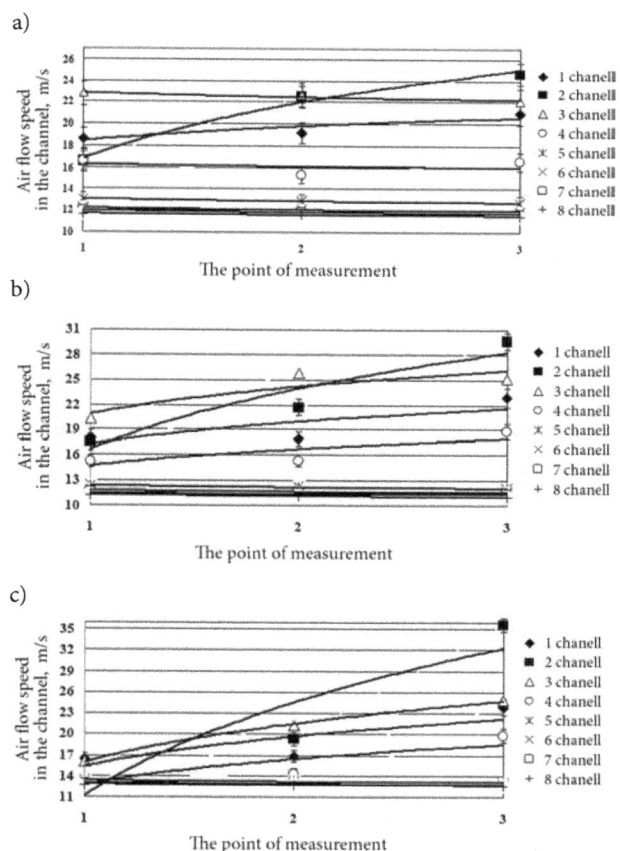

Fig. 6. Distribution of airflow velocities in 8-channel cyclone's channels when quarter-rings with 5° opening angle of plates in holes are used: a) peripheral airflow volume is by 50% smaller than transit airflow volume (25/75); b) peripheral airflow volume is equal to transit airflow volume (50/50 position); c) peripheral airflow volume is by 50% larger than transit air volume (75/25 position)

biggest aerodynamic resistance, 1605 Pa, was determined when the airflow was regulated with curvilinear semi-rings at 75/25 distribution ratio. The smallest aerodynamic resistance, 1330 Pa, was found when the airflow was distributed with semi-rings at 25/75 ratio. The aerodynamic resistance obtained in 8-channel cyclone at even peripheral and transit airflow distribution (50/50) reaches 1550 Pa. It can be assumed that due to the peripheral and transit airflow volume distribution ratio of 75/25, a bigger part of the airflow is directed to the previous channel towards the device's periphery and therefore air flow pressure in the device's axis and outlet airflow duct increases.

Figure 8 presents the relationship between the aerodynamic resistance and airflow distribution ratio when the airflows in the 8-channel cyclone are regulated by quarter-rings without holes. It has been determined that the highest aerodynamic resistance at 16 m/s average flow velocity is created at 75/25 flow distribution ratio and reaches 1550 Pa. It can be assumed that when a reciprocating flow is larger, the flow makes greater swirl movements in the cyclone's external channels than in the internal ones, which at the same time causes bigger losses of the flow energy resulting in an increase in aerodynamic resistance.

The smallest aerodynamic resistance was obtained when the peripheral and transit flows were distributed at 25/75 ratio. This produces a reverse effect of flow movement within the cyclone's channels, i.e. a bigger part of the airflow enters internal channels, thus generating bigger velocity in the outgoing flow duct because of smaller flow energy losses.

The results of tests on aerodynamic resistance in the cyclone when quarter-rings with 5° opening angle of plates in holes are used are presented in Figure 9.

While using quarter-rings with 5° opening angle of plates in holes it has been determined that the values of aerodynamic resistance are lower compared to cases when semi-rings or solid quarter-rings were used. It can be stated that the holes in quarter-rings enable the airflow to move along a trajectory of lower resistance and return to the previous channel, i.e. fewer obstacles are created to the airflow movement. The aerodynamic resistance determined at 25/75 flow distribution ratio reaches 1330 Pa, at 50/50 – 1370, 75/25 – 1450 Pa.

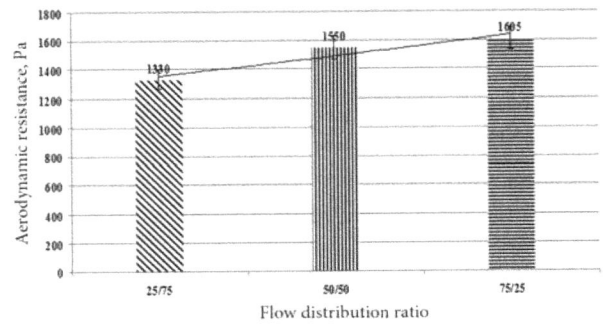

Fig. 7. Relationship between aerodynamic resistance and airflow distribution ratio in one-level cylindrical 8-channel cyclone by using curvilinear semi-rings

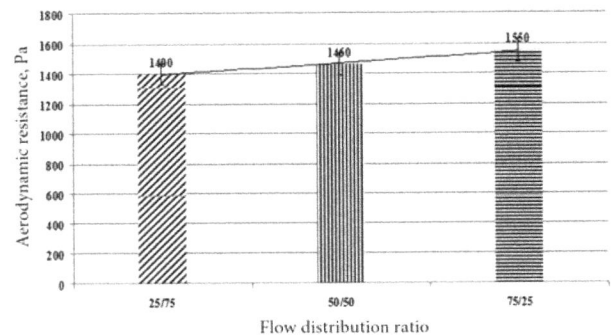

Fig. 8. Relationship between aerodynamic resistance and airflow distribution ratio in one-level 8-channel cyclone when quarter-rings without holes are used

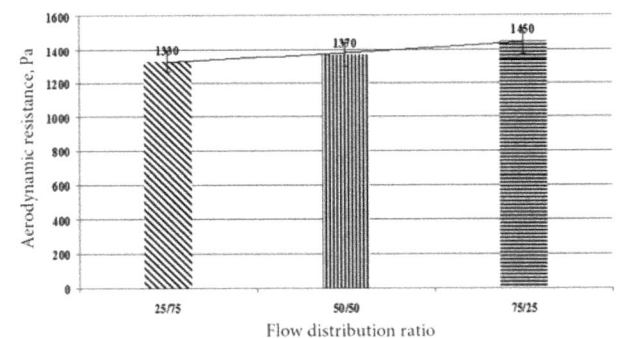

Fig. 9. Relationships between aerodynamic resistance and airflow distribution ratio in one-level 8-channel cyclone when quarter-rings with 5° opening angle of plates in holes are used

Conclusions

1. When peripheral and transit airflow volumes were regulated with semi-rings, the highest value of airflow velocity, 20.0–25.2 m/s, in an 8-channel air purifier was determined in the last measuring point of channel 1. Velocity increases due to a significant increase in the channel's geometry, i.e. channel 1 is wider at the beginning of flow entry into cyclone than at the end and flow velocity increases by maintaining a uniform volume of the passing airflow.

2. When the peripheral and transit flows in the cyclone were regulated with semi-rings, the smallest values of airflow velocity were identified in channel 2, as the channel's volume was the highest due to the arrangement of curvilinear semi-rings in the descending order of radii. Compared to velocity in channel 1, velocity change accounts for up to 18.9–21.2%. In channels 3 to 8 velocity decreases by 3–5% from the beginning towards the end of the channel.

3. When quarter-rings without holes and with 5° opening angle of plates in holes were used, the highest airflow velocity in 8-channel cyclone was determined in the last measuring point of channel 2 reaching 24.8–31.9 m/s. Velocity increases due to the quarter-ring arrangement leading to a sudden decrease in the channel's volume at the end of the last channel.

4. It has been noticed that after holes with 5° opening angle of their plates were made in quarter-rings, airflow velocity in channels 1 to 4 rose by up to 1 m/s, while in internal channels (channels 5 to 8) respectively fell (up to 7%). Due to the holes with plates opened at 5° angle, part of the airflow is additionally returned and therefore velocity in external channels increases.

5. The biggest aerodynamic resistance, 1605 Pa, was determined when the airflow was regulated with curvilinear semi-rings at 75/25 distribution ratio. It can be assumed that due to the peripheral and transit airflow volume distribution ratio of 75/25, a bigger part of the airflow is directed to the previous channel towards the device's periphery and therefore air flow pressure in the device's axis and outlet airflow duct increases. The smallest aerodynamic resistance, 1330 Pa, was recorded when the airflow was distributed with semi-rings at 25/75 ratio.

6. As regards the use of quarter-rings without holes and with 5° opening angle of plates in holes, a bigger aerodynamic resistance of the system is achieved using quarter-rings without holes, i.e. 0° opening of plates and, depending on a flow distribution ratio, reaches 1400 to 1550 Pa. When quarter-rings with 5° opening angle of plates in holes are used the resistance ranges between 1330 and 1450 Pa. When holes are made, the reciprocating airflow increases and for this reason outgoing airflow velocity and aerodynamic resistance decrease.

References

Altmeyer, S.; Mathieu, V.; Jullemier, S.; Contal, P.; Midoux, N.; Rode, S.; Leclers, J. P. 2004. Comparision of different models of cyclone prediction performance for various operating conditions using a general software, *Chemical Engineering and Processing* 43: 511–522. http://dx.doi.org/10.1016/S0255-2701(03)00079-5

Baltrėnas, P.; Vaitiekūnas, P.; Jakštonienė, I.; Konoverskytė, S. 2012. Study of gas-solid flow in a multichannel cyclone, *Journal of Environmental Engineering and Landscape Management* 20(2): 129–137. http://dx.doi.org/10.3846/16486897.2011.645825

Bernardo, S.; Mori, M.; Peres, A. P.; Dionisio, R. P. 2006. 3-D Computational fluid dynamics for gas and gas-particle flows in a cyclone with different inlet section angles, *Powder Technology* 162(3): 190–200. http://dx.doi.org/10.1016/j.powtec.2005.11.007

Burov, A. A.; Burov, A. I.; Gamolič, V. A. 2007. Kontinualnaia model zapylennogo krivolineinogo techeniia gaza, *Trudy Odesskogo politekhnicheskogo universiteta* 1(27): 235–237. Odessa.

Burov, A. A.; Burov, A. I.; Silin, A. V.; Tcabiev, O. N. 2005. Tcentrobezhnaia ochistka promyshlennykh vybrosov v atmosferu, *Ekologiia dovkillia ta bezpeka zhittediialnosti* 6: 44–51.

Burov, A. A.; Burov, A. I.; Karamushko, A. V. 2011. Povitriana techiia u krivoliniinomu kanali, *Ekologiia dovkillia ta bezpeka zhittediialnosti* 12: 174–177.

Elsayed, K.; Lacor, C. 2010. Optimization of the cyclone separator geometry for minimum pressure drop using mathematical models and CFD simulations, *Chemical Engineering Science* 65(22): 6048–6058. http://dx.doi.org/10.1016/j.ces.2010.08.042

Elsayed, K.; Lacor, C. 2011. The effect of cyclone inlet dimensions on the flow pattern and performance, *Applied Mathematical Modelling* 35(4): 1952–1968. http://dx.doi.org/10.1016/j.apm.2010.11.007

Gujun, W.; Guogang, S.; Xiaohu, X.; Mingxian, S. 2008. Solids concentration simulation of different size particles in a cyclone separator, *Powder Technology* 183: 94–104. http://dx.doi.org/10.1016/j.powtec.2007.11.019

Gimbun, J.; Chuah, T. G.; Fakhru'l-Razi, A.; Choong, T. S. Y. 2005. The influence of temperature and inlet velocity on cyclone pressure drop: a CFD study, *Chemical Engineering Progress* 44: 7–12. http://dx.doi.org/10.1016/j.cep.2004.03.005

Hu, L. Y.; Zhou, L. X.; Zhang, J.; Shi, M. X. 2005. Studies on strongly swirling flows in the full space of volute cyclone separator, *AIChE Journal* 51(3): 740–749. http://dx.doi.org/10.1002/aic.10354

Jakštonienė, I.; Serebryansky, D. A.; Vaitiekūnas, P. 2011. Experimental research on the work of centrifugal filter when eliminating solid particles from clinker cooling system, in *The 8th International Conference „Environmental Engineering": selected papers*, 19–20 May 2011, Vilnius, Lithuania, vol. 1. Vilnius: Technika, 134–138. ISBN 978-9955-28-263-1.

Kaya, F.; Karagoz, I. 2008. Performance analysis of numerical schemes in highly swirling turbulent flows in cyclones, *Current Science* 94(10): 1273–1278.

LAND 27-98/M-07. 1998. *Stacionarūs atmosferos teršalų šaltiniai. Dujų srauto greičio ir tūrio debito ortakyje matavimas* [online], [cited 09 October 2014]. Available from Internet: http://www.litlex.lt/scripts/sarasas2.dll?Tekstas=1&Id=28387

Pushnov, A.; Berengarten, M. 2011. Ecological aspects of industrial cooling towers exploitation and it's influence to environment, *Journal of Environmental Engineering and Landscape Management* 19(2): 158–166. http://dx.doi.org/10.3846/16486897.2011.583390

Serebrensky, D. A. 2004. *Povyshenie effektivnosti gazoochistki teplovykh energeticheskikh ustanovok*. Odesskii natcionalnyi politekhnicheskii universitet, Odessa. 155 c.

Serebrensky, D. A. 2011. *Vysokoeffektivnyi tcentrobezhnyi filtr dlia ochistki gazov*. Institut tekhnicheskoi teplofiziki Natcionalnoi akademii nauk Ukrainy, Kiev. 8 c.

Vaitiekūnas, P.; Jakštonienė, I. 2010. Analysis of numerical modelling of turbulence in a conical reverse-flow cyclone, *Journal of Environmental Engineering and Landscape Management* 18(4): 321–328. http://dx.doi.org/10.3846/jeelm.2010.37

Veriankaitė, L.; Šaulienė, I.; Bukantis, A. 2011. Evaluation of meteorological parameters influence upon pollen spread in the atmosphere, *Journal of Environmental Engineering and Landscape Management* 19(1): 5–11. http://dx.doi.org/10.3846/16486897.2011.557252

Žilvinas VENCKUS. Junior Researcher, Dept of Institute of Environmental Protection, Vilnius Gediminas Technical University (VGTU). Master of Science (Environmental Engineering) Vilnius Gediminas Technical University (VGTU), Bachelor of Science (Environmental Engineering), ŠU, 2009. Research Interests: environmental protection, clean technology, environmental modeling.

Albertas VENSLOVAS. Dr (since 2014), Dept of Institute of Environmental Protection, Vilnius Gediminas Technical University (VGTU). Master of Science (Environmental Engineering), VGTU, 2008. Bachelor of Science (Environmental Engineering), LŽŪU, 2006. Research Interests: environmental protection, environmental modeling.

Mantas PRANSKEVIČIUS. Dr (since 2012), Dept of Institute of Environmental Protection, Vilnius Gediminas Technical University (VGTU). Master of Science (Environmental Engineering), VGTU, 2008. Bachelor of Science (Environmental Engineering), ŠU, 2006. Research Interests: environmental protection, environmental modelling, air pollution.

11

GEOTECHNICAL CHARACTERISTICS OF EFFLUENT CONTAMINATED COHESIVE SOILS

Muhammad Imran KHAN[a, b], Muhammad IRFAN[a], Mubashir AZIZ[c], Ammad Hassan KHAN[d]

[a] Department of Civil Engineering, University of Engineering and Technology, Lahore, Pakistan
[b] Department of Civil Engineering and Applied Mechanics, McGill University, Montreal, Canada
[c] Department of Civil Engineering, Al Imam Mohammad Ibn Saud Islamic University, Riyadh, Saudi Arabia
[d] Department of Transportation Engineering and Planning,
University of Engineering and Technology, Lahore, Pakistan

Abstract. In developing countries like Pakistan, raw industrial effluents are usually disposed-off directly into open lands or in water bodies resulting in soil contamination. Leachate formation due to rainfalls in openly dumped solid waste also adds to soil contamination. In this study, engineering behavior of soils contaminated by two industrial effluents, one from paper industry (acidic) and another from textile industry (basic), has been investigated. Laboratory testing revealed significant effects of effluent contamination on engineering behavior of tested soils. Liquid limit, plasticity index, optimum moisture content and compression index of tested soils were found to increase with effluent contaminant, indicating a deterioration in the engineering behavior of soils. Whereas maximum dry density, undrained shear strength and coefficient of consolidation of the contaminated soils showed a decreasing trend. The dilapidation in engineering characteristics of soils due to the addition of industrial effluents could pose serious threats to existing and future foundations in terms of loss of bearing capacity and increase in settlement.

Keywords: soil contamination, industrial waste, engineering behavior, effluent waste, leachate.

Introduction

Industries play a vital role in the development of any country. However, the industrial waste, if released to the environment untreated, may result in several environmental and health hazards. In developed countries, reinforcement of environmental laws and ample resources ensure proper treatment of industrial waste before releasing it to the environment. However, the industrial effluents are often discharged untreated which contaminate water channels and ultimately the soil through seepage.

Industrial sector of Pakistan contributes about 24% of GDP and a huge volume of effluent waste is being generated from various industries. This contaminated effluent waste may carry poisonous or harmful substances like chromium, sulfides, free chlorine, lead, chromium etc. (Bond et al. 1973). Although environmental regulations, outlining the necessary treatment for industrial effluents exist in the country, yet their implementation in true sense remains distant from reality. Unfortunately, most of the waste from industries is discharged to the local environment untreated. Besides their effects on the quality of drinking water, which always remains the prime concern, effluents also affect the engineering properties of surrounding soils through seepage. Previous work (Adebisi, Fayemiwo 2010; Gratchev, Towhata 2009; Jia et al. 2011; Khamehchiyan et al. 2007; Naeini, Jahanfar 2011; Nazir 2011; Olgun, Yıldız 2010; Patel 2011; Reddy et al. 2011; Sunil et al. 2006, 2009) has shown that the index and engineering properties of contaminated soils tend to alter due to chemical reactions between the soil minerals and the contaminant. Soil contamination may also be detrimental to foundation material (Sunil et al. 2006, 2009). Rapid urbanization in various metropolitans of Pakistan is forcing the developers to construct at places which have been used as dumping sites of such industrial effluents. Special considerations, with reference to the modifications in engineering behavior of such soils, are required to be taken into account for design and construction of foundations on such soils.

Corresponding author: Muhammad Irfan
E-mail: mirfan1@msn.com

The main objective of this study is therefore to explore the effects of industrial effluents on the geotechnical properties of local cohesive soils. Considering the widespread paper and textile industry in Pakistan, effluents from these industries are employed in this study. Contaminated soil specimens were prepared by adding various proportions of industrial effluents in the virgin soil samples and their effects on various engineering properties of contaminated soils are investigated.

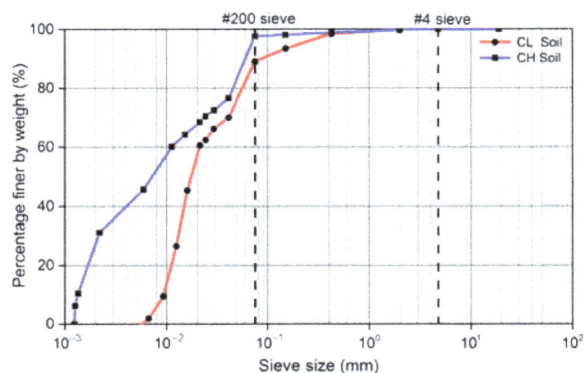

Fig. 1. Grain size distribution of D.G Khan (CL) and Nandipur (CH) soil samples

Fig. 2. Photographs (a) Soil sample from D.G. Khan (CL), (b) Soil sample from Nandipur (CH)

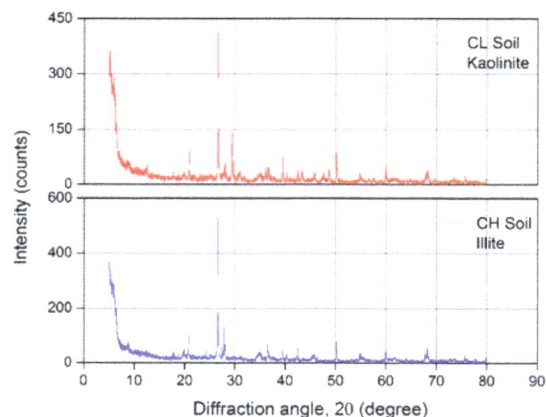

Fig. 3. X-ray diffraction analysis of soil samples

1. Materials

1.1. Soil samples

Soil samples were collected from D.G. Khan and Nandipur (Pakistan). Grain size distribution of both soils are shown in Figure 1 and were classified as CL / A-6(10) and CH / A-7-6 (20) as per USCS / AASHTO soil classification systems, respectively. Actual photographs of both soil samples are shown in Figure 2. Soil samples were collected from test pits at 0.6 m depth. The samples were not collected from natural ground surface in order to avoid the presence of any organic matter in the form of roots, vegetation, leaves etc. X-ray diffraction (XRD) analysis of soil samples (Fig. 3) revealed the presence of kaolinite in CL soil and illite in CH soil as dominant mineral. Physical and Chemical Properties of soil samples are summarized in Table 1.

Table 1. Physical and chemical properties of the soil samples tested as per respective ASTM standard

Description	Units	D.G. Khan Soil	Nandipur Soil
Natural moisture content, NMC	%	6	10
Fine contents (Silt + Clay)	%	89	98
Liquid limit, w_L	%	37	58
Plastic limit, w_P	%	21	27
Plasticity index, I_P	%	16	31
Specific gravity	–	2.645	2.717
USCS soil classification	–	CL	CH
AASHTO soil classification	–	A-6 (10)	A-7-6 (20)
Maximum dry unit weight, γ_{dmax}	kN/m³	18.41	17.18
Optimum moisture content, OMC	%	12.82	17.13
pH value	–	7.5	7
Electrical conductivity, EC	μ Simen	849	228
Sulphate content, SO_4	%	0.04	0.02
Chloride content, Cl	%	0.123	0.015
Organic matter content	%	0.140	0.135
Calcium content, Ca	%	0.014	0.007
Magnesium content, Mg	%	0.004	0.002
Total dissolved solids, TDS	%	0.011	0.005
XRD analysis (dominant clay mineral)	–	Kaolinite	Illite

1.2. Industrial effluents

The main objective of this study was to investigate the effects of effluents from paper industry (Fig. 4 (a)) and textile industry (Fig. 4 (b)) on engineering properties of

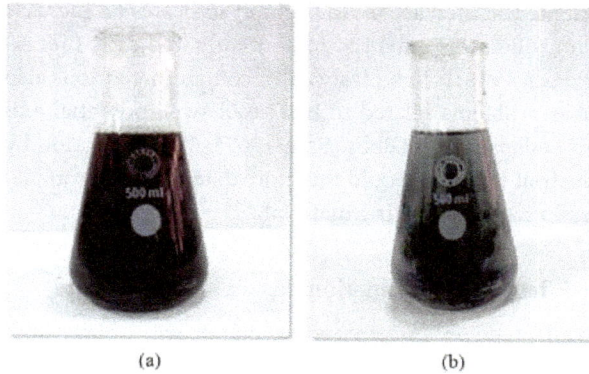

Fig. 4. (a) Effluent from paper industry (acidic in nature), (b) Effluent from textile industry (basic in nature)

cohesive soils. Effluents from paper industry are typically observed to be acidic (Kumar 2005), while the effluents from textile industry are commonly basic (Choudhury 2006). The representative industrial effluents were collected from Century Paper Mill, Lahore and Denim Textile, Lahore (Pakistan). The collected effluent samples were subjected to chemical examination, the summary of which as well as the respective national standards according to NEQS (2000) are presented in Table 2.

Table 2. Chemical properties of effluent samples

Description	Units	Textile Industry Effluent	Paper Industry Effluent	NEQS (2000)
Calcium content, Ca	%	0.002	0.012	–*
Magnesium content, Mg	%	0.001	0.031	–*
Chloride content, Cl	%	0.098	0.125	0.10
Sulphate content, SO$_4$	%	0.012	0.016	0.10
pH	–	11	3	6–9
Total dissolved solids, TDS	%	0.145	0.015	0.35

Note: * National guidelines do not exist.

2. Experimental program

A systematic procedure was established for the preparation of contaminated soil samples. Cohesive soil samples were first oven dried and pulverized. Industrial effluents were then mixed with soil samples in specified proportions of 0, 5, 10, 15 and 20% by dry weight of soil. The soil-effluent mixture was left for 48 hours to bring the moisture in equilibrium before laboratory testing. The contaminated soil samples were then air dried and sieved through 4.75 mm sieve. By adding different proportions

of each effluent in each soil type, a total of 16 contaminated samples were prepared. A systematic nomenclature was adopted to represent the contaminated soil samples. The first two letters represent the soil type, either low plastic (CL) or high plastic (CH). Third letter represents the source of effluent (paper industry (P) or textile industry (T)) and the digits at the end represent the percentage of effluent in each soil sample e.g. "CHP5" represents a high plastic clay sample with 5% contamination from paper industry. The contaminated soil samples plus the two original uncontaminated soils were then subjected to various tests in order to explore the effects of effluent contamination. Soil properties evaluated as part of this study included specific gravity, Atterberg limits, modified compaction, one-dimensional consolidation, unconfined compression, pH and electrical conductivity. All tests were performed according to the relevant ASTM standard.

3. Results and discussions

3.1. Effect of contamination on Atterberg's limits and soil classification

The addition of acidic (paper industry) and basic (textile industry) contaminants increased the plasticity of soil as shown in Figures 5 and 6. The liquid limit of CL soil increased by 13.5% and 10.8% with 20% acidic and basic contamination respectively. However, the corresponding increase in liquid limit of CH soil was not as pronounced and the increase in liquid limit was limited to 6.9% and 5.2% for acidic and basic contamination addition respectively.

Similarly, the plasticity index of both the soils showed an increasing trend with the addition of contaminants. The plasticity index of CL soil increased by 37.5% with the mixing of 20% acidic and basic contamination. However, the corresponding increase in plasticity index of CH soils was not as pronounced and the increase in plasticity index was limited to 22.6% and 19.4% for acidic and basic contaminants respectively.

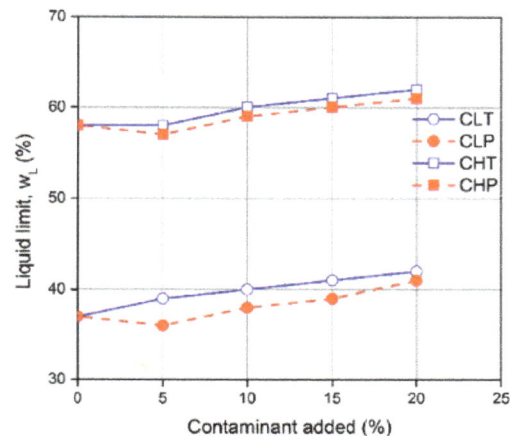

Fig. 5. Effect of contamination on liquid limit

With effluent contamination, an increase in liquid limit is followed by a corresponding increase in plasticity index. Therefore, soil classification, as shown in Figure 7 does not change due to contamination.

The increase in Atterberg limits of soil is mainly because of the chemical action between soil particles and

effluent. The increase in liquid limit indicates an increase in consolidation potential of contaminated soils (Terzaghi, Peck 1948). High plasticity of contaminated soils also causes problems related to increased swell potential and high collapsibility (Gibbs, Bara 1967). Contamination by industrial effluents would therefore deteriorate the quality of soil as an engineering material.

3.2. Effect of contamination on soil specific gravity

The effect of contaminants on specific gravity of both soils is summarized in Figure 8. In general, the specific gravity of CH soils was observed to decrease slightly with the addition of contaminants. Quantitatively, the specific gravity of CH soil decreased by 1.4% and 2.3% with the addition of acidic and basic contaminants, respectively. However, the specific gravity of CL soil was unaffected by the addition of contaminants. This behavior can be associated with both the effluents having nearly the same specific gravity as CL soil.

3.3. pH value of contaminated soil

The industrial effluents used in this study were recovered from a textile and a paper industry. The effluent recovered from textile industry was basic in nature whereas the effluent from paper industry was acidic in nature. The overall pH of soil samples was affected accordingly with the addition of these effluents. Effect of contaminant on pH is summarized in Figure 9. The pH of CL and CH soil samples increased by 5.3% and 5.7% with the addition of 20% textile waste effluent. On the contrary, the pH of both the soil samples decreased with the addition of effluent from paper industry because of its acidic nature. The observed decrease in pH of CL and CH soil samples with 20% contamination of paper effluent was 5.3% and 7.1% respectively. This behavior is consistent with the findings of Sunil *et al.* (2009).

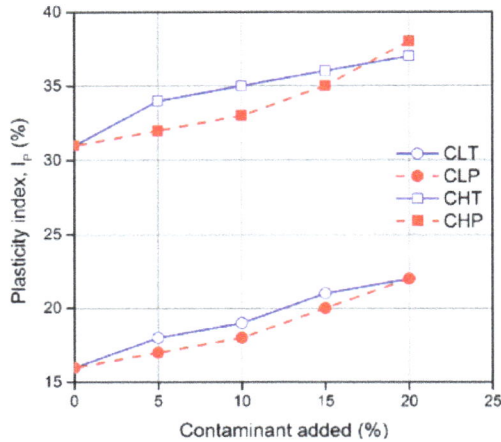

Fig. 6. Effect of contamination on plasticity index

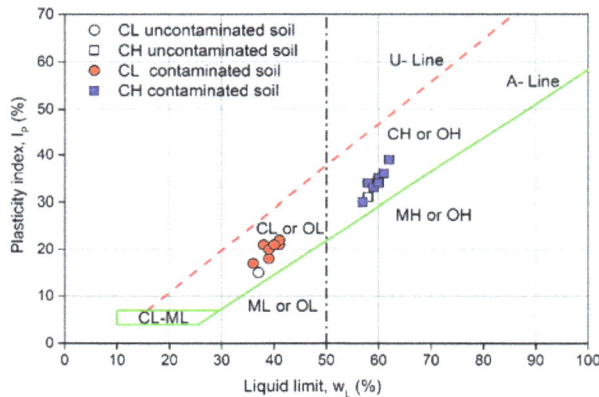

Fig. 7. Effect of contamination on soil classification

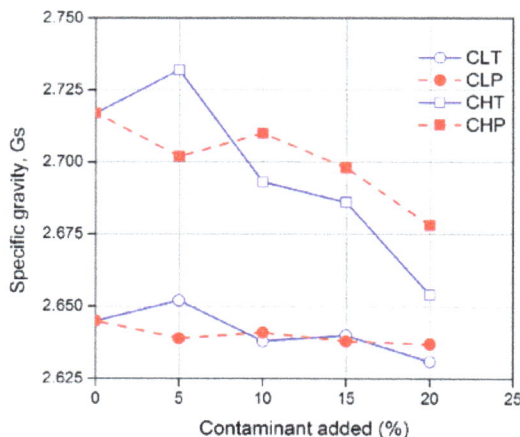

Fig. 8. Variation of specific gravity with concentration of contaminant

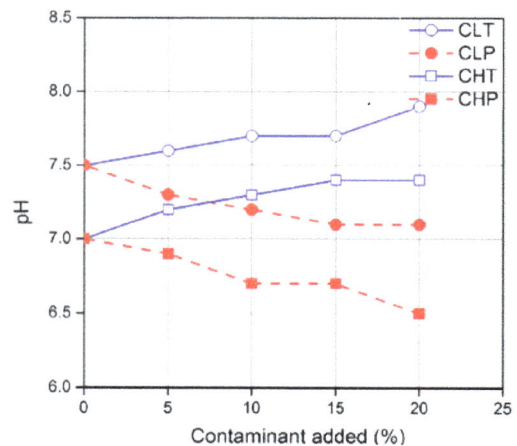

Fig. 9. Variation of pH with concentration of contaminant

3.4. Effect of contamination on electrical conductivity

The effect of contamination addition on the electrical conductivity of both soils is illustrated in Figure 10. The electrical conductivity of both the soil samples increased with the addition of contaminants. Quantitatively, the specific gravity of CL soil increased by 1.2% and 1.4% with the addition of acidic and basic contaminants respectively. However the corresponding increase in electrical conductivity of CH soils was more pronounced and it was 4.0% and 5.8% for acidic and basic contamination addition respectively. The increase in electrical conductivity of contaminated soils compared to the parent soil, may be associated to the presence of more free ions in the acidic/basic contaminated soils.

3.5. Compaction characteristics of contaminated soils

The effect of contamination addition on the compaction characteristics of both soils is represented in Figures 11 and 12. In general, the optimum moisture content was observed to increase by around 10% for CL soil and around 7.5% for CH soil with the addition of contaminants. On contrary, maximum dry unit weights of both cohesive soils were found to decrease with the addition of contaminants. The decrease in maximum dry unit weights of both the soils with the addition of 20% acidic/basic effluent was quite identical at around 4%.

The variation in the compaction characteristics of cohesive soils because of contaminant addition can be explained on the basis of soil plasticity. Optimum moisture content of cohesive soils increases whereas the maximum dry unit weight obtained through compaction tests decreases with plasticity index of soil (Berawala, Solanki 2010; Pandian et al. 1997; Sridharan, Nagaraj 2005). As discussed previously in Section 4.1, increase in contamination concentration makes the soil more plastic, thereby leading to an increase in optimum moisture content and a decrease in maximum dry unit weight. From engineering applications perspective, this means a high water demand to attain optimum moisture in the field; which, in general, increases the project cost and is typically undesirable. In other words, soil with high contaminant concentration would be difficult to compact and would yield a lower unit weight compared to uncontaminated soil under the same compactive effort and moisture conditions.

3.6. Effect of contamination on unconfined compressive strength

Addition of contamination to cohesive soils was found to adversely affect shear strength as illustrated in Figure 13. In general, unconfined compressive strength of both CL and CH soils showed a slight decrease on the addition of effluent contamination. Unconfined compressive strength of CL and CH soils decreased by around 5% and 6%,

respectively with contamination. The reduction of strength due to contamination is attributed to possible breakage of internal bonds (Umesha et al. 2012). Typical photographs of specimens before and after the unconfined compression test are shown in Figure 14.

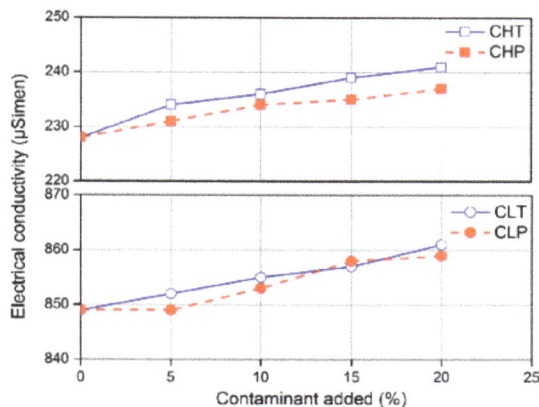

Fig. 10. Effect of contamination on electrical conductivity of soil

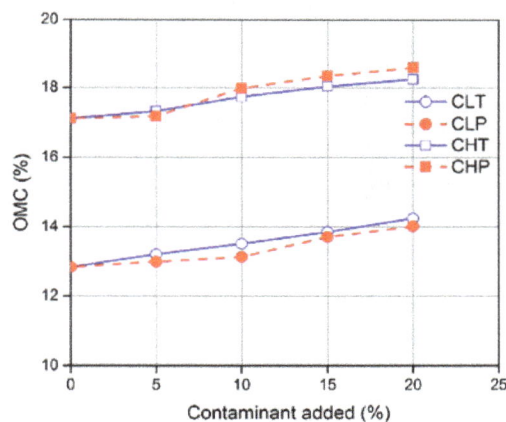

Fig. 11. Variation of OMC with effluent contamination

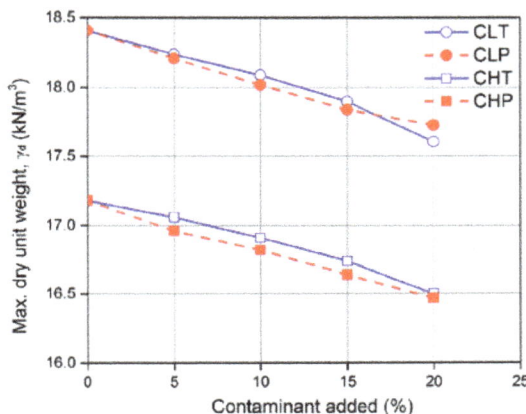

Fig. 12. Variation of maximum dry unit weight with effluent contamination

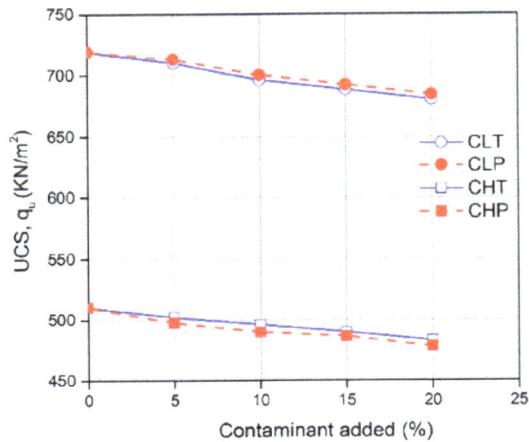

Fig. 13. Effect of contamination on unconfined compressive strength

(a)   (b)

Fig. 14. Typical unconfined compression test specimens (a) before test (b) after test

3.7. Consolidation characteristics of contaminated soils

The consolidation characteristics of industrial effluent contaminated soils were evaluated by performing one-dimensional consolidation test on original, and contaminated soil specimens. The effect of effluent contamination on the rate of consolidation (represented by coefficient of consolidation, C_v), and the magnitude of consolidation (represented by compression index, C_c) are illustrated in Figures 15 and 16 respectively.

With the addition of acidic (paper industry) and basic (textile industry) contaminants, C_v of CL soil decreased by 40.2% and 50.0%, and that of CH soil decreased by 50.8% and 60.9% respectively. Time required for the consolidation of contaminated soil would therefore be greater than that required for the completion of same degree of consolidation in the parent soil. C_v is an indirect measure of soil permeability as well. Decrease in C_v also indicates a reduction in the hydraulic conductivity of contaminated soil. Suspended solids in the industrial waste typically have very small particle size. These particles might clog the inter-particle space in parent soil sample thereby reducing the hydraulic conductivity.

The compression index (C_c), represented in Figure 15, demonstrates an increase in the magnitude of consolidation with contaminant concentration. The increase in C_c of CL soil ranged from 7.5% to 11.3% whereas the corresponding increase for CH soil was from 9.4% to 11.1%, for 20% addition of acidic/basic contaminant. This indicates the consolidation potential of contaminated soils to be higher than uncontaminated soils. The same behavior could also be deduced from the increase in soil plasticity with effluent contamination (Fig. 5), since C_c is a direct function of soil plasticity (Terzaghi, Peck 1948).

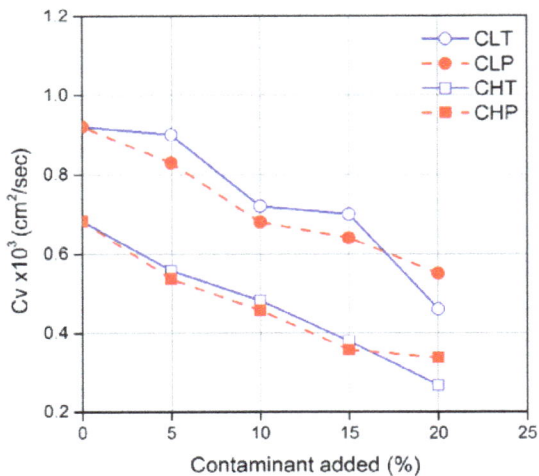

Fig. 15. Variation of Cv with the addition of acidic and basic contaminants

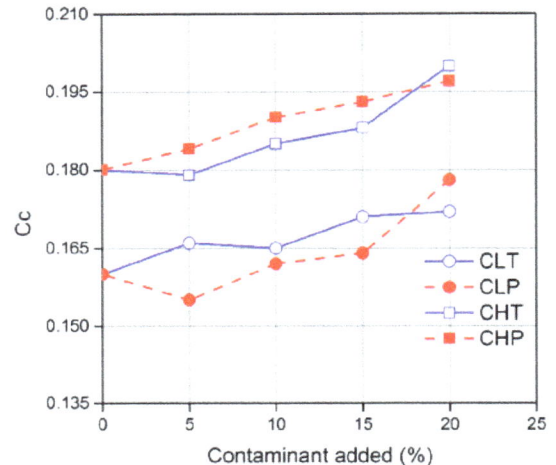

Fig. 16. Variation of Cc with the addition of acidic and basic contaminants

Conclusions

The main objective of this research was to study the effects of industrial contamination on local cohesive soils. The test results showed a significant change in the geotechnical properties of contaminated soils when compared with the parent soil samples. The conclusions drawn from this research work are as follows:

- The increase in liquid limit and plasticity index of cohesive soils is mainly attributed towards an increase in specific surface area of soil due to contaminant addition which leads to high adsorption of water affecting the Atterberg's limits values.

- Addition of acidic and basic contaminants vary the pH of soil accordingly. Due to the presence of more free ions in contaminated soils, electrical conductivity is also increased.

- Due to chemical reaction between contaminant and soil particles, soil becomes more plastic in nature which causes a decrease in maximum dry unit weight and an increase in optimum moisture content. A corresponding decrease in unconfined compressive strength of contaminated soils was also observed.

- The chemical reaction between effluents and soil particles also affect the consolidation characteristics of soil. Contamination makes the soil highly compressible as indicated by an increase in compression index. However, a decrease in the coefficient of consolidation of contaminated soils indicate a decrease in the permeability of contaminated soil. This implies that contaminated soils would take longer to accomplish a certain degree of consolidation.

- In general, the soil properties have been found to deteriorate because of being contaminated by industrial effluents. Special considerations would therefore be required for constructing on such soils. The loss of bearing capacity indicated by a decrease in unconfined compressive strength, and an expected increase in the magnitude of consolidation settlement should be carefully considered while designing foundations on such soils. The increased chemical aggressivity of contaminated soils can also be a concern for foundation concrete.

Acknowledgements

The University of Engineering and Technology, Lahore and Institute of Environmental Engineering and Research, Lahore (Pakistan) are gratefully acknowledged for the financial and technical assistance.

References

Adebisi, S.; Fayemiwo, K. 2010. Pollution of Ibadan soil by industrial effluents, *New York Science Journal* 3(10): 37–41.

Berawala, K.; Solanki, C. 2010. Empirical correlations of expansive soils parameters for the Surat Region, in *Indian Geotechnical Conference*, 16–18 December 2010, Mumbai, India.

Bond, R. G.; Straub, C. P.; Prober, R. 1973. *Water supply and treatment*. Ohaio: CRC Press.

Choudhury, A. K. R. 2006. *Textile preparation and dyeing*. Enfield: Science Publishers.

Gibbs, H. J.; Bara, J. P. 1967. Stability problems of collapsing soil, *Journal of Soil Mechanics* & *Foundations Division, ASCE* 93(4): 577–594.

Gratchev, I.; Towhata, I. 2009. Effects of acidic contamination on the geotechnical properties of marine soils in Japan, in *ISOPE-2009 Osaka: 19th International Offshore (Ocean) and Polar Engineering Conference*, 21–26 June 2009, Osaka, Japan.

Jia, Y.; Wu, Q.; Shang, H.; Yang, Z. N.; Shan, H. 2011. The influence of oil contamination on the geotechnical properties of coastal sediments in the Yellow River Delta, China, *Bulletin of Engineering Geology and the Environment* 70(3): 517–525. https://doi.org/10.1007/s10064-011-0349-8

Khamehchiyan, M.; Charkhabi, A. H.; Tajik, M. 2007. Effects of crude oil contamination on geotechnical properties of clayey and sandy soils, *Engineering Geology* 89(3): 220–229. https://doi.org/10.1016/j.enggeo.2006.10.009

Kumar, A. 2005. *Concepts of biophysics*. New Delhi: APH Publishing Corporation.

Naeini, S.; Jahanfar, M. 2011. Effect of salt solution and plasticity index on undrain shear strength of clays, *World Academy of Science, Engineering and Technology* 49: 982–986.

Nazir, A. K. 2011. Effect of motor oil contamination on geotechnical properties of over consolidated clay, *Alexandria Engineering Journal* 50(4): 331–335. https://doi.org/10.1016/j.aej.2011.05.002

NEQS. 2000. *National environmental quality standards*. Environmental Protection Agency, Government of Pakistan, Islamabad.

Olgun, M.; Yildiz, M. 2010. Effect of organic fluids on the geotechnical behavior of a highly plastic clayey soil, *Applied Clay Science* 48(4): 615–621. https://doi.org/10.1016/j.clay.2010.03.015

Pandian, N.; Nagaraj, T.; Manoj, M. 1997. Re-examination of compaction characteristics of fine-grained soils, *Geotechnique* 47(2): 363–366. https://doi.org/10.1680/geot.1997.47.2.363

Patel, A. 2011. Study of Geotechnical properties of black cotton soil contaminated by castor oil and stabilization of contaminated soil by sawdust, in *National Conference on Recent Trends in Engineering* & *Technology*, 13–14 May 2011, Gujrat, India.

Reddy, K. R.; Hettiarachchi, H.; Gangathulasi, J.; Bogner, J. E. 2011. Geotechnical properties of municipal solid waste at different phases of biodegradation, *Waste Management* 31(11): 2275–2286.

Sridharan, A.; Nagaraj, H. 2005. Plastic limit and compaction characteristics of finegrained soils, *Proceedings of the ICE-Ground Improvement* 9(1): 17–22. https://doi.org/10.1680/grim.2005.9.1.17

Sunil, B.; Nayak, S.; Shrihari, S. 2006. Effect of pH on the geotechnical properties of laterite, *Engineering Geology* 85(1): 197–203. https://doi.org/10.1016/j.enggeo.2005.09.039

Sunil, B.; Shrihari, S.; Nayak, S. 2009. Shear strength characteristics and chemical characteristics of leachate-contaminated lateritic soil, *Engineering Geology* 106(1): 20–25.

https://doi.org/10.1016/j.enggeo.2008.12.011

Terzaghi, C.; Peck, R. B. 1948. *Soil mechanics in engineering practice.* New York: John Wiley & Sons.

Umesha, T. S.; Dinesh, S. V.; Sivapullaiah, P. V. 2012. Effects of acids on geotechnical properties of black cotton soil, *International Journal of Geology* 6(3): 69–76.

Muhammad Imran KHAN. MSc Geotechnical Engineering. PhD Scholar, Department of Civil Engineering and Applied Mechanics, Montreal, Canada. Lecturer, Civil Engineering Department, University of Engineering and Technology, Lahore. Research Interest: Contaminated Soils, Laboratory testing of soils, Foundation Engineering, Soil Structure Interaction. No. of publications: 0. No. of attended conferences: 2.

Muhammad IRFAN. PhD Geotechnical Engineering. Assistant Professor, Civil Engineering Department, University of Engineering and Technology, Lahore.
Research Interest: Finite element modeling of soil-structure interaction, Unsaturated soil mechanics, Rain-induced landslides, Landslide early warning, In-situ soil characterization, Soil dynamics, Foundation Engineering. No. of publications: 37. No. of attended conferences: 18.

Mubashir AZIZ. PhD Geotechnical Engineering. Assistant Professor, College of Engineering, Al Imam Mohammad Ibn Saud Islamic University, Riyadh.
Research Interest: Soil Mechanics, Engineering Geology, Geotechnical Earthquake Engineering. No. of publications: 30. No. of attended conferences: 18.

Ammad Hassan KHAN. PhD Geotechnical Engineering. Professor & Chairman, Transportation Engineering Department, University of Engineering and Technology, Lahore.
Research Interest: Geotechnical Investigation, In-situ and Laboratory Testing, Transportation Engineering Materials, Advances and Trends in Geotechnical Engineering. No. of publications: 18. No. of attended conferences: 11.

12

BIOINDICATION OF TRACE METAL POLLUTION IN THE ATMOSPHERE OF BAKU CITY USING *LIGUSTRUM JAPONICUM, OLEA EUROPEA,* AND *PYRACANTHA COCCINEA* LEAVES

Naglaa Youssef[a,b], Bernd Markert[c], Elshad Gurbanov[a], Haciyeva Sevnic[d],
Simone Wünschmann[c]

[a]*Department of Botany, Faculty of Biology, Baku State University, Baku, Azerbaijan*
[b]*Department of Botany, Faculty of Sciences, Sohag University, Sohag, Egypt*
[c]*Environmental Institute of Scientific Neworks (EISN-Institute), Fliederweg 17, 49733 Haren, Germany*
[d]*Department of Ecological Chemistry, Faculty of Chemistry, Baku State University, Baku, Azerbaijan*

Abstract. The leaves of *Ligustrum japonicum (Oleaceae), Olea europaea* (Oleaceae), and *Pyracantha coccinea* (Rosaceae) were evaluated with the aim of using them as bioindicators for trace metal contamination in Baku city, Azerbaijan, one of the most highly polluted cities worldwide. These species of trees are the most abundant in urban and rural areas of Azerbaijan, because of high tolerance against climatic influences due to their modesty and adaptability. Concentrations of Cd, Cr, Cu, Fe, and Pb were determined in the leaves by AAS method. The samples were collected at three locations with different degrees of trace metal pollution (industrial, high traffic, and reference [botanical garden] site). The highest element concentrations were detected at sites of high traffic. Up to 70 times higher Pb concentrations could be found in the leaves of the trees that reflect the known Pb problem around Baku. The results presented give a first impression of a correlation between the degree of trace metal contamination in the environment and the trace metal concentration in the leaves of *L. japonicum* and *O. europaea.*

Keywords: air pollution, trace metals, environmental monitoring, leaves, bioindicators, Baku, Azerbaijan.

Reference to this paper should be made as follows: Youssef, N.; Markert, B.; Gurbanov, E.; Sevnic, H.; Wünschmann, S. 2014. Bioindication of trace metal pollution in the atmosphere of Baku city using *Ligustrum japonicum, Olea europea,* and *Pyracantha coccinea* leaves, *Journal of Environmental Engineering and Landscape Management* 22(1): 14–20. http://dx.doi.org/10.3846/16486897.2013.804828

Introduction

Air pollution is an offensive risk and can be a genuine health hazard to humans as well as to animals, plants, and microorganisms. Plant tissues have shown to behave as an effective indicator of atmospheric pollution (e.g. Markert *et al.* 2003; Wolterbeek *et al.* 2010). Vegetation is an acceptable indicator of pollution impact to its vicinity, because plants have the ability to accumulate or reject trace metals so that their metal levels correlate to those in the air. The effect observed is a time-averaged result, which will be – in most cases – more reliable than that obtained from direct determination of the pollutant concentrations in air for a short period. Hence, analyzing plant tissues can often produce better results in terms of sensitivity and reproducibility (Lau, Luk 2001). Some plant species are especially useful as biological indicators to assess

air pollution because of their global distribution (e.g. Kardel *et al.* 2010). A lot of plant species have already been applied as bioindicator species (Baycu *et al.* 2006; Celik *et al.* 2005; Fraenzle *et al.* 2012; Markert *et al.* 2003; 2008; Mingorance, Oliva 2006; Youssef 2012). The contaminants can deposit on and in the surface of leaves (Salma *et al.* 2001) and increase their harmful effects on human health (Temesi *et al.* 2003). Accumulation of heavy metals causes chronic damage to ecosystems and must be carefully observed and monitored taking into account uptake, movement, and effects of the contaminants on both the environment and its biota (Fraenzle *et al.* 2012; Markert *et al.* 2008; Mingorance, Oliva 2006; Morselli *et al.* 2004).

The aim of the present study is to investigate the pollution levels of Cd, Cr, Cu, Fe, and Pb using leaves of plant species, such as *Ligustrum japonicum, Olea*

Corresponding author. Simone Wünschmann
E-mail: wuenschmann@eisn-institute.de

europaea, and *Pyracantha coccinea* from three different sites in and around Baku (the capital of Azerbaijan). Baku has been discussed as one of the highest polluted cities of the world (Luck 2008), being polluted by emissions of oil drillings. Ismailov and Akhoun-Zade (1999) reported that especially during 1991–1994, serious environmental problems were caused by petro-chemical industry, power plants, metallurgical, and building materials industries situated in Baku. Impor-tant contaminants have been hydrocarbons through 1,665,000 tons/y of emission from oil refineries in Baku. After Sawidis *et al.* (2011) the environmental lead pollution is directly related to the density of traffic. The main part of the total vehicle fleet up to 65% is located in the city of Baku, which significantly affects the ecological background of the city.

The results of this study should answer the scientific question whether plant leaves of *L. japonicum,* *O. europaea,* and *P. coccinea* can be used as bioindica-tors for environmental pollution observation and con-trol in Baku city. The three species under investigation meet the requirements of bioindicators: wide distribu-tion, high abundance, accumulation of pollutants, easy to collect, simple identification and cultivation, and analytical fitness. The studied elements are common pollutants in the air, especially in industrial cities as Baku characterized by large petroleum industry and by high density of traffic (Babayeva 2002). The results are especially of interest as preliminary baseline data for environmental air pollution of Baku derived from tree leaves.

1. Materials and methods

1.1. Sampling of plant leaves

Leaves of the evergreen trees *L. japonicum* (*Oleaceae*), *P. coccinea* (*Rosaceae*), and *O. europaea* (*Oleaceae*) were selected in three different areas in and around Baku (Fig. 1): (1) industrial zone of Baku (Absheron peninsula), (2) region of Baku airport with high traffic, and (3) botanical garden of Baku State University as reference site (10 km distance to pollution sources). From each sampling point, leaves of five trees per species (all in between 30 and 40 years) were taken from April to June 2012. Leaves of four different branches of each tree were sampled at a height of 1.5–2 m above the ground. From each branch three shoots were chosen. The sampling procedure followed the method after Sawidis *et al.* (2011). Care was given to avoid the collection of leaves characterized by insect infestation, presence of honeydew, bird dropping, pesticide treatment, chlorosis or necrosis, coarse, and anomalous dust cover. The collected leaves (about 30 g per sample) were placed in paper envelopes immedi-ately after collection and dried in the laboratory.

1.2. Analytical procedure

After drying and homogenization, the samples were placed in polythene bags and stored in a refrigerator at 4 °C. One gram of each sample was digested in an open quartz tube and 10 ml of concentrated HNO_3 (Merck) was added to each sample and the mixture was left at room temperature overnight. In the further course, the

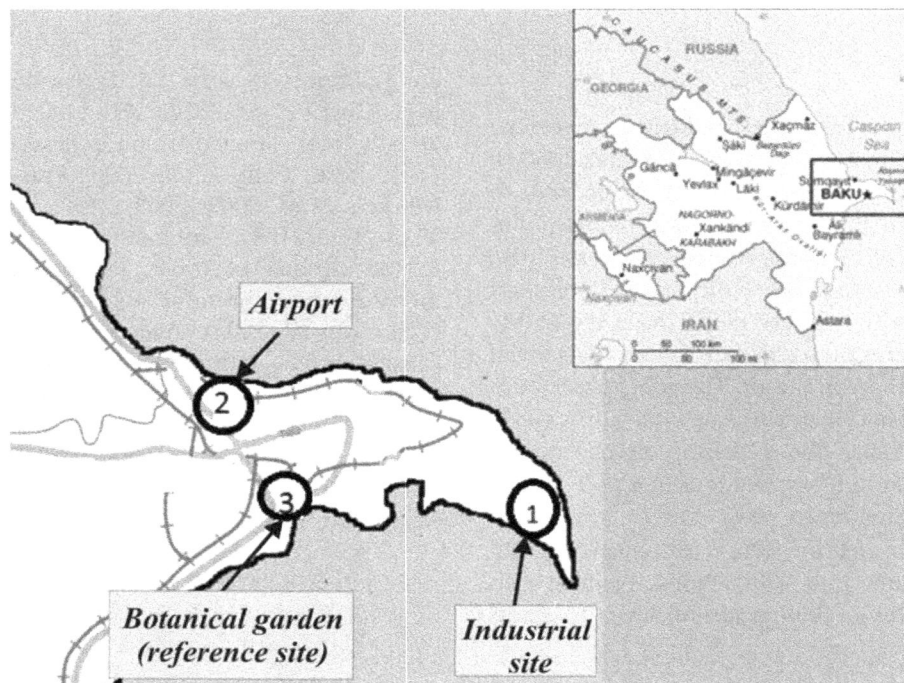

Fig. 1. Sampling locations of tree leaves in Baku, Azerbaijan

samples were heated at 50 °C for 2 h and subsequently heated at 160 °C for 4 h. The solution was filtered through a Whatman type 589/2 filter and the filtrate was diluted to 25 ml volume with double de-ionized water. These final solutions were analyzed for trace metal concentrations using an Atomic Absorption Spectrophotometer 850 (flame atomization, Perkin Elmer) at the Analysis Center of the Geological Institute at Azerbaijan, National Academy of Science. The quality controls of instrumental measurements were performed by using the National Standard of Azerbaijan BZ No.7–95, *O. europaea*. Statistical analyses were done by using Excel, Origin 6, and SPSS 11.5.

2. Results and discussion

The detected concentrations of Cd, Cr, Cu, Fe, and Pb in tree leaves from each sampling area are summarized in Table 1. Element specific concentrations are represented in Figures 2–6.

2.1. Cadmium

The leaves of *L. japonicum* showed the highest Cd amount up to 0.514 ± 0.13 ppm dry wt at site 2 in relation to the other tree species (traffic area, see Fig. 2). In general the Cd distribution pattern demonstrates higher Cd concentrations at the sampling sites 1 (industry) and 2 (traffic) in relation to the reference site (botanical garden). Compared to the average reference concentration for Cd in plants (0.03–0.5 ppm dry wt, Markert 1997) the Cd concentrations found do not cause pathological effects to plants.

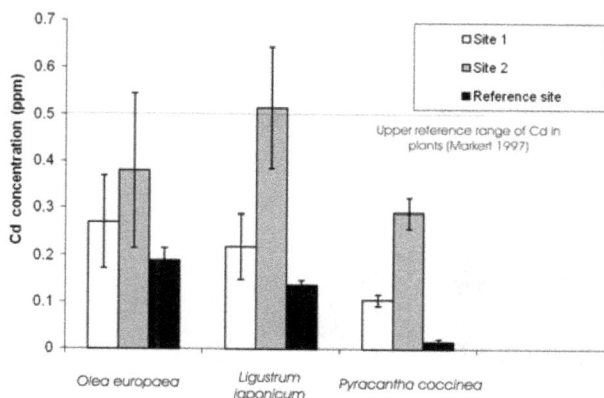

Fig. 2. Cadmium concentrations (ppm dry wt) in the leaves of *Olea europaea*, *Ligustrum japonicum*, and *Pyracantha coccinea* from different studied sites (site 1: industry; site 2: traffic; reference site: botanical garden)

2.2. Chromium

The leaves of *O. europaea* and *L. japonicum* represented much higher levels of 1.6–9 ppm dry wt in relation to the average reference concentration for Cr in plants (0.2–1 ppm dry wt, Markert 1997). However, the Cr concentrations in the leaves of *P. coccinea* taken from all three sampling sites matched the reference range of Cr contents in plants (up to 1 ppm dry wt).

2.3. Copper

The leaves of *O. europaea* showed the highest Cu amounts of up to 84 ppm dry wt at the sampling site 2 (traffic) and up to 73 ppm dry wt at sampling site 1 (industry) (Fig. 4). After Markert (1997) the average reference concentrations of Cu in plants differ between 2 and 20 mg/l. These values correspond with the Cu

Table 1. Comparison of Cd, Cr, Cu, Fe, and Pb concentrations in leaves of *Olea europaea*, *Ligustrum japonicum*, and *Pyracantha coccinea* from different studied sites

Plant species	Site	Cd, ppm ± sd dry wt	Cr, ppm ± sd dry wt	Cu, ppm ± sd dry wt	Fe, ppm ± sd dry wt	Pb, ppm ± sd dry wt
Olea europaea	Site 1: industrial zone	0.27 ± 0.10	9 ± 3	73 ± 14	273 ± 86	221 ± 31
	Site 2: high traffic (airport)	0.38 ± 0.16	7 ± 2	84 ± 18	188 ± 55	282 ± 40
	Site 3: reference site, botanical garden	0.19 ± 0.03	4 ± 1	21 ± 4	165 ± 66	70 ± 13
Ligustrum japonicum	Site 1: industrial zone	0.22 ± 0.07	4 ± 0.70	33 ± 9	232 ± 55	257 ± 24
	Site 2: high traffic (airport)	0.54 ± 0.13	4 ± 1	45 ± 10	250 ± 33	341 ± 73
	Site 3: reference site, botanical garden	0.14 ± 0.01	1.63 ± 0.35	6 ± 1	170 ± 31	99 ± 19
Pyracantha coccinea	Site 1: industrial zone	0.10 ± 0.01	0.47 ± 0.15	14 ± 3	98 ± 24	60 ± 11
	Site 2: high traffic (airport)	0.29 ± 0.03	0.39 ± 0.13	21 ± 5	122 ± 19	106 ± 13
	Site 3: reference site, botanical garden	0.02 ± 0.01	0.23 ± 0.02	3 ± 0.79	135 ± 23	53 ± 9

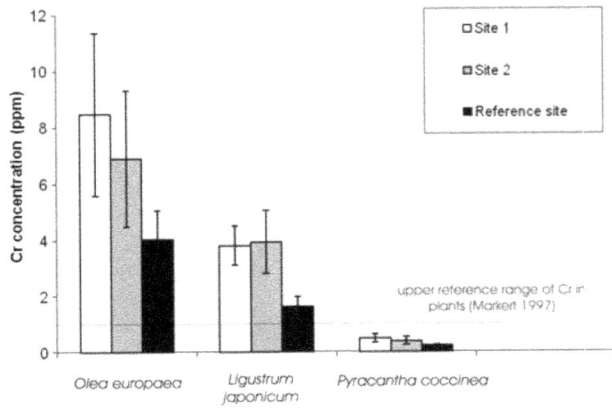

Fig. 3. Chromium concentrations (ppm dry wt) in the leaves of *Olea europaea*, *Ligustrum japonicum*, and *Pyracantha coccinea* from different studied sites (site 1: industry; site 2: traffic; reference site: botanical garden)

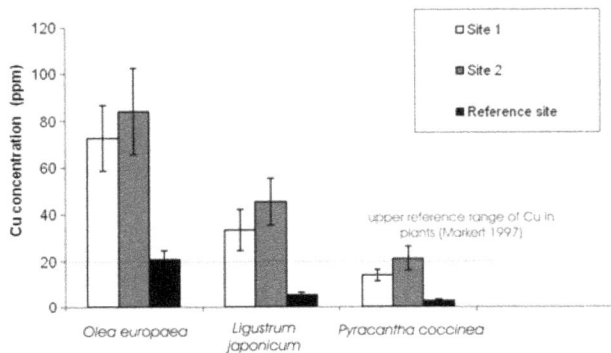

Fig. 4. Copper concentrations (ppm dry wt) in leaves of *Olea europaea*, *Ligustrum japonicum*, and *Pyracantha coccinea* from different studied sites (site 1: industry; site 2: traffic; reference site: botanical garden)

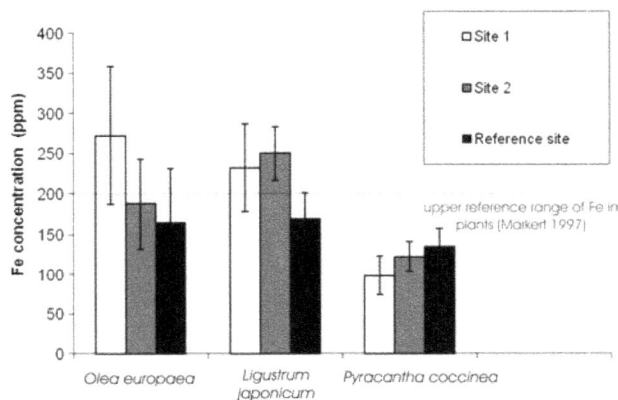

Fig. 5. Iron concentrations (ppm dry wt) in the leaves of *Olea europaea*, *Ligustrum japonicum*, and *Pyracantha coccinea* from different studied sites (site 1: industry; site 2: traffic; reference site: botanical garden)

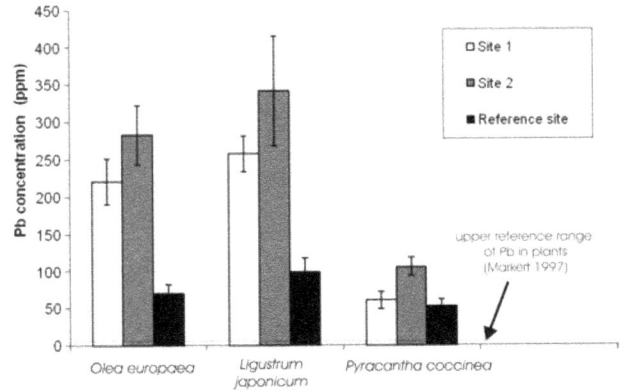

Fig. 6. Lead concentrations (ppm dry wt) in the leaves of *Olea europaea*, *Ligustrum japonicum*, and *Pyracantha coccinea* from different studied sites (site 1: industry; site 2: traffic; reference site: botanical garden)

concentrations evaluated in all three leave species collected from the reference site (botanical garden; up to 21 ppm dry wt) and with the Cu concentrations measured in the leaves of *P. coccinea* at all studied places.

The Cu concentrations in the leaves of *O. europaea* and *L. japonicum* show up to four times higher Cu concentrations for the sampling points 1 and 2 compared to the "normal" values between 2 and 20 mg/l.

2.4. Iron

The leaves of *O. europaea* and *L. japonicum* showed higher Fe concentrations up to 273 ppm dry weight related to the leaves of *P. coccinea* (up to 135 ppm dry wt at the reference site, Fig. 5). According to Markert (1997) the average reference concentration for Fe in plants ranges from 5 to 200 ppm dry wt. These values corresponded with the Fe concentrations found in the leaves of *P. coccinea* at all samplings sites and with the Fe concentrations found in the leaves of *O. europaea* and *L. japonicum* collected from the reference site (botanical garden). The leaves of *O. europaea* taken from the sampling site no. 1 and the leaves of *L. japonicum* taken from the sampling sites 1 and 2 represented up to 30% higher Fe concentrations.

2.5. Lead

All evaluated Pb concentrations in the leaves of *O. europaea*, *L. japonicum*, and *P. coccinea* exceeded to a considerable degree the average reference concentration in plants of 0.1–5 ppm dry wt (Markert 1997). Figure 6 demonstrates that the Pb concentrations of the reference sites (botanical garden) were lower compared to site 1 (industry) and 2 (traffic). But these concentrations are much higher than the "normal" average value of 0.1–5 ppm dry wt given by Markert (1997). Values up to around 350 ppm dry wt in the leaves of *L. japonicum* could be detected at the sampling site 2 (traffic). Figure 6 also demonstrates that the leaves of the tree species *O. euopaea* and *L. japonicum* accumulate Pb in much higher concentrations compared to the leaves of *P. coccinea*.

2.6. Comparison of all chemical elements to each other

In summary, as expected, both sites (industry (1) and high traffic (2)) show higher element concentrations in comparison to the reference site (botanical garden site (3)). In case of Cd concentrations, all values measured at the sampling sites seem to be within a tolerable range. After Markert (1997) the average reference concentration of Cd in plants differs between 0.03 and 0.5 ppm dry wt. In the present study, the highest Cd concentration could be detected with 0.514 ppm dry wt in the leaves of *L. japonicum*. Based on these results, there is probably no significant Cd pollution in the urban region of Baku. The same could be found for Fe concentrations in the leaves studied. Related to the "normal" average reference concentration of Fe in plants after Markert (1997) with up to 200 ppm dry wt, the highest Fe concentration in the present study showed a value of 270 ppm dry wt. So the Fe concentrations in the leaves of the trees *O. europaea* and *L. japonicum* are slightly increased, but not significantly. The Fe concentrations in the leaves of *P. coccinea* are absolutely in an acceptable range.

For the elements Cr, Cu, and especially for Pb the situation is quite different. It is noticeable that the Cr and Cu concentrations in the leaves of *P. coccinea* are quite acceptable according to all studied places. Related to the Cr and Cu concentrations in the leaves of the tree species *O. europaea* and *L. japonicum*, with exception for the Cu concentration in leaves taken from the reference sites, there are higher values compared to the given average reference values of Cr and Cu in plants after Markert (1997): up to nine times higher for Cr concentrations and up to four times higher for Cu concentrations. The Pb concentrations detected in all leaves samples reflect a very serious Pb emission problem in and around the region of Baku. Even the lowest Pb concentration measured in the leaves of *P. coccinea* at the reference site (botanical garden) showed a 10 times higher Pb concentration compared to the "normal" average reference concentration of Pb in plants given by Markert (1997). Especially, the more Pb contaminated sample regions site 1 (industry) and site 2 (traffic) showed up to 70 times higher Pb concentrations. The reason for the higher lead values might be the greater use of cars and buses fuelled with leaded gasoline. Industrial and metallurgical processes, as well as the combustion of diesel oil produce the largest emissions of lead. The city's past as a Soviet industrial center has left it as one of the most polluted cities in the world (Luck 2008). At

site 1, industrial facilities (chemical, pharmaceutical, metallic, petroleum industry) are randomly distributed in the central parts of the region. They represent, together with city traffic and coal power stations, the sources of various types of pollutants (Mitrovic *et al.* 2008; Pavlovic *et al.* 2007). It is known that the main sources of copper and lead pollution are the steel industry and coal combustion (Anagnostatou 2008). These results give an indication that the accumulation of trace metals depend on the traffic, industrial activities, and urbanization levels (Markert *et al.* 2011; Mingorance, Oliva 2006).

Additionally, the results show that the concentrations of all studied elements in *P. coccinea* tissue were comparable to the given "normal" averaged reference ranged values for plants after Markert (1997) with the exception of Pb. These results give the conclusion that the tree species *P. coccinea* is not applicable for bioindication of air pollution. Unlike that situation, the tree species *O. europaea* and *L. japonicum* demonstrate the effects of urbanization on elemental concentrations in leaves, but the results may also depend on the morphological anatomical parameters of leaves (Kardel *et al.* 2010).

Nevertheless, there are some general trends in the data which, notwithstanding effects of selective uptake of metal ions by plants, can be used to show that there are different sources of emission/pollution which contribute to metal contents of the leaves, that is, Pb, and Cr get there from other sources than Fe does (which is quite reasonable), possibly via particles of different sizes and origins or part via soil/ground water, part through atmospheric deposition (total digestion of leaves prior to analysis or complete burning in F-AAS covers both pathways to the leaf). Of course, the three plants differ in both biochemical fractionation behavior (which unfortunately is not yet numerically specified for these three species) when obtaining metal ions from soil and in relative impacts of atmospheric deposition likewise (the leaves are differently thick and "sticky").

Conclusions

Air pollution with trace metals is a matter of great interest, especially in urban areas (Markert *et al.* 2011). Monitoring of air quality using plants has been widely applied to detect and to monitor the level, distribution and effect of pollution (Gajic *et al.* 2009; Markert *et al.* 2003; Mingorance, Oliva 2006). Although biomonitoring of air quality using plants has been practiced for years in many European countries, it has still not been applied at a satisfactory level, due to different and even opposing results. Trees are very efficient at trapping atmospheric particles, and they play a special role in reducing the level of fine, "high risk" respirable particulates, which have the potential to cause serious human health problems. The present primary results on trace metal concentrations in leaves of tree species in and around Baku show that especially, the leaves of *O. europaea* and *L. japonicum* can be probably used as bioindicators, because of their ability to accumulate the essential elements Cu, Cr, and Fe. But of highest interest are the Pb concentrations in the environment. Pb is a nonessential element and can lead to toxic effects to humans, animals, and plants alike. So, the present study showed that the leaves especially of *O. europaea* and *L. japonicum* reflect in an excellent way the Pb status of the environment of Baku.

After Sawidis *et al.* (2011) the environmental lead pollution is directly related to the density of traffic. The main part of the total vehicle fleet up to 65% is located in the city of Baku, which significantly affects the ecological background of the city. In addition to the traffic density, industrial activities also tend to increase the concentration. The samples obtained from an industrial area, as well as from urban roadsides, which encounter the highest human activity and vehicular density, had the highest accumulation of the metal concentrations. Most of this contamination can clearly be traced back to motor vehicle traffic emissions.

In general, because the leaves of *O. europaea* and *L. japonicum* as bioindicators are reflecting the environmental status it would be enriching to join global monitoring programs of air pollution.

In conclusion,

(1) the leaves of *O. europaea* and *L. japonicum* have been used for the first time as possible bioindicators for controlling the air pollution in Baku city; they reflect the environmental status;

(2) the leaves of *P. coccinea* are not suitable for bioindication.

Acknowledgements

This work is a result of cooperation between Baku State University and Analysis Center of the Geological Institute at Azerbaijan, National Academy of Sciences. PD Dr. Stefan Fraenzle is thanked for improving the English of this article.

References

Anagnostatou, A. V. 2008. *Assessment of heavy metals in central Athens and suburbs using plant material*. Dissertation. University of Surrey, UK.

Babayeva, S. 2002. Transport and pedestrian communications in the residential quarters of Baku central zone, *Urbanizm* 2(3): 55–58.

Baycu, G.; Tolunay, D.; Özden, H.; Günebakan, S. 2006. Ecophysiological and seasonal variations in Cd, Pb, Zn, and Ni concentrations in the leaves of urban deciduous trees in Istanbul, *Environmental Pollution* 143: 545–554.

Celik, A.; Kartal, A. A.; Akdogan, A.; Kaska, Y. 2005. Determining the heavy metal pollution in Denizli

(Turkey) by using *Robinio pseudoacacia* L., *Environment International* 31: 105–112.

Fraenzle, S.; Markert, B.; Wuenschmann, S. 2012. *Introduction to environmental engineering*. New York, Tokyo: Wiley-VCH, Weinheim.

Gajic, G.; Mitrovic, M.; Pavlovic, P.; Stevanovic, B.; Djurdjevic, L.; Kostic, O. 2009. An assessment of the tolerance of *Ligustrum ovalifolium* Hassk to traffic-generated Pb using physiological and biochemical markers, *Ecotoxicology and Environmental Safety* 72: 1090–1101.

Ismailov, T.; Akhoun-Zade, F. 1999. *Environmental information system in Azerbaijan, assessment report for establishing a national environmental and natural resource information network Compatible with the UNEP/GRID* [online], [cited March 2003]. Available from Internet: http://www.grida.no/enrin/htmls/azer.htm

Kardel, F.; Wuyts, K.; Babanezhad, M.; Vitharana, U. W. A.; Wuytack, T.; Potters, G.; Samson, R. 2010. Assessing urban habitat quality based on specific leaf area and stomatal characteristics of *Plantago lanceolata* L., *Environmental Pollution* 158: 788–794.

Lau, O. W.; Luk, S. F. 2001. Leaves of *Bauhinia blakeana* as indicators of atmospheric pollution in Hong Kong, *Atmospheric Environment* 35: 3113–3120.

Luck, T. 2008. *The world's dirtiest cities* [online], Forbes [cited November 2010]. Available from Internet: http://www.forbes.com/2008/02/26/pollution-baku-oil-biz-logistics-cx_tl_0226dirtycities.html

Markert, B. 1997. *Instrumental element and multi-element analysis of plant samples. Methods and applications.* Weinheim: Wiley-VCH.

Markert, B.; Breure, A.; Zechmeister, H. (Eds.). 2003. *Bioindicators and biomonitors. Principles, concepts and applications.* Amsterdam: Elsevier.

Markert, B.; Wuenschmann, S.; Fraenzle, S.; Wappelhorst, O.; Weckert, V.; Breulmann, G.; Djingova, R.; Herpin, U.; Lieth, H.; Schroeder, W.; Siewers, U.; Steinnes, E.; Wolterbeek, B.; Zechmeister, H. 2008. On the road from environmental biomonitoring to human health aspects, monitoring atmospheric heavy metal deposition by epiphytic/epigeic plants: present status and future needs, *Environmental Pollution* 32(4): 486–498.

Markert, B.; Wuenschmann, S.; Fraenzle, S.; Figueiredo, A. M.; Ribeiro, A. P.; Wang, M. 2011. Bioindication of atmospheric trace metals – with special reference to megacities, *Environmental Pollution* 159: 1991–1995.

Mingorance, M. D.; Oliva, S. R. 2006. Heavy metals content in *N. oleander* leaves as urban pollution assessment, *Environmental Monitoring and Assessment* 119: 57–68.

Mitrovic, M.; Pavlovic, P.; Lakusic, D.; Djurdjevic, L.; Stevanovic, B.; Kostic, O.; Gajic, G. 2008. The potential of *Festuca rubra* and *Calamagrostis epigejos* for the revegetation of fly ash deposits, *Science of Total Environment* 407(1): 338–347.

Morselli, L.; Brusoni, B.; Passarini, F.; Bernardi, E.; Francaviglia, R.; Gataleta, L. 2004. Heavy metal monitoring at a Mediterranean natural ecosystem of Central Italy. Trends in different environmental matrixes, *Environmental International* 30(2): 173–81.

Pavlovic, P.; Mitrovic, M.; Djurdjevic, L.; Gajic, G.; Kostic, O.; Bojovic, S. 2007. The ecological potential of *Spiraea vanhouttei* (Briot.) Zabel for urban (the city of Belgrade) and fly ash deposit (Obrenovac) landscaping in Serbia, *Polish Journal of Environmental Studies* 16: 427–431.

Salma, I.; Maenhaut, W.; Zemplén-Papp, É.; Záray, G. 2001. Comprehensive characterization of atmospheric aerosols in Budapest, Hungary: physicochemical properties of inorganic species, *Atmospheric Environment* 35: 4367–4378.

Sawidis, T.; Breuste, J.; Mitrovic, M.; Pavlovic, P.; Tsigaridas, K. 2011. Trees as bioindicators of heavy metal pollution in three European cities, *Environmental Pollution* 159: 3560–3570.

Temesi, D.; Molnár, Á; Mészáros, E.; Feczkó, T. 2003. Seasonal and diurnal variation in the size distribution of fine carbonaceous particles over rural Hungary, *Atmospheric Environment* 37: 139–146.

Wolterbeek, B.; Sarmento, S.; Verburg, T. 2010. Is there a future for biomonitoring of element air pollution? A review focused on a larger-scaled health-related (epidementological) context, *Journal of Radioanalytical and Nuclear Chemistry* 24: 926–934.

Youssef, N. A. 2012. Environmental monitoring of trace elements in leaves of *Ligustrum japonicum* L. by X-ray fluorescence (EDRF), in *Proc. of the International Scientific Conference on Innovation Problems of Modern Biology*, Baku, Azerbaijan.

Naglaa YOUSSEF. MSc (Botany, 2009). Pre-Master Diploma in Plant Ecology (October 2005). BSc Honours (Botany, June 2003), Botany Department, Faculty of Science, Sohag University, Egypt. Currently, Assistant Lecturer (2009) and Demonstrator (full-time; April 2005) in the Botany Department, Faculty of Science, Sohag University, Egypt. Research interests: plant ecology, plant community; air pollution; biomonitoring; biodiversity information systems; plant ecophysiology and plant responses to climate changes.

Bernd MARKERT. Professor, Dr Habil, Natural Scientist. Studies in Chemistry and Biology, Ludwig Maximilian University of Munich, Germany. PhD in 1986 and Habilitation for Ecology in 1993 at the University of Osnabrueck, Germany. Former Head of the International Graduate School, Germany, and Professor for Environmental High Technology. Member of the "Environmental Institute of Scientific Networks" (EISN-Institute.de), Haren, Germany. Publications: author/coauthor of about 400 scientific papers and 25 scientific books were authored/coauthored, edited or coedited. Research interests: biochemistry of trace substances in the water/soil/plant system; instrumental analysis of chemical elements; developing the "Biological System of the Elements"; eco- and human-toxicological aspects of hazardous substances; pollution control by use of bioindicators and technologies for waste management, environmental restoration, and remedial action on soils; different interdisciplinary working fields in between natural, economic and social sciences; developing of an Ethical Consensus as smart method of conflict management which provides integrative solutions of problems; modelling Dialogic Education Processes (DEP) as a future principle of communication; attempt on integration of religious and scientific laws in relation to each other under specific consideration of mental and psychological influences or effects.

Elshad GURBANOV. Professor, Dr Biological Science, Chairman of the Botany Department of Biology, Baku University. Graduated in 1979 at the Faculty of Biology, Baku State University. During 1982–1992 worked as a Teacher and an Associate Professor at the chair of Botany, Baku State University; since 1992 a Chairman of Botany, Chair of Biology Faculty, Baku State University. Member of Central Election Commission from 2000 to 2002. Teaching lessons: systematization of supreme plants, ecology of plants, geobotany, bases of botany, and local flora. Author of 150 scientific publications, 2 monographs, and 5 books. Research interests: systematic of supreme plants; geobotany; and plant ecology.

Haciyeva SEVNIC. Dr, Chemistry Professor, the Head of Ecological Chemistry Sub Department, Baku State University. She worked at the Chemistry Department (1983), Baku State University, and graduated in 1988 with honours diploma. During 1989–1993 she was a postgraduate student of the Institute of Chemical Problem NASA. In 2003, won a prize of Y. Mammadaliev. During 2006–2007, won a prize and a honorary title of "Head of sub department of the year". From 1995 to 2004 was Senior Lecturer of the Chemistry Department. Since 2004 she is the Head of Ecological Chemistry Sub Department. Teaching lessons: general ecology, ecological chemistry, ecology and protection of the environment, and physical chemical parameters of the environment. Author of over 200 scientific publications: 70 articles, 6 books, 12 patents, 3 PhD. Research interests: investigation of chelate sorbents by spectrophotometric method for analyzing inorganic compounds; techniques of determination of ions by spectrophotometric methods; and monitoring the environment.

Simone WÜNSCHMANN. PhD, Natural Scientist and a former Scientific Assistant at the International Graduate School, Germany, Department of Environmental High Technology, Working Group for Human and Ecotoxicology. Diploma Engineer for Ecology and Environmental Protection, University of Applied Sciences, Germany. PhD in Environmental Sciences, University of Vechta, Germany. Member of the "Environmental Institute of Scientific Networks" (EISN-Institute.de), Haren, Germany. Publications: author/co-author of about 40 scientific papers and 4 scientific books. Participant at more than 50 international conferences. Research interests: pollution control, human and ecotoxicology; ecology and environmental protection; and environmental engineering with an emphasis on renewable energy.

13

BIOSORPTION OF COPPER FROM SYNTHETIC WATERS BY USING TOBACCO LEAF: EQUILIBRIUM, KINETIC AND THERMODYNAMIC TESTS

Mehmet ÇEKIM[a], Sayiter YILDIZ[b], Turgay DERE[c]

[a]*Cumhuriyet University, Institute of Science, 58140 Sivas, Turkey*
[b]*Cumhuriyet University, Engineering Faculty, Department of Environmental Engineering, 58140 Sivas, Turkey*
[c]*Adıyaman University, Engineering Faculty, Department of Environmental Engineering,*
02040 Adıyaman, Turkey

Abstract. In this study, biosorption of Cu^{2+} ions on to tobacco leaves was investigated. The optimum conditions for biosorption of Cu^{2+} ions onto tobacco leaves were determined: pH – 4.0, temperature – 20 °C, shaking rate – 200 rpm, biosorbent dose – 0.4 g, and initial Cu^{2+} ion concentration – 25 mg/L. The state of equilibrium lasted for 60 minutes. COD, TN and TP analyses were performed to determine the negative impacts of biosorbent on the system. Compliance of equilibrium data of tobacco biosorption of Cu^{2+} ions to Langmuir, Freundlich, Temkin and D–R isotherm models was also investigated. High correlations were achieved in all four isotherm models. Within the scope of kinetic tests, it was observed that biosorption of Cu^{2+} ions with tobacco biosorbent complied with the pseudo second-order kinetic rate constant. FTIR, SEM and EDX analyses were carried out to investigate the surface characteristics and chemical structure of tobacco biosorbent and absorption of Cu^{2+} ions were observed. It was concluded that tobacco leaves are a highly efficient (90.72%) sorbent that can remove Cu^{2+} ions from wastewater.

Keywords: biosorption, heavy metal, tobacco, copper, isotherm, kinetic.

Introduction

Heavy metals are major water pollutants. One of such heavy metals is copper, which can be found as Cu (0), Cu (+) and Cu (2+) in wastewater of various industries, such as metal cleaning and coating, paper, fertilizers etc. Cu (2+) can easily bind to organic and inorganic substances depending on ambient pH levels, which makes it especially dangerous (Hasan *et al.* 2008). In humans, excessive copper intake may cause serious mucosa irritation, problems of the central nervous system, such as depression, and damage of internal organs, such as liver. The threshold Cu (2+) concentration allowed in drinking water by the World Health Organization is 1.5 mg/L (Kalavathy *et al.* 2005).

The traditional methods used to remove heavy metals from aqueous ambient are; chemical precipitation, ion exchange, adsorption through active carbon, reverse osmosis, filtration and membrane technologies (Liu *et al.* 2004). Ambient metals may not be removed fully through these traditional methods. These techniques have also various disadvantages such as high cost equipment and monitoring system requirements, excessive chemical and energy demands, toxic sludge and other waste material creation (Horsfall *et al.* 2003). Therefore, biotechnological approaches are prominent in recent years as alternative treatment methods (Davis *et al.* 2003).

Biosorption is a physicochemical process that occurs naturally in certain biomass which allows it to passively concentrate and bind contaminants onto its cellular structure (Churchill *et al.* 1995). In other words, biosorption is a process that can be explained by adsorption principles and is used to remove heavy metal ions, radioactive elements and dyes through inactive biomass (Özer *et al.* 2009).

In biosorption, physicochemical principles of metal bindings onto cellular structure are composed of complexation, coordination, chelate formation, ion exchange, adsorption and inorganic micro precipitation processes. In general, ion exchange is the vastly observed mechanism (Volesky 2001). Biosorption process has various advantages such as low operational costs, relative low waste sludge volumes, easy-use of waste materials, seaweeds, natural

Corresponding author: Sayiter Yıldız
E-mail: sayildiz@cumhuriyet.edu.tr

substances or raw biosorbents obtained from the other industrial activities (fermentation wastes), recycle of the metals over biosorbent and waste material (Wang, Chen 2006).

In this study, removal of Cu^{2+} ions from synthetic wastewater by using tobacco leaf through biosorption in an intermittent system was investigated. The effects of ambient pH, initial metal concentration, initial biosorbent concentration, duration of contact, shaking rate and temperature parameters on biosorption were investigated through isotherm, kinetic and thermodynamic assessments. Variations in chemical oxygen demand (COD), total nitrogen (TN) and toplam phosphorus (TP) values of tobacco-leaf biosorption system were also investigated. Additionally, Fourier Transform Infrared Spectroscopy (FT-IR), Scanning Electron Microscope (SEM) and Energy Dispersive X-Ray (EDX) analyses were also performed over experimental data.

1. Material and methods

1.1. Biosorbent preparation

Tobacco leaves to be used in biosorption tests were supplied from tobacco fields of Pınarbaşı town of Adıyaman Province. The leaves were washed with distilled water until they totally lose their color and dried in an oven at 105°C for 24 hours. Dried leaves were then grinded and made active through keeping inside 1% H_2SO_4 solution for 24 hours. Activated absorbent washed and filtrated again with distilled water to remove the acid content and dried at 105 °C for 24 hours. The resultant dried tobacco leaves sieved through 0.30 mm sieve and made ready for experiments.

1.2. Preparation of copper solution

Cu^{2+} solution was prepared by using sufficient (3.929 g) amount of copper sulphate ($CuSO_4 \cdot 5H_2O$) chemical as to have a solution at 1 L in volume and 1000 mg/L in concentration. Different metal ion concentrations were prepared by making relevant dilutions from the stock solution.

1.3. Experimental setup and method of analysis

Biosorption tests were carried in in 250 mL Erlenmeyer flasks in an intermittent system by using 100 mL copper solution and biosorbent. Wise Shake (SHO–2D) shaking incubator able to operate at constant temperature and shaking rate was used in biosorption tests. The time at which biosorbent added to copper solution is assumed to be $t = 0$ and analyses were performed at certain intervals. Free Cu^{2+} ions were read in samples to find out the absorbance at 478 nm wave length and analyzed in a Hach-Lange DR-6000 UV spectrophotometer. During the tests, pH values were read by Thermo Orion – STARA2145

brand pH meter. The pH levels were adjusted separately with concentrated and diluted H_2SO_4 and NaOH. Chemical oxygen demand (COD), total nitrogen (TN) and total phosphorus (TP) analyses were performed by using Hach-Lange instant kits. Biosorption capacity (q_e) and percent removal efficiency (yield) were calculated by using the following equations:

$$q_e = \frac{V\left(C_o - C_e\right)}{X}; \tag{1}$$

$$\left(E\right)\% = \frac{\left(C_o - C_e\right)}{C_o} \cdot 100, \tag{2}$$

where: q_e (mg/g) is the concentration of the substances bound over adsorbent; X (g) is the amount of adsorbent used in tests; V (mL) is the solution volume; C_o (mg/L) is the initial concentration of the solution; C_e (mg/L) is the final concentration of the solution.

Compliance of the experimental data with Langmuir, Freundlich, Temkin and D–R isotherm models were investigated. In Langmuir isotherm, adsorption linearly increase with the initial concentration of the adsorbed material (Aksu, Akpınar 2000). Langmuir isotherm equation is provided below (Eq. 3):

$$q_e = \frac{Q_{max} K_a C_e}{1 + K_a C_e}, \tag{3}$$

where: q_e is the amount of adsorbed material over a unit adsorbent (mg/g), Q_{max} is Langmuir isotherm constant expressing the maximum amount of material to be adsorbed over the sorbent under operational conditions (mg/g), K_a is Langmuir constant expressing adsorption energy (L/mg), C_e is the concentration of material left within the solution following the adsorption (mg/L).

Dimensionless R_L (dispersion) coefficient is calculated to find out the availability of biosorption (Eq. 4) and a value between 0 and 1 indicates the availability of the adsorption (Başıbüyük, Forster 2003):

$$R_L = \frac{1}{1 + K_a C_o}. \tag{4}$$

Freundlich isotherm assumes that the amount of adsorbed material increase with the increasing concentration of adsorbed material in solution (Weber 1972). Freundlich isotherm model is expressed by Equation (5):

$$q_e = K_F C_e^{\left(\frac{1}{n}\right)}, \tag{5}$$

where K_F and n are constants depending on temperature, adsorbent and adsorbed material. K_F is called as adsorption capacity and the term $1/n$ is called as heterogeneity factor. This isotherm can successfully be used to define heterogeneous systems (Ghoreishi, Haghighi 2003).

According to linearized Freundlich equation (Eq. 6), a graph is drawn with ln Ce versus ln q_e. The term $1/n$ is derived from the slope and $\ln K_F$ is taken from the interception point on y-axis:

$$\ln q_e = \ln K_F + (1/n)\ln C_e. \tag{6}$$

Temkin model was developed by considering adsorption temperature of entire molecules within the layer and linearly decrease because of the area influenced by adsorbent interactions (Kılıç et al. 2011). The equation expressing Temkin isotherm is provided in Eq. (7) (Allen et al. 2004):

$$q_e = \left(\frac{RT}{b}\right)\ln\left(A \cdot C_e\right), \tag{7}$$

where: R is gas constant (J/mol K), T is temperature (K), and A is Toth constant (dm³/g). When this equation is linearized and the term (RT/b) in this equation is expressed as B, Equation (8) is obtained. B is related to adsorption temperature of Temkin isotherm constant (Allen et al. 2003):

$$q_e = B\ln A + B\ln C_e. \tag{8}$$

The constants obtained from Langmuir and Freundlich isotherms do not provide any information about the physical and chemical characteristics of biosorption. However, average adsorption energy (E) calculated with D–R isotherm developed by Dubinin and Radushkevich (Dubinin, Radushkevich 1947) provide information about physical and chemical characteristics of the sorption (Ceyhan, Baybaş 2001). D–R isotherm is expressed by Equation (9) provided below (Sarı et al. 2008; Malik et al. 2005):

$$\ln q_e = \ln q_{max} - \beta\varepsilon^2. \tag{9}$$

The amount adsorbed (q_e; mol/g) is related to maximum sorption capacity (q_{D-R}; mol/g), average adsorption energy (E) and relevant activity coefficient (β; mol²/J²) and is a function of Polanyi potential (ε). The Polanyi potential (ε) is calculated by using Equation (10) (Sarı et al. 2008):

$$\varepsilon = RTln\left(1 + \frac{1}{C_e}\right). \tag{10}$$

Maximum holding capacity q_{D-R} and β can be calculated from the $q_e - \varepsilon^2$ graph of the model drawn by the experimental data. Adsorption energy (E; kJ/mol) is calculated by using the Equation (11) (Sarı et al. 2008):

$$E = \frac{1}{\sqrt{-2\beta}}. \tag{11}$$

1.4. Biosorption kinetics

Kinetic tests were performed at pH levels of 2–3–4–5 for 1–60 minutes to determine the relations of Cu²⁺ ion biosorption onto tobacco sorbent with duration of contact at different pH levels. Compliance of the resultant data with the pseudo first-order and second-order kinetic models

was assessed. The kinetic models and expressions used in these tests are provided below:

Pseudo First-Order Kinetic Model: Pseudo first-order rate equation was derived by Lagergren and rate equation is provided below (Ho, Chiang 2001):

$$\log\left(q_e - q_t\right) = \log\left(q_e\right) - \frac{k_1}{2.303} \cdot t, \tag{12}$$

where: k_1 is rate constant (1/min), q_e is adsorption capacity (mg/g), q_t is adsorbed material concentration at time t (mg/g).

When the graph of $\log(q_e - q_t)$ versus t is drawn according to Eq. (12), the slope of the line provides $k_1/2.303$ and y-axis interception point provides $\log(q_e)$. Relevant calculations are then performed to find k_1 and q_e values.

Pseudo Second-Order Rate Equation: The pseudo second-order rate equation developed by Ho in 1995 expressed independency of the rate from adsorbent concentration and dependency to adsorption capacity at solid phase and time (Ho, Wang 2004). Pseudo second-order rate equation is provided in Eq. (13) (Ip et al. 2010):

$$\frac{t}{q_t} = \frac{k_1}{k_2\left(q_e\right)^2} + \frac{1}{q_e} \cdot t, \tag{13}$$

where: k_2 is pseudo second-order rate constant (g/mg·min), q_e is the amount of adsorbed material at equilibrium (mg/g), q_t is the material concentration adsorbed at time t (mg/g). When the graph of (t/q_t) vs t is drawn according to equation, slope of the line provides $1/q_e$ and y-axis interception point provides $1/k(q_e)^2$ value.

Inter-Particle Diffusion (Weber-Morris) Model: in case the diffusion mechanism is not explained clearly by the pseudo first and second-order equations, kinetic results are tried to be explained by inter-particle diffusion model (Kılıç et al. 2011). Inter-particle diffusion rate equation is provided in Eq. (14) (Ghasemi et al. 2012):

$$q_t = k_{id}(t)0.5 + C, \tag{14}$$

where: k_{id} is inter-particle diffusion constant (mg·g⁻¹·min⁻⁰·⁵), t is time (minute), C is the thickness of boundary layer in adsorption duration (Polat, Aslan 2014). When the graph of qt versus $t^{0.5}$ is drawn, the slope of the line provides k_{id} adsorption constant.

1.5. Thermodynamic tests

Thermodynamics allows the researchers to gain knowledge about thermal events accompanying chemical reactions, thermal characteristics of reaction materials, especially about entropy and enthalpy, general criteria about volition of reactions and equilibrium (Pahlavanzadeh *et al.* 2010). Thermodynamic parameters are provided in Eqs 15–16:

$$\Delta G^0 = \Delta H^0 - T\Delta S^0, \qquad (15)$$

where: ΔG^0: Gibbs free energy (kJ/mol), ΔH^0 is enthalpy exchange (kJ/mol), ΔS^0 is entropy exchange (kJ/mol K), T is absolute temperature (Kelvin):

$$\ln K_c = \frac{\Delta G^0}{R} + \frac{\Delta H^0}{RT}. \qquad (16)$$

Gibbs free energy value of adsorption process carried out at a certain temperature is calculated with K_c by using the equation $\Delta G^0 = -RT \ln K_c$ ($R = 8.314$ J/mol K). Then the graph of $\ln K_c$ versus $1/T$ is drawn and ΔH^0 and ΔS^0 can be calculated by using the slope and interception point. K_c is the ratio of adsorbed Cu^{2+} concentration to residual Cu^{2+} concentration in solution.

Sorption experiments were performed in triplicate and the average values of samples were presented. Also, blank samples (without Cu^{2+}) were used to compare the results through all batch procedures. Data presented are the mean values from the experiments, standard deviation ($\leq 2\%$) and error bars are indicated in figures.

2. Results and discussion

2.1. Biosorbent characterization

FT-IR analyses

In several previous studies, raw biomass and metal-loaded biomass spectrums of Fourier Transform InfraRed (FT–IR) spectrum analyses were compared to explain biosorption process between biomass and metal ions (Pradhan *et al.* 2007). FT–IR spectrums of tobacco leaves were taken to investigate the changes in functional groups before and after the biosorption of Cu^{2+} ions (Fig. 1).

It was observed that wavelength absorbances and intensities of two spectrums obtained before and after biosorption process were different. Absorbances of $C = O$ strains of carboxyl groups over 1614.41 cm^{-1} and 1434.56 cm^{-1}, assumed to play a role especially in binding Cu^{2+} ions, were respectively observed as 1614.48 cm^{-1} and 1429.52 cm^{-1} after biosorption process. Such a case indicated the existence of broad-band hydroxyl groups (–OH) observed between 3000–3750 cm^{-1} (Yazıcı 2007). Right at this strain point, the absorbance value of 3356.75 cm^{-1} before the biosorption changed to 3291.26 cm^{-1} after the biosorption. The other bands around 2925 and 2893 cm^{-1} indicate symmetric strain vibrations in alkyl (–CH$_3$ or –CH^{2-}) groups (Majumdar *et al.* 2008). At this point, the absorbance value of 2920.92 cm^{-1} changed as 2918.08 cm^{-1}.

The increases and decreases (especially the decreases) in FT-IR spectrums after the adsorption process are considered as the indications of participation of these active groups in adsorption process.

SEM (Scanning Electron Microscope) and EDX analyses

SEM (Scanning Electron Microscope) was used to take SEM images of the sorbent before and after the biosorption process and to determine whether or not Cu2+ ions are adsorbed onto tobacco biosorbent (Fig. 2).

Color coding of the elements within green rectangular in SEM images (Fig. 2) are presented in Figure 3. As it can be seen from Figure 3, Cu^{2+} metal was adsorbed onto biosorbent surface with a red color code.

EDX (Energy dispersive X-ray) detector of SEM device was used to make element analyses of tobacco biosorbent before and after the biosorption and different colors

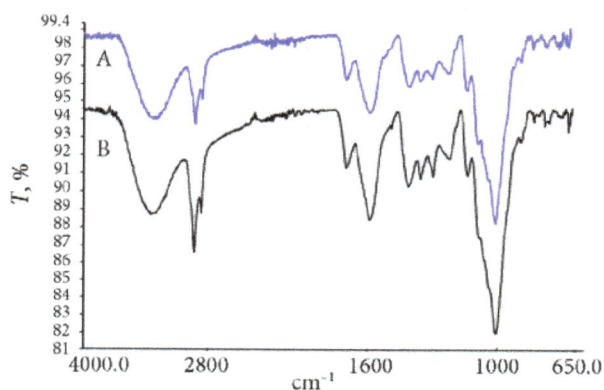

Fig. 1. (a) FT-IR spectrum of tobacco leaf after biosorption and (b) FT-IR spectrum of tobacco leaf before biosorption

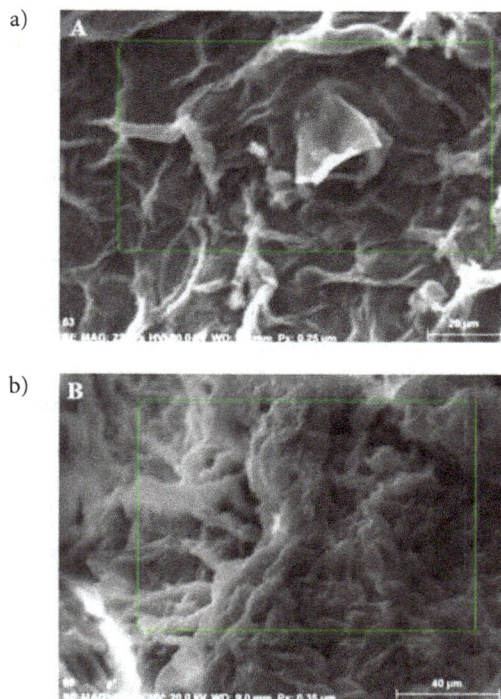

Fig. 2. (a) SEM images of tobacco leaf before biosorption (b) SEM images of tobacco leaf after biosorption

a)

b)

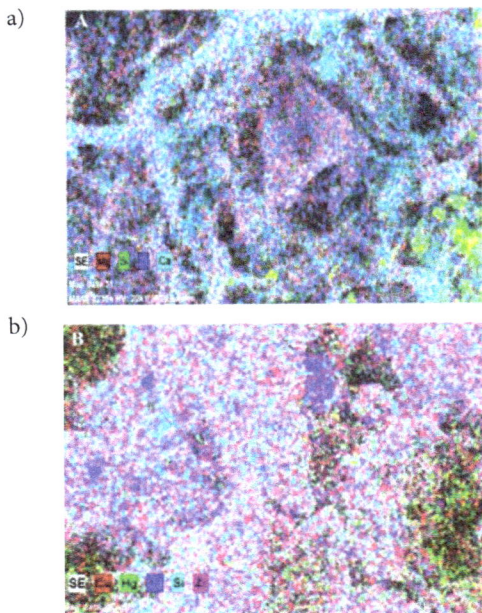

Fig. 3. (a) Element distributions of EDX analysis region before biosorption (b) Element distributions of EDX analysis region after biosorption

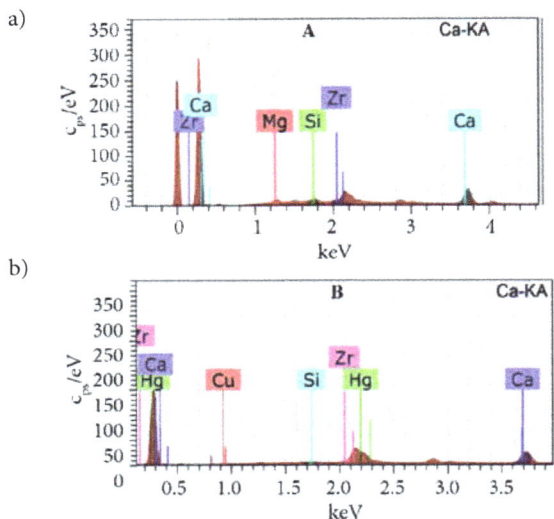

a)

b)

Fig. 4. (a) EDX results of tobacco leaf before biosorption (b) EDX results of tobacco leaf after biosorption

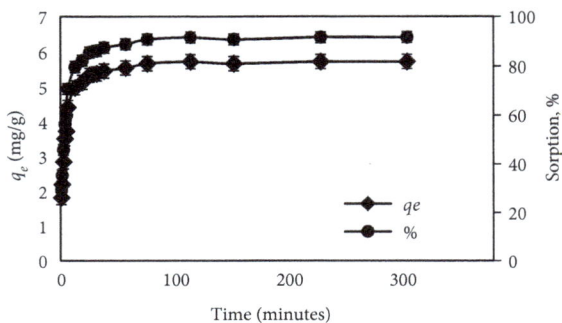

Fig. 5. Variation of biosorption capacity and yield with contact time

were coded (Fig. 4). Figures revealed that copper did not exist in EDX analysis before biosorption. However, EDX analyses after biosorption revealed that copper ions were adsorbed.

2.2. Biosorption equilibrium duration

Initial concentration (Co) was selected as 25 mg/L to determine the equilibrium duration of Cu^{2+} biosorption. Other ambient conditions were set as; biosorbent dose $(X) = 0.4$ g, shaking rate = 200 rpm, pH = 4. Resultant outcomes are presented in Figure 5.

Results revealed a rapid Cu^{2+} ion binding onto biosorbent during the initial 5 minutes (70%). The highest biosorbent capacity was achieved in 60 minutes. Later on, significant changes were not observed. Therefore, equilibrium duration was determined to be 60 minutes.

2.3. Effects of pH on Cu+2 biosorption

Previous biosorption tests revealed that ambient pH had significant impacts on biosorption capacity (Al-Rub *et al.* 2004; Keskinkan *et al.* 2004; Gong *et al.* 2005; Lodeiro *et al.* 2005). Tests were carried out in this study with pH values in between 2–5 in 0.5 unit increments.

While the yield was 43% at pH of 2, the value increased to 86% at pH of 3. The sorption capacity (q_e) was 5.58 at pH of 3.5, whereas the value was 5.72 at pH of 5.

COD analyses were carried out to determine the amount of organic pollutants provided to the ambient with the changing pH levels. Increased COD levels were observed when the pH of the ambient is around 2. However, there were not significant changes in COD levels after the pH level of 2.5. Results revealed that tobacco sorbent increased COD level of the ambient (239–363 mg/L).

2.4. Effects of copper concentration on biosorption

There are several studies in literature indicating increasing biosorption capacities with increasing initial metal concentrations (Gülnaz *et al.* 2005). Similarly, increased biosorption capacities (q_e) were observed in the present study with increasing initial Cu^{2+} concentrations (Fig. 6a). The highest q_e value (10.66 mg/g) was observed at initial Cu^{2+} concentration of 50 mg/L. The highest removal efficiency at Cu^{2+} concentrations of 10–15–20 mg/L was observed as 92%. Although the highest q_e value was achieved at 50 mg/L, the lowest removal efficiency (85%) was also observed at this concentration.

2.5. Effects of biosorbent quantity

Significant effects of biosorbent quantity on biosorption capacity and yield were reported in previous biosorption studies (Al-Qodah 2006; Ju *et al.* 2008). The tests were carried out in this study with biosorbent quantities of 1–8 g/L

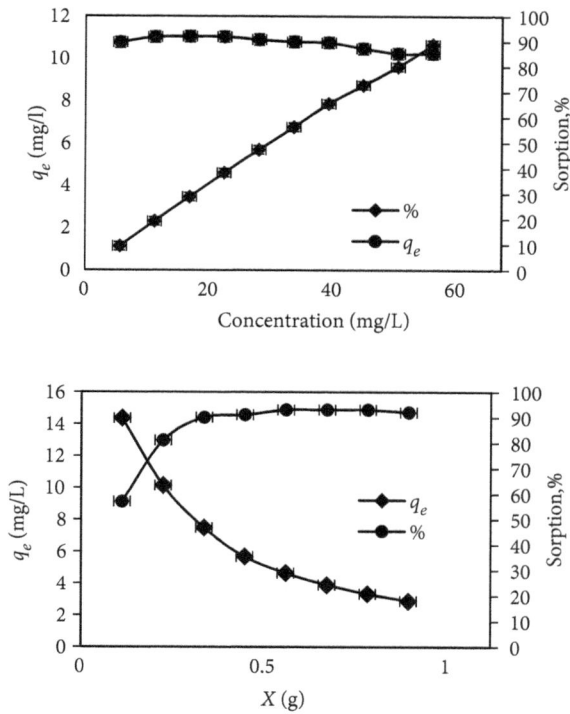

Fig. 6. (a) Effects of initial Cu^{2+} concentrations on biosorption capacity and yield (b) Effects of biosorbent quantity on biosorption capacity and yield

to determine the effects of biosorbent quantity on Cu^{2+} ion biosorption onto tobacco sorbent (Fig. 6b). The highest yield at 5-6-7 g/L sorbent rates was observed as 93%. Percent removal at 1 g/L biosorbent rate was determined to be 57% and biosorbent capacity was determined to be 14.36 mg/g.

Effects of changing biosorbent quantities on COD, TN and TP levels of the solution were also investigated in this study. COD, TN and TP values at 8 g/L biosorbent quantity were respectively observed as 399 mg/L, 5.37 mg/L and 5.33 mg/L. Same values at 1 g/L biosorbent were respectively determined to be 88 mg/L, 1.66 mg/L and 0.771 mg/L. Results revealed increasing COD, TN and TP levels with increasing biosorbent quantities.

2.6. Effects of shaking rate on biosorption

The yields varied between 90-91% and q_e values varied between 5.60-5.70 mg/g at shaking rates between 100-300 rpm. It was determined that shaking rates over 100 rpm did not have significant impacts on biosorption process (Ho, McKay 1999).

2.7. Effects of temperature on biosorption

Tests were carried out at temperatures in between 20-50 °C to determine the effects of temperature on biosorption. While biosorption capacity was 5.74 mg/g at 20 °C, the value was 5.55 mg/g at 50 °C. Biosorption

capacity (q_e) varied between 5.55-5.74 mg/g and yield values varied between 89-91% at temperatures in between 20-50 °C. While some previous studies indicated linear increases in biosorption capacities with increasing temperatures (Polat, Aslan 2014; Martins et al. 2004), others indicated decreasing biosorption capacities with increasing temperatures (Aksu 2001; Cruz et al. 2004). Some others indicated that temperature did not have any significant effects on biosorption process (Topal et al. 2011; Güler, Sarıoğlu 2013). Similarly, significant effects of temperature on biosorption system were not observed in this study.

COD analyses were performed to assess the effects of temperature on organic pollutants released to ambient. Results revealed that temperature did not have any significant impacts of COD release to ambient. COD levels varied between 208-235 mg/L with varying temperatures.

2.8. Isotherm models

The data obtained at different initial Cu^{+2} concentrations were fitted to Langmuir, Freundlich, Temkin and D-R (Dubinin-Radushkevich) models. The coefficients for these models are provided in Table 1. Q_{max} indicates the maximum biosorption capacity of tobacco (mg/g). K_a is Langmuir constant (L/mg) and indicates affinity of adsorbent related to biosorption enthalpy to biosorbent and the strength of the tie between them. K_F expresses biosorption capacity (L/g), n expresses intensity of biosorption. High K_F and n values indicate higher affinity of biosorbent to biosorption and desired levels of biosorption process.

Table 1. Sorption parameters and correlation coefficients for different models

Model	Equation	Sorption Parameters	
Langmuir	$1/q_e = (1/Q_{max}K_a)$ $(1/C_e) + (1/Q_{max})$	R^2	0.995
		K_a (L/mg)	0.213
		Q_{max} (mg/g)	17.182
		R_L	0.158
Freundlich	$\ln q_e = \ln K_F + (1/n)\cdot$ $\ln C_e$	R^2	0.979
		K_F (L/g)	3.361
		n	1.700
Temkin	$q_e = B \ln A +$ $B \ln C_e$	R^2	0.991
		A (L/g)	1.994
		B	3.824
D-R	$\ln q_e = \ln q_{max} - \beta\varepsilon^2$	R^2	0.937
		β (mol^2/j^2)	-0.499
		q_{max} (mg/g)	9.606
		E (kj/mol)	1.001

As it can be seen from Table 1, correlation coefficients were relatively high in all models. The highest value was observed in Langmuir isotherm with $R^2 = 0.995$. Maximum biosorbent capacity of Langmuir isotherm model was 17.182 mg/g, dimensionless R_L coefficient was calculated as 0.158. This R_L value indicated that tobacco is

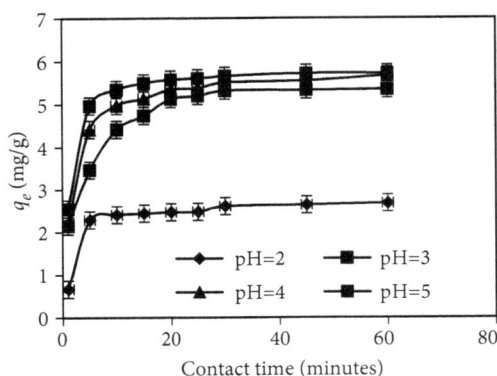

Fig. 7. Variation of biosorption capacity with contact durations at different pH levels

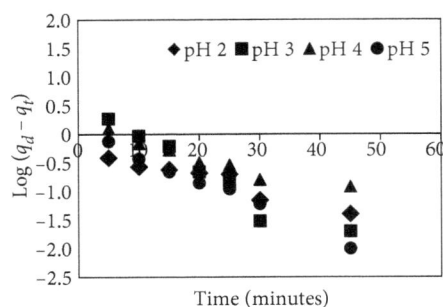

Fig. 8. Graphs of pseudo first – order kinetic models for biosorption of Cu^{2+} ions onto tobacco

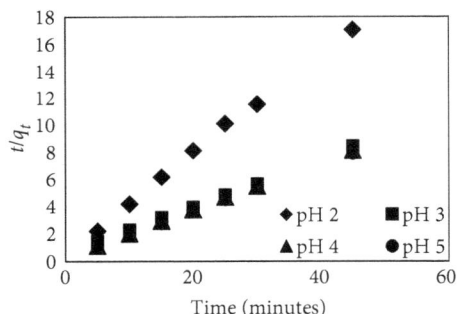

Fig. 9. Graphs of pseudo second-order kinetic models for biosorption of Cu^{2+} ions onto tobacco

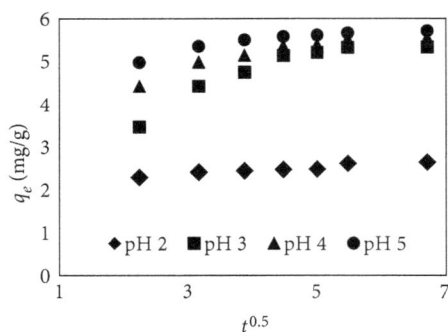

Fig. 10. Graphs of Weber-Morrisi for biosorption of Cu^{2+} ions onto tobacco

an available sorbent for biosorption of Cu^{2+} ions (Doğan, Alkan 2003).

It was asserted that entire molecules at adsorption layer of Temkin isotherm decreased linearly with adsorption temperature. In this isotherm model, effects of indirect adsorbed/adsorbed interactions on adsorption isotherms are investigated (Temkin, Phyzev 1940). The highest R^2 (0.991) value was also observed in Temkin isotherm model and B Temkin isotherm constant related to adsorption temperature was calculated as 3.824.

In D–R model, adsorption energy (E) was calculated as 1.001 kj/mol, q_{max} as 9.606 mg/g and β value as -0.499 mol^2/j^2. Of these values, adsorption energies (E) between 8–16 kj/mol indicated that sorption mainly realized through ion exchange mechanism which is a chemical process. E values lower than 8 kj/mol indicates physical interactions as the binding mechanisms (Sawalha et al. 2006). In the present study, binding mechanisms may be explained by physical interactions ($E = 1.001$ kj/mol $<$ 8 kj/mol).

2.9. Kinetic constants

Kinetic tests were carried out at pH levels of 2–3–4–5 and for durations of 1–60 minutes (Fig. 7). Compliance of data with pseudo first-order (Fig. 8) and pseudo second-order (Fig. 9) kinetic models and inter-particle diffusion model (Fig. 10) were assessed. The pseudo first and second order rate constants (k_1 and k_2), amount of Cu^{2+} adsorbed by a unit biosorbent at equilibrium (q_d, teo), initial sorption rate (h; mg/g.dk), inter-particle (Weber-Morris model) diffusion rate constant k_i (mg/g·dk0.5) are provided in Table 2.

Previous studies indicated relatively low correlations and high errors for the pseudo first and second-order kinetic models (Ho, McKay 1999). As it can be seen from the Table 2, almost identical values of theoretical and experimental q_d values for the pseudo first and second-order kinetic models indicated a well compliance. While the greatest k_2 value was observed at pH of 2, the lowest k_2 value was obtained from pH of 3. The highest R^2 and sorption rate values at pH 4 and 5 were respectively observed as 0.9999–0.9999 and 3.5442–6.4465 mg/g·dk.

2.10. Calculation of thermodynamic parameters

The graph of $\ln K_c$ versus $1/T$ for different temperatures is presented in Figure 11 and calculated thermodynamic parameters are provided in Table 3.

A negative Gibbs free energy indicates spontaneous biosorption, a positive value indicates nonspontaneous biosorption and a value of zero indicates equilibrium biosorption (Aksu 2002). In Table 3, Gibss energy of Cu^{2+} ions (ΔG^0) was negative (–). Negative (–) ΔG^0 indicated spontaneous biosorption.

A positive biosorption entalphy (ΔH^0) value indicates that the system received energy from an external source, in other words the system was endothermic (Ho 2003). Negative ΔS^0 value indicates metal ion affinity to biosorbent at solid-solution interspace or a more regular state of metal ion over biosorbent surface (Padmavathy 2008).

Comparisons of Cu^{2+} biosorption studies carried out with different sorbents are provided in Table 4.

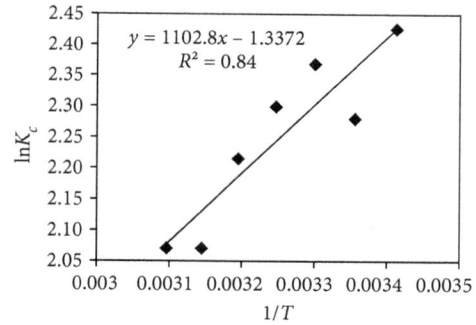

Fig. 11. Van t'Hoff graph of Cu^{2+} biosorption for temperatures in between 20–50 °C

Table 2. Kinetic constant for biosorption of Cu^{2+} ions onto tobacco

| pH | $q_{d,den}$ | Pseudo-first-order | | |
		k_1	$q_{d,teo}$ (mg/g)	R^2
2	2.68	0.058	0.558	0.911
3	5.35	0.123	2.979	0.930
4	5.67	0.059	1.276	0.927
5	5.72	0.103	1.166	0.990

| pH | Pseudo-second-order | | | |
	k_2	$q_{d,den}$ (mg/g)	h (mg/g·dk)	R^2
2	0.262	2.701	1.879	0.998
3	0.060	5.744	1.718	0.998
4	0.110	5.754	3.544	0.999
5	0.197	5.821	6.447	0.999

| pH | Inter-Particle Diffusion (Weber-Morris) | | |
	k_i (mg/g·dk$^{0.5}$)	C	R^2
2	0.077	2.137	0.930
3	0.407	3.004	0.806
4	0.244	4.113	0.850
5	0.155	4.799	0.819

Table 3. K_c, ΔG^0, ΔH^0 and ΔS^0 values of Cu^{2+} ions at different temperatures

Temperature (°C)	K_c	ΔG^0 (kj/mol)	ΔH^0 (kj/mol)	ΔS^0 (kj/molK)
20	11.3153	−5.91011		
25	9.7759	−5.64866		
30	10.6822	−5.9668		
35	9.9649	−6.14479	9.1687	−0.0111
40	9.1626	−5.76439		
45	7.9286	−5.47402		
50	7.9285	−5.56009		

Table 4. Comparison of biosorption studies carried out with various biosorbent/adsorbents

	q_e mg/g	X (g/L)	Cu^{2+} (mg/L)	pH	References
Waste eggshell	6.48	0.5–5	15–35	2–5	Polat, Aslan (2014)
Lobaria pulmonaria (L.)	65.3	1–50	5–100	2–6	Kılıç et al. (2014)
Sawdust	3.89	4	10–200	2–6	Putra et al. (2014)
Eggshell	34.48	4	10–200	2–6	Putra et al. (2014)
Sugar cane	3.65	4	10–200	2–6	Putra et al. (2014)
Eichhornia crassipes	27.7	0.3	2–25	2.5–6	Komy et al. (2013)
Waste sludge	19.34	0.1–1	25–300	2–7	Güler, Sarıoğlu (2013)
S. cerevisiae (maya)	8.25	0.1–1	25–300	2–7	Güler, Sarıoğlu (2013)
Rhizopus oryzae	34.2	0.1	10–300	3–3,5	Fu et al. (2012)
Grass	58.34	0.5–2	10–100	1–8	Hossain et al. (2012)
Mangrov shell	6.95	1–3	10–100	2–10	Rozaini et al. (2010)
Lentils shell	15.62	0.5–0.8	100–500	2–6	Aydın et al. (2008)
Wheat shell	11.0	0.5–0.8	100–500	2–6	Aydın et al. (2008)
Brass shell	3.60	0.5–0.8	100–500	2–6	Aydın et al. (2008)
Tobacco leaf	14.36	1	25	4	This study

Conclusions

The present study revealed that tobacco sorbent could reliably be used to remove Cu^{2+} ions from aqueous solutions. The following conclusions can be drawn from the present study:

1. COD, TN and TP analyses were carried out at different biosorbent quantities and increasing values were observed with increasing biosorbent quantities.

2. A significant change was not observed in biosorbent capacity after a shaking rate of 100 rpm.

3. The amount of COD released to ambient with temperature was also investigated and significant effects of temperature on COD release were not observed.

4. The data on Cu^{2+} ion biosorption onto tobacco were applied to Langmuir, Freundlich, Temkin and D–R (Dubinin-Radushkevich) isotherm models. The resultant correlations coefficients were relatively high in all models.

5. Kinetic tests revealed that pseudo second-order kinetic model might better define the biosorption system.

It was concluded in this study that tobacco biosorbent had high Cu^{2+} removal efficiency (90.72%) but it resulted in organic-originated pollution over biosorption system.

References

Aksu, Z.; Akpınar, D. 2000. Modelling of simultaneous biosorption of phenol and nickel(II) onto dried aerobic activated sludge, *Separation and Purification Technology* 21: 87–99. http://dx.doi.org/10.1016/S1383-5866(00)00194-5

Aksu, Z. 2001. Equilibrium and kinetic modelling of cadmium (II) biosorption by C.vulgaris in a batch systems: effect of temperature, *Seperation and Purification Technology* 21: 285–294. http://dx.doi.org/10.1016/S1383-5866(00)00212-4

Aksu, Z. 2002. Determination of the equilibrium, kinetic and thermodynamic parametres of the biosorption of nickel(II) ions onto Chlorella vulgaris, *Process Biochemistry* 38: 89–99. http://dx.doi.org/10.1016/S0032-9592(02)00051-1

Allen, S. J.; McKay, G.; Porter, J. F. 2004. Adsorption isotherm models for basic dye adsorption by peat in single and binary component systems, *Journal of Colloid and Interface Science* 280: 322–333. http://dx.doi.org/10.1016/j.jcis.2004.08.078

Allen, S. J.; Gan, Q; Matthews, R.; Johnson, P. A. 2003. Comparison of optimised isotherm models for basic dye adsorption by kuzdu', *Bioresour Technology* 88(2): 143–52. http://dx.doi.org/10.1016/S0960-8524(02)00281-x

Al-Rub, F. A. A.; El-Naas, M. H.; Benyahia, F.; Ashour, I. 2004. Biosorption of nickel on blank alginate beads, free and immobilized algal cells, *Process Biochemistry* 39: 1767–1773. http://dx.doi.org/10.1016/j.procbio.2003.08.002

Al-Qodah, Z. 2006. Biosorption of heavy metal ions from aqueous solution by activated sludge, *Desalination* 196: 164–176. http://dx.doi.org/10.1016/j.desal.2005.12.012

Aydın, H.; Buluta, Y.; Yerlikaya, Ç. 2008. Removal of copper (II) from aqueous solution by adsorption onto low-cost adsorbents, *Journal of Environmental Management* 87: 37–45. http://dx.doi.org/10.1016/j.jenvman.2007.01.005

Başıbüyük, M.; Forster, C. F. 2003. An examination of adsorption characteristics of basic dye on to live activated sludge system, *Process Biochemistry* 38: 1311–1316. http://dx.doi.org/10.1016/S0032-9592(02)00327-8

Ceyhan, Ö.; Baybaş, D. 2001. Adsorption of some textile dyes by Hexadecyltrimethylammonium bentonite, *Turk Journal Chemistry* 25: 193–200.

Churchill, S. A.; Walters, J. V.; Churchill, P. F. 1995. Sorption of heavy metals by prepared bacterial cell surfaces, *Journal of Environmental Engineering* 121(10): 706–711. http://dx.doi.org/10.1061/(ASCE)0733-9372(1995)121:10(706)

Cruz, C. V. C; Da Costa, A. C. A.; Henriques, C. A.; Luna, A. S. 2004. Kinetic modelling and equilibrium studies during cadmium biosorption by dead Sargassum sp. biomass, *Bioresource Technology* 91: 249–257. http://dx.doi.org/10.1016/S0960-8524(03)00194-9

Davis, T. A.; Volesky, B.; Mucci, A. A. 2003. Review of the biochemistry of heavy metal biosorption by brown algae, *Water Research* 37: 4311–4330. http://dx.doi.org/10.1016/S0043-1354(03)00293-8

Doğan, M.; Alkan, M. 2003. Removal of methly violet from aqueous solution by perlite, *Journal of Colloid and Interface Science* 267(1): 32–41. http://dx.doi.org/10.1016/S0021-9797(03)00579-4

Dubinin, M. M.; Radushkevich, L. V. 1947. Equation of the characteristic curve of activated charcoal, *Chemistry Zentralblatt* 1: 875.

Fu, Y. Q.; Li, S.; Zhu, H. Y.; Jiang, R.; Yin, L. F. 2012. Biosorption of copper (II) from aqueous solution by mycelial pellets of Rhizopus oryzae, *African Journal of Biotechnology* 11(6): 1403–1411.

Ghasemi, Z.; Seif, A.; Ahmadi, T. S.; Zargar, B.; Rashidi, F.; Rouzbahani, G. M. 2012. Thermodynamic and kinetic studies for the adsorption of Hg(II) by nano-TiO_2 from aqueous solution, *Advanced Powder Technology* 23(2): 148–156. http://dx.doi.org/10.1016/j.apt.2011.01.004

Ghoreishi, S. M.; Haghighi, R. 2003. Chemical catalytic reaction and biological oxidation for treatment of non-biodegradable textile effluent, *Chemical Engineering Journal* 95: 163–152. http://dx.doi.org/10.1016/S1385-8947(03)00100-1

Gong, R.; Ding, Y. D.; Liu, H.; Chen, Q.; Liu, Z. 2005. Lead biosorption by intact and pretreated spirulina maxima biomass, *Chemosphere* 58: 125–130. http://dx.doi.org/10.1016/j.chemosphere.2004.08.055

Güler, U. A.; Sarioglu, M. C. 2013. Mono and binary component biosorption of Cu(II), Ni(II), and Methylene Blue onto raw and pretreated S. cerevisiae: equilibrium and kinetics, *Desalination and Water Treatment* 52 (25–27): 4871–4888. http://dx.doi.org/10.1080/19443994.2013.810359

Gülnaz, O.; Saygideger, S.; Kusvuran, E. 2005. Study of Cu(II) biosorption by dried activated sludge: effect of physicochemical environment and kinetics study, *Journal of Hazardous Materials* B120: 193–200. http://dx.doi.org/10.1016/j.jhazmat.2005.01.003

Hasan, S.; Ghosh, T. K.; Viswanath, D. S.; Boddu, V. M. 2008. Dispersion of chitosan on perlite for enhancement of copper (II) adsorption capacity, *Journal of Hazardous Materials* 152: 826–837. http://dx.doi.org/10.1016/j.jhazmat.2007.07.078

Ho, Y. S.; McKay, G. 1999. Pseudo-second order model for sorption processes, *Process Biochemistry* 34: 451–465. http://dx.doi.org/10.1016/S0032-9592(98)00112-5

Ho, Y. S.; Chiang, C. C. 2001. Sorption studies of acid dye by mixed sorbents, *Adsorption* 7: 139–147. http://dx.doi.org/10.1023/A:1011652224816

Ho, Y. S. 2003. Removal of copper ions from aqueous solution by tree fern, *Water Research* 37: 2323–2330. http://dx.doi.org/10.1016/S0043-1354(03)00002-2

Ho, Y. S.; Wang, C. C. 2004. Pseudo-ishotherms for the sorption of cadmium ions onto tree fern, *Process Biochemistry* 39: 759–763. http://dx.doi.org/10.1016/S0032-9592(03)00184-5

Horsfall, M. Jnr.; Abıa, A. A.; Spıff, A. I. 2003. Removal of Cu (II) and Zn (II) ions from wastewater by Cassava (Manihot esculenta Cranz) waste biomass, *African Journal of Biotechnology* 2: 360–364. http://dx.doi.org/10.5897/AJB2003.000-1074

Hossain, M. A.; Ngo, H. H.; Guo, W. S.; Setiati, T. 2012. Adsorption and desorption of copper(II) ions onto garden grass, *Bioresource Technology* 121: 386–395. http://dx.doi.org/10.1016/j.biortech.2012.06.119

Ip, A. W. M.; Barford J. P.; McKay, G. 2010. A comparative study on the kinetics and mechanism of removal of reactive Black 5 by adsorption onto activated carbons and bone char, *Chemical Engineering Journal* 157: 434–442. http://dx.doi.org/10.1016/j.cej.2009.12.003

Ju, D. J.; Byun, I. G.; Park, J. J; Lee, C. H.; Ahn, G. H.; Park, T. J. 2008. Biosorption of a reactive dye (Rhodamine-B) from an aqueous solution using dried biomass of activated sludge, *Bioresource Technology* 99: 7971–7975. http://dx.doi.org/10.1016/j.biortech.2008.03.061

Kalavathy, M. H.; Karthikeyan, T.; Rajgopal, S.; Miranda, L. R. 2005. Kinetic and isotherm studies of Cu(II) adsorption onto H3PO4-activated rubber wood sawdust, *Journal of Colloid and Interface Science* 292: 354–362. http://dx.doi.org/10.1016/j.jcis.2005.05.087

Keskinkan, O.; Göksu, M. Z. L.; Başıbüyük, M.; Forster, C. F. 2004. Heavy metal adsorption properties of submerged aquatic plant (Ceratophyllum demersum), *Bioresource Technology* 92: 197–200. http://dx.doi.org/10.1016/j.biortech.2003.07.011

Kılıç, M.; Apaydın-Varol, E.; Pütün, A. E. 2011. Adsorptive removal of phenol from aqueous solutions on activated carbon prepared from tobacco residues: equilibrium, kinetics and thermodynamics, *Journal of Hazardous Materials* 189: 397–403. http://dx.doi.org/10.1016/j.jhazmat.2011.02.051

Kılıç, Z.; Atakol, O.; Aras, S.; Duman, D. C.; Çelikkol, P.; Emregül, E. 2014. Evaluation of different isotherm models, kinetic, thermodynamic, and copper biosorption efficiency of Lobaria pulmonaria (L.) Hoffm., *Journal of the Air & Waste Management Association* 64(1): 115–123. http://dx.doi.org/10.1080/10962247.2013.831383

Komy, Z. R.; Abdelraheem, W. H.; Ismail, N. M. 2013. Biosorption of Cu^{2+} by Eichhornia crassipes: physicochemical characterization, biosorption modeling and mechanism, *Journal of King Saud University – Science* 25: 47–56.

Liu, H.; Chen, B.; Lana, Y.; Chenga, Y. 2004. Biosorption of Zn(II) and Cu(II) by the indigenous Thiobacillus thiooxidans, *Chemistry Engineering Journal* 97: 195–201. http://dx.doi.org/10.1016/S1385-8947(03)00210-9

Lodeiro, P.; Cordero, B.; Barriada, J. L.; Herrero, R.; Sastre de Vicente, M. E. 2005. Biosorption of cadmium by biomass of brown marine macroalgae, *Bioresource Technology* 96: 1796–1803. http://dx.doi.org/10.1016/j.biortech.2005.01.002

Majumdar, S. S.; Das, S. K.; Saha, T.; Panda, G. C; Bandyopadhyoy, T.; Guha, A. K. 2008. Adsorption behavior of copper ions on Mucor rouxii biomass through microscopic and FTIR analysis, *Colloids and Surfaces B: Biointerfaces* 63: 138–145. http://dx.doi.org/10.1016/j.colsurfb.2007.11.022

Malik, U. R.; Hasany, S. M.; Subhani, M. S. 2005. Sorptive potential of sunflower stem for Cr(III) ions from aqueous solutions and its kinetic and thermodynamic profile, *Talanta* 66: 166–173. http://dx.doi.org/10.1016/j.talanta.2004.11.013

Martins, R. J. E.; Pardo, R.; Boaventura, R. A. R. 2004. Cadmium (II) and zinc (II) adsorption by the aquatic moss Fontinalis antipyretica: effect of temperature, pH and water hardness, *Water Research* 38: 693–699. http://dx.doi.org/10.1016/j.watres.2003.10.013

Özer, A.; Gürbüz, G.; Çalımlı, A.; Körbahtii B. K. 2009. Biosorption of copper(II) ions on Enteromorpha prolifera: Application of response surface methodology (RSM), *Chemical Engineering Journal* 146: 377–387. http://dx.doi.org/10.1016/j.cej.2008.06.041

Padmavathy, V. 2008. Biosorption of nickel(II) ions by baker's yeast: Kinetic, thermodynamic and desorption studies, *Bioresource Technology* 99: 3100–3109. http://dx.doi.org/10.1016/j.biortech.2007.05.070

Pahlavanzadeh, H.; Keshtkar, A. R.; Safdari, J.; Abadi, Z. 2010. Biosorption of nickel(II) from aqueous solution by brown algae: equilibrium, dynamic and thermodynamic studies, *Journal of Hazardous Materials* 175: 304–310. http://dx.doi.org/10.1016/j.jhazmat.2009.10.004

Polat, A.; Aslan, S. 2014. Kinetic and isotherm study of cupper adsorption from aqueous solution using waste eggshell, *Journal of Environmental Engineering and Landscape Management* 22(2): 132–140. http://dx.doi.org/10.3846/16486897.2013.865631

Pradhan, S.; Singh, S.; Rai, L. C. 2007. Characterization of various functional groups present in the capsule of Microcystis and study of their role in biosorption of Fe, Ni and Cr, *Bioresource Technology* 98: 595–601. http://dx.doi.org/10.1016/j.biortech.2006.02.041

Putra, W. P.; Kamari, A.; Yusoff, S. N. M.; Ishak, C. F.; Mohamed, A.; Hashim, N.; Isa, I. M. 2014. Biosorption of Cu(II), Pb(II) and Zn(II) ions from aqueous solutions using selected waste materials: adsorption and characterisation studies, *Journal of Encapsulation and Adsorption Sciences* 4: 25–35. http://dx.doi.org/10.4236/jeas.2014.41004

Rozaini, C. A.; Jain, K.; Oo, C. W.; Tan, K. W.; Tan, L. S.; Azraa, A.; Tong, K. S. 2010. Optimization of nickel and copper ions removal by modified mangrove barks, *International Journal of Chemical Engineering and Applications* 1(1): 84–89. http://dx.doi.org/10.7763/IJCEA.2010.V1.14

Sarı, A.; Mendil, D.; Tuzen, M.; Soylak, M. 2008. Biosorption of Cd(II) and Cr(III) from aqueous solution by moss (Hylocomium splendens) biomass: equilibrium, kinetic and thermodynamic studies, *Chemical Engineering Journal* 144: 1–9. http://dx.doi.org/10.1016/j.cej.2007.12.020

Sawalha, M. F.; Videa, J. R. P.; Gonz´alez, J. R.; Gardea-Torresdey, J.L. 2006. Biosorption of Cd(II), Cr(III), and Cr(VI) by Saltbush (Atriplex canescens) biomass: thermodynamic and isotherm studies, *Journal Colloid Interface Science* 300: 100–104. http://dx.doi.org/10.1016/j.jcis.2006.03.029

Temkin, M. J.; Phyzev, V. 1940. Recent modifications to Langmuir isotherms, *ActaPhysiochim USSR* 12: 217–222.

Topal, M.; Arslan Topal, E. I.; Aslan, S. 2011. Removal of Cu(II) from aqueous solutions by using lemon peel, *Erciyes University of the Journal Institute of Science and Technology* 27(3): 265–270.

Volesky, B. 2001. Detoxification of metal-bearing effluents biosorption for the next century, *Hydrometallurgy* 59: 203–216. http://dx.doi.org/10.1016/S0304-386X(00)00160-2

Wang, J.; Chen, C. 2006. Biosorption of heavy metals by Saccharomyces cerevisiae: a review, *Biotechnology Advances* 24: 427–451. http://dx.doi.org/10.1016/j.biotechadv.2006.03.001

Weber, J. R. 1972. *Physicochemical processes for water quality control.* USA: Willey Interscience.

Yazıcı, H. 2007. *The investigation of the biosorption of Cr and Cu²⁺ ions from aqueous solutions by marrubium globosum ssp. globosum plant*: M. Sc. Thesis. Isparta: Süleyman Demirel University Graduate School of Applied and Natural Sciences Department of Environmental Engineering. 140 p.

Mehmet ÇEKIM. He is a Master Student at the Department of Environmental Engineering, Cumhuriyet University, Sivas. Research interests include biosorption and wastewater treatment.

Sayiter YILDIZ, Dr Lecturer at the Department of Environmental Engineering, Cumhuriyet University, Sivas. His research interests include heavy metal removal, sewage sludge, sorption.

Turgay DERE, Dr Lecturer at the Department of Environmental Engineering, Adiyaman University, Adiyaman. His research interests include heavy metal removal, sorption.

14

DEVELOPMENT OF A PERFORMANCE THRESHOLD APPROACH FOR IDENTIFYING THE MANAGEMENT OPTIONS FOR STABILISATION/SOLIDIFICATION OF LEAD POLLUTED SOILS

Pietro Paolo Falciglia[a], Abir Al-Tabbaa[b], Federico G. A. Vagliasindi[a]

[a]*Department of Civil and Environmental Engineering, University of Catania,*
Viale A. Doria, 6-95125 Catania, Italy
[b]*Department of Engineering, University of Cambridge, Trumpington Street, Cambridge CB2 1PZ, UK*

Abstract. Two soils spiked with lead at different rates were stabilised/solidified using Portland cement and fly ash at different soil:binder ratios, and tested for their setting time, unconfined compressive strength, leachability and durability. A performance threshold approach was used in order to identify optimal management options for the products of the S/S treatment. Results show that soil texture, percentage of binders and lead concentration play an important part in the treatment, significantly influencing the performance of the resulting products in terms of curing, compressive strength and durability. Pb soil concentrations higher than 15000 mg kg^{-1} were found to heavily reduce the applicability of the treatment requiring the maximum amount of binder in order to satisfy the performance criteria. The performance of sandy soils was shown to be limited by setting time and UCS features due to the retardation of the hydration reactions and also by its leaching behaviour, whereas for silt-clayey soils the critical parameter is the mechanical resistance.

Keywords: cement, fly ash, lead (Pb), management options, soil contamination, stabilisation/solidification (S/S).

Reference to this paper should be made as follows: Falciglia, P. P.; Al-Tabbaa, A.; Vagliasindi, F. G. A. 2014. Development of a performance threshold approach for identifying the management options for stabilisation/solidification of lead polluted soils, *Journal of Environmental Engineering and Landscape Management* 22(02): 85–95.

Introduction

Lead (Pb)-based compounds have been a major source of environmental contamination in the past few decades where lead has been reported to affect human health and it is considered as a possible cause of human cancer (Tang, Yang 2012).

Treatment options for Pb-polluted soils have been reported to be either expensive (soil washing, Torres *et al.* 2012), require intensive energy consumption (electrokinetic remediation, Ryu *et al.* 2011), or time consuming, as in the case of biological methods such as bioleaching (Cheng *et al.* 2009) or phytoremediation (Bech *et al.* 2012), or environmentally unsustainable, as in the case of disposal in landfills (Harbottle *et al.* 2007).

Recent studies (Leonard, Stegemann 2010) have also shown the possibility of using stabilization/solidification (S/S) with hydraulic binders as a possible treatment option, with the potential of using the S/S product for useful purposes. S/S has been widely used due to its versatility, efficiency, time and costs to dispose of low-level radioactive and hazardous wastes, as well as to remedy metal or radionuclides contaminated soils (Falciglia *et al.* 2012).

Stabilisation/solidification (S/S) is a treatment process by which contaminated soils, sediments or waste materials are mixed with a binder and specific additives with the aims of reducing the mobility of the toxic contaminants by increasing the pH and fully or partially binding the contaminants in the solid matrix (stabilisation), and of improving the physical properties (strength, compressibility, permeability and durability) of the final treatment products (solidification) (Antemir *et al.* 2010; Polettini *et al.* 2001).

Corresponding author: Pietro P. Falciglia
E-mail: ppfalci@dica.unict.it

Binder-based stabilisation/solidification (S/S) techniques are used for Pb-polluted soil treatment because they ensure physical and chemical stabilisation by significantly reducing the mobility and the solubility of Pb in soil (Yin et al. 2006), requiring minimal input of energy, resulting in potential applications in concrete works (Al-Ansary, Al-Tabbaa 2007; Harbottle et al. 2007).

For Pb-polluted soil treatment, most applications of S/S are cement-based, and rely on Portland cement (PC) as the primary binder (Svensson and Allard 2008). PC is a heterogeneous mixture of five mineral phases: 50–70% alite (C_3S), 20–30% belite (C_2S), 5–12% alluminate (C_3A), 5–12% ferrite (C_4AF) and 2% gypsum. More recently PC has been combined in blends with other minerals such as lime, blast-furnace slag, clays and fly ash. Fly ash (FA), also known as pulverised fuel ash (PFA), is generated from coal fired power plants and, due to its low cost, has been widely used in PC manufacturing and as a partial substitute for PC as stabilising agent for Pb in S/S treatments. If FA contains more than 20% CaO, it is classified as self-cementing Class C FA, while Class F FA generally contains less than 10% CaO. FA presents mechanical features slightly lower than PC (especially Class F), influencing the performance of the S/S treated matrices if used instead of PC as a binder, but it can be successfully used for soil stabilisation or in other civil construction applications (Moon, Dermatas 2007).

The hydration of PC is a sequence of overlapping chemical reactions between dry binder compounds and water, leading to continuous cement paste stiffening and hardening. The early behaviour of hydrating PC is governed by reactions of aluminate phases while the setting and the early strength development behaviour is mostly dependent on the hydration of silicates, particularly alite. The formation of hydration products and the development of micro-structural features depend on solution processes and interfacial and solid-state reactions. The hydration products of PC are mainly made up by 20–25% Ca(OH)$_2$ (CH), 60–70% calcium silicate hydrate gel (CSH) and 5–15% other phases including grains of still-unhydrated cement (Chen et al. 2009).

In particular, using Class F FA, even if it does not readily exhibit self cementing characteristics, upon PC or lime addition, pozzolanic reactions take place leading to the formation of calcium silicate hydrate gel (CSH) (Dermatas, Meng 2003).

In Pb polluted soils treated with PC and/or FA, two possible mechanisms may be responsible for the immobilisation of the contaminant during the hydration phases (Moon, Dermatas 2007). One mechanism may be precipitation resulting from the formation of lead silicate oxide. Another may be inclusion, either by physical encapsulation and or by chemical inclusion. Physical encapsulation can be achieved by creating a solidified monolith, while chemical inclusion can be achieved through the incorporation of lead in binder hydration products, such as CSH gel phases, which play such an important role in the retention of metal. The main mechanisms that determine Pb immobilization in the solid matrix, similarly to other heavy metals, are: sorption on clay and pozzolanic reaction products (Moon, Dermatas 2007) and addition (Eq. (1)) or substitution (Eq. (2)) reactions with CSH (Ouki, Hills 2002):

$$CSH + Pb \rightarrow Pb - CSH; \qquad (1)$$

$$CSH + Pb \rightarrow Pb - CSH + Ca^{2+}. \qquad (2)$$

Several studies on S/S treatment have been performed to better understand the fundamentals of heavy metal immobilisation and leaching (Kundu, Gupta 2008) or to investigate the performance of PC or PC-FA-based S/S treatments of soils polluted by lead. Work carried out by Jing et al. (2004) on leachability of sandy soil contaminated by Pb at rate of 3800 mg kg^{-1}, treated using PC, lime and lime-FA mixtures, showed that S/S treatments significantly reduced Pb leaching (below the regulatory concentration) in all performed tests and that Pb concentration in the leachate is mainly controlled by the leachate pH. Moon and Darmatas (2007) investigated the effectiveness of a FA-based S/S treatment of two soils polluted by Pb at concentration of 459 and 3530 mg kg^{-1} by means of semi-dynamic leaching tests. Results showed a reduction of up to 98.5% in the Pb release upon addition of 25% FA and that soil contaminant concentration influenced Pb level in leachate. Moreover, other literature findings (Dermatas, Meng 2003; Qian et al. 2008; Yin et al. 2006) reported that, in a PC and/or FA based treatment, the percentage of binder used, curing age and Pb contamination level influenced the setting time, compressive strength, leachability and chemical and crystalline structure of the treated soils.

Indeed, it was found that the hydration of cement, and materials such as FA, can be highly modified by heavy metal concentration due to coating around binder grains (Malviya, Chaudhary 2006) and that altered features of hydration products can result in a change of characteristics of the matrices treated by S/S in terms of mechanical strength, durability, curing time, permeability and leachability (Jing et al. 2004; Ouki, Hills 2002). In particular, in the presence of PC and FA, Pb, if present at high concentrations, may form, depending on the pH value, hydroxide and silicate phases characterized by a low solubility such as leadhillite (lead carbonate sulfate hydroxide, $Pb_4SO_4(CO_3)_2(OH)_2$), lead carbonate hydroxide hydrate ($3PbCO_3 \cdot 2Pb(OH)_2 \cdot H_2O$) (Lee 2007) and lead silicate (Pb_2SiO_4, Pb_3SiO_5) (Moon, Dermatas 2007) that reduce the permeability of the matrices and increase the setting times.

Therefore, PC and FA based S/S techniques have been documented as appropriate for Pb-contaminated soil treatment but the Pb concentration level has been reported to possibly significantly influence the effectiveness of the treatment. However the correlation between metal contamination level variation and the performance of S/S treated matrices has not been investigated in any detail and in particular, the limits for the application of S/S techniques to treat high Pb concentration polluted soils are not clear.

Hence, the general objective of this work was to better understand the potential of S/S for the treatment of Pb polluted soils and in particular the effects of Pb concentration and soil:binder ratio on the physical and mechanical properties of soil treated with PC and FA which were assessed analysing the setting time, unconfined compressive strength (UCS), leachability and durability of the treatment products.

1. Materials and methods

1.1. Soil, contaminants and binders

A sandy soil (soil A) and a silty-clay soil (soil B), with properties shown in Table 1, were used for the experiments. Selected soils were spiked with Pb at different rates (C) (1000, 2000, 4000, 8000, 15000 and 25000 mg kg^{-1}), by adding a known quantity of a contaminant solution containing deionized water and reagent grade lead (II) nitrate, purchased from Merck KGaA (Darmstadt, Germany). This nitrate form of Pb was chosen due to its high solubility, representing a "worst-case" scenario (Jing *et al.* 2004) and because it is a main pollutant contained in industrial wastes (Gervais, Ouki 2002).

Table 1. Characteristics of the soils

Parameter	soil A	soil B
Texture	sandy	silty-clay
Sand (silica s. 75–350 μm) [%]	80	20
Silt (silica flour 10–75 μm) [%]	10	56
Clay (kaolin <75 μm) [%]	10	24
pH (L:S of 10)	8.73	8.39
Organic matter [%]	2.79	2.98
Total Organic Carbon [%]	1.67	1.75
Bulk density [g cm^{-3}]	1.42	1.31
Surface area [m^2 g^{-1}]	3.33	14.1
Moisture content [%]	14	25.5

After the contamination procedure, the soils were kept in a closed vessel and stored in a dark room at 4 °C for 1 month then five samples for each contamination level were collected and analysed for Pb content before S/S treatments. Pb content was performed by ICP-OES (Perkin Elmer Optima 4300 with Dual View).

The cement used was purchased as type CEM I Portland cement (PC) CEM 11/B-LL 32.5R from Italcementi

S.p.A. (Italy). Class F fly ash (FA) was obtained from Buzzi Unicem S.p.A. (Italy). The chemical and physical properties of PC and FA are presented in Table 2.

Table 2. Characteristics of the binders

Parameter	PC (type I)	FA
Properties		
Moisture content [%]	0.1	0.4
Bulk density [g cm^{-3}]	1.2	1.5
pH (L/S of 10)	13.1	12.4
Surface area [m^2 g^{-1}]	7.8	30.0
Chemical composition		
SiO$_2$ [% w/w dry]	23.5	42.4
Fe$_2$O$_3$ [% w/w dry]	3.7	2.4
Al$_2$O$_3$ [% w/w dry]	4.9	37.4
CaO [% w/w dry]	66.3	8.3
MgO [% w/w dry]	1.6	1.9

1.2. Binder systems and S/S sample production

The S/S treatment was performed by mixing control or spiked soil (S) samples with a binder mixture (B) of PC and FA (PC:FA 1:1) at three different S:B ratios (3.3:1, 4.0:1, 5.0:1), applied wet using a water (W) to dry binder (DB) ratio of 0.42:1 as summarised in Table 3.

Table 3. Details of the soil-binder mixes

PC:FA	W:DB	S:B
1:1	0.42:1	5.0:1
1:1	0.42:1	4.0:1
1:1	0.42:1	3.3:1

The mixing was performed by means of a food mixer for 15 min to a homogeneous consistency and the treated soil samples were then cast and compacted into cylindrical moulds (100.0 mm in height and 50.0 mm in diameter) in accordance with the ASTM D1557-91 standard. After 1 day, the samples were demoulded then cured for 28 days in sealed sample bags at a temperature of 20±2 °C and a relative humidity of 95±3% prior to UCS and durability testing.

1.3. Testing protocol

To verify the effectiveness of the S/S treatment, it is necessary to assess the characteristics of the treatment products and compare them with specific performance criteria. It is appropriate to establish a testing regime that addresses the relevant issues for the management scenario of the treatment products (e.g. disposal or utilisation) being considered (Perera *et al.* 2004). The testing protocol on control and soils contaminated at different C included: (1) setting time; (2) UCS; (3) leaching; and (4) wet-dry and freeze-thaw durability values.

The initial and final setting times of the mixtures were determined by using the ASTM C191-82 method. The setting time of a cementitious mixture is referred to as the period from which water is introduced into the mixture system to the onset of hardening. The initial setting time occurred when the Vicat needle 1.00 mm in diameter penetrated the mortar mixture to a point of 25±1 mm, while final setting time occurred when the needle did not visibly sink into the paste.

UCS test relates to the mechanical resistance of the S/S products. UCS values were measured according to ASTM test method D1633 by applying a vertical load axially at a constant strain rate of 0.5 MPa s^{-1} using a Laumas Electronics CTS compressive strength testing apparatus until failure of the cylindrical specimen.

Durability test methods are applied to analyse the long-term performance of the S/S products and in particular the resistance of the material to repeated cycles of weathering. Cured test specimen were subjected to twelve wet/dry (W/D) and freeze/thaw (F/T) cycles according to ASTM D4843 and ASTM D4842 methods, respectively. Specifically, for W/D test each cycle consisted of a period of 5 hours submerged under water and 42 hours in an oven under low-temperature drying condition (71 °C), while for the F/T test, each cycle consisted of a 24-hours freezing period at –20 °C and a 24-hours thawing period in water.

Pb^{2+} leaching behaviour of the products was investigated applying the EN 12457-2 test and the results were compared with E.U. landfill acceptance criteria (Council Decision 2003/33/EC). For the EN 12457-2 test, after the UCS test, coarse particles were separated using a 4.0 mm sieve. Ninety grams of particle samples were placed into a 1000 mL polypropylene plastic bottle containing 900 mL of deionised water (pH 6.81) (L:S weight ratio 10:1). The suspension was shaken in a rotary shaker for a period of 24 hours at 10 rpm and 20 °C. After extraction, the final pH of the leachate was measured and the liquid was separated from the solids by filtration through a 0.45 μm glass fibre filter. The filtered leachates were then preserved for Pb content, which was performed by ICP-OES (Perkin Elmer Optima 4300 with Dual View).

All tests were carried out in triplicate and mean values are shown.

1.4. Quality criteria

In preparation for full-scale treatment, and to evaluate the effectiveness of the S/S treatment and the degree to which the S/S objectives were met, some specific quality performance criteria were defined. Performance criteria are also usually developed in conjunction with the objectives of

the treatment and the management scenario of the end material (Perera *et al.* 2004).

The criteria refer to some physical and chemical properties of the S/S solids, measured at 28 days, considering the specific methods adopted for the experiments. The quality criteria were extracted from regulatory limits proposed by US Environmental Protection Agency (US EPA) and United Kingdom Environmental Agency (UK EA) (Stegemann, Cote 1990) and are summarized in Table 4.

Table 4. Quality criteria for S/S materials

		Range/ value	Regulatory
Physical properties			
Setting time (h)		5–72	US EPA (Perera *et al.* 2004)
Initial setting time (h)		2–8	BS EN 196-3:2005 US EPA (Perera *et al.* 2004)
Final setting time (h)		<24	BS EN 196-3:2005 US EPA (Perera *et al.* 2004)
UCS (MPa)	Landfill disposal	0.35	USEPA/530-SW-016 US EPA (Perera *et al.* 2004)
UCS (MPa)	Landfill disposal	1.00	BS EN 196-3:2005
UCS (MPa)	Sanitary landfill disposal or construction application	3.45	USEPA/530-SW-016; WTC (1991) US EPA (Perera *et al.* 2004)
UCS (MPa)	Mortar manufacturing	20	UK waste disposal regulatory (Yin *et al.* 2006)
Weight loss after 12 cycles of durability test (%)		<30	WTC (1991); US EPA (Perera *et al.* 2004), Stegemann and Cote (1990)
Chemical properties			
Leachate Pb (mg kg^{-1})	Inert waste landfill disposal	0.5	EU landfill acceptance criteria (2003/33/EC) US EPA (Perera *et al.* 2004)
Leachate Pb (mg kg^{-1})	Non hazardous waste landfill disposal	10	EU landfill acceptance criteria (2003/33/EC) US EPA (Perera *et al.* 2004)
Leachate Pb (mg kg^{-1})	Hazardous waste landfill disposal	50	EU landfill acceptance criteria (2003/33/EC) US EPA (Perera *et al.* 2004)

2. Results

2.1. Testing protocol

2.1.1. Setting times

Figures 1 (a) and 1 (b) show the effects of Pb soil concentration (C) on the initial and final setting times of the S/S treated soils for a S:B ratio of 3.3:1, 4.0:1 and 5.0:1 and for soils A and B, respectively. Results indicate that, especially for soil A, C and S:B ratio significantly influenced the setting times of the S/S treated soils. In particular, setting times increased for both soils, with increasing C and decreased with decreasing the S:B ratio. Higher setting times were observed for soil A, where the presence of Pb at a S:B ratio of 5.0 strongly delayed the hydration reactions and significantly lengthened setting time values (up to 240 h) for C higher than 15000 mg kg^{-1}. This specific behaviour suggests that when a Pb:B threshold ratio is exceeded, a significant increase in the setting time occurs.

These results are consistent with the literature findings where Pb has been reported to suppress cement hydration and lengthen the setting times due to the precipitation of protective coatings of gelatinous hydroxide around the cement grain surface (Chen *et al.* 2009). Moreover, results are in agreement with experimental findings obtained by other authors: Gervais and Ouki (2002) found that the initial setting time for a PC system doped with Pb as nitrate salt at a Pb:B ratio of 0.01 and a W:B ratio of 0.45 is delayed to 30 h, whereas, for a PC S/S treated soil with an S:B ratio of 1, Yin *et al.* (2006) reported final setting times of 4, 8, 20 and 22 h for a C of 500, 5000, 25000 and 50000 mg kg^{-1}, respectively.

Considering the less rigorous US EPA acceptance quality criteria of 72 h for final setting time, inadequate values were obtained for soil A at a S:B ratio of 5.0 for any Pb contamination level and at a S:B ratio of 4.0 only for a C of 25000 mg kg^{-1}. Adequate values were obtained for soil B for all the experimental conditions tested. If the BS EN criteria is considered (initial setting time <8 h, final setting time <24 h), adequate values were obtained only for soil B contaminated at the maximum level of 2000 mg kg^{-1} for all the S:B ratios investigated.

2.1.2. UCS

Results of compressive strength (UCS) tested at 28 days of curing for control and treated polluted soils (soil A and B) are presented as a function of C in Fig. 2.

Results were observed to have an average margin of error of ±7%. Results showed that C highly influenced UCS and the same trend between UCS and C was observed for both tested soils. Specifically, for all the S:B ratios tested, an increase of UCS was observed for a C = 1000 mg kg^{-1} (up to 13000 KPa) respect to the control samples, followed by a decrease with increasing C for the highest values. As expected, UCS of the samples increased

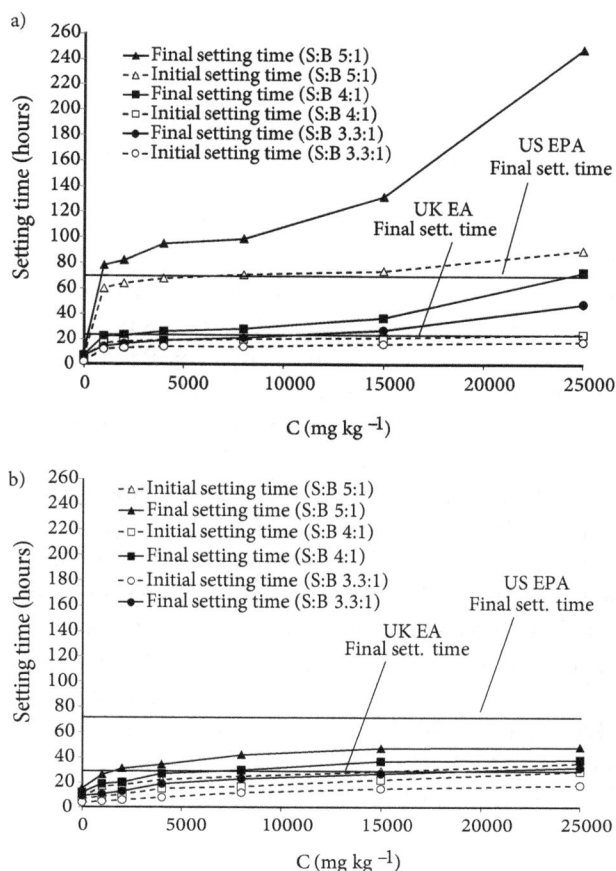

Fig. 1. Setting times of S/S treated soils vs Pb soil concentration for soil A (a) and soil B (b) (S:B ratio of 3.3, 4.0 and 5.0)

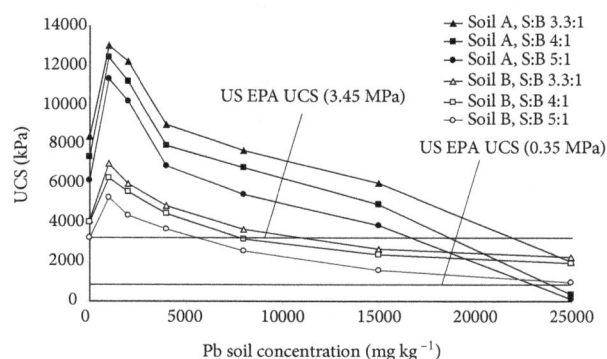

Fig. 2. Effect of Pb soil concentration on the UCS development of S/S soils for soil A and B (S:B ratio of 3.3, 4.0 and 5.0)

with decreasing S:B ratio, and highest values were measured for sandy soil. Overall, excluding the data referred to a C = 25000 mg kg^{-1}, a difference of between 3000 and 6000 KPa was recorded between UCS of soil A and B for all the S:B ratios. For the samples spiked at a C = 25000 mg kg^{-1}, a drastic UCS decrease was observed only for the sandy soil. Therefore, it is clear that the presence of low C (<4000 mg kg^{-1}) results in an improvement of the mechanic characteristics of the S/S treated soils respect to the control samples, whereas for higher C, UCS decreased, doing so more rapidly for sandy soil up to values lower than 100 KPa.

This specific behaviour is probably due to the lengthening of the hydration reaction that was observed to be more consistent for the sandy soils. As a matter of fact, the better strength performances of the sandy soils are hindered by the higher setting times recorded, especially for high values of C. But, for the lowest C, it seems that the presence of Pb in the structure of the S/S treated soils gave them an improvement in the mechanical features that are not weakened by a significant retardation of the hydration reactions. A previous study (Yin et al. 2006) on S/S treated soils reported an increase of UCS with increasing the Pb contamination level in soil. Authors observed for a matrix of PC and soil (S:B 2:1) cured at 28 days an UCS of 34, 37 and 47 MPa for a contamination level of 0, 500 and 25000 mg kg^{-1}, respectively. UCS of 34 MPa was also found by Gervais and Ouki (2002) but for a 100% PC solidified previously doped with Pb at a 0.01 Pb:B ratio.

In terms of quality acceptance, considering the US EPA criteria, insufficient strength values were obtained only for soil A considering a S:B ratio equal to or higher than 4.0 and the maximum C (25000 mg kg^{-1}) in the cases

of S/S treated soils landfill disposal (UCS > 0.35 MPa). If the more restrictive limit for construction application (UCS > 3.45 MPa) is considered, sufficient UCS values were reached for C equal to or lower than 15000 mg kg^{-1} for sandy soil and 4000 mg kg^{-1} for silty-clay soil. Higher C silty-clay soils (up to 8000 mg kg^{-1}) could be successfully treated using a S:B ratio of 3.3 or changing the composition of the binder (i.e. increasing the PC content).

2.1.3. Leachability

The effect of C on the Pb leachability expressed as mg kg^{-1} of solid material and on pH values of leachates resulting from EN 12457-2 test are shown in Fig. 3.

Results were observed to have an average margin of error of ±4%. For all the single batch extractions low values of leached Pb were observed for C values up to 8000 mg kg^{-1}, while for higher C a significant increase was observed in soil A and a slight increase in soil B. This specific behaviour corresponded to pH values which were above 9.8 for soil B but decreased down to a value of 8.0 for soil A at the highest C, indicating the higher buffering capacity of the treated samples of soil B compared to those of soil A. As expected, an increase of leached Pb and a decrease of pH values were observed with increasing S:B ratio. Generally, for most alkaline materials the leaching concentration of Pb decreases for a pH reduction from 10 to 8, but in this case a different trend was observed, probably due to the higher concentration tested that determined an increase in the leaching value even though the reduction in pH would have yielded a lower concentration.

Furthermore, the theory for which leaching of contaminants such as Pb is reduced with increasing the percentage of fine texture soil and/or humidity (i.e. soil B) seems to be confirmed. Indeed, an increase of the specific surface area of the S/S matrices and, therefore an increased adsorption of contaminant onto the clay fraction of the soil may produce a decrease in the amount of coating on the cement grains allowing the cement hydration. Moreover, for soil B, the presence of kaolin, that is known to be a good Pb^{2+} adsorbent (Jiang et al. 2009), especially for high pH values, improved the adsorption phenomena reducing Pb leaching. This phenomenon is more relevant for Pb concentration equal to or higher than 8000 mg kg^{-1}. For both types of soil, the good results obtained, in terms of leachability, are also probably due the presence of a low concentration of organic matter. This condition is in fact known to play an important role in the increase of the immobilization phenomena of Pb (Janoš et al. 2010).

Comparing the obtained results with E.U. landfill acceptance criteria (Council Decision 2003/33/EC), the leached Pb for soil B (all treatments) was below the limit of 0.5 mg kg^{-1} for inert waste landfill disposal. For soil A, the limit of 0.5 mg kg^{-1} was respected only for C equal

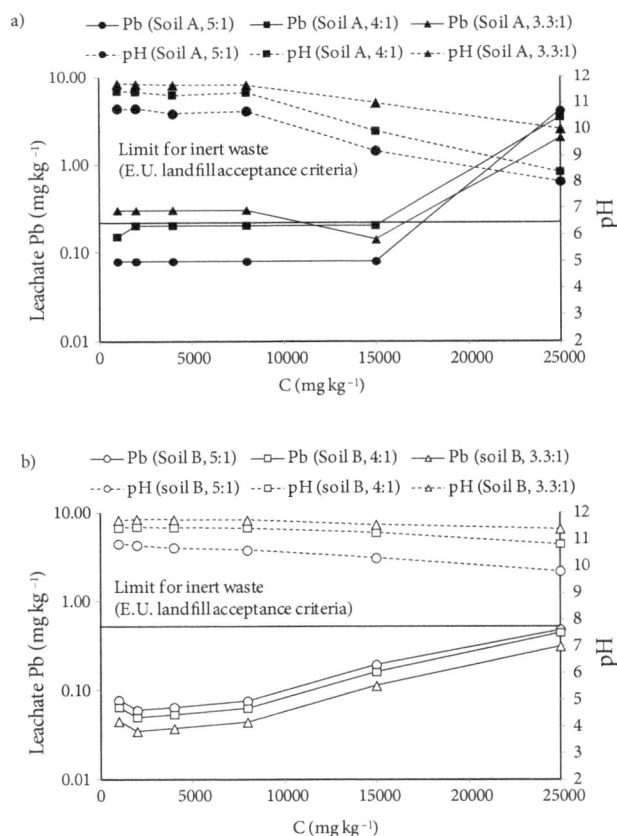

Fig. 3. Effect of Pb soil concentration on the leachability of lead and pH of S/S treated soils for soil A (a) and soil B (b) (S:B ratio of 3.3, 4.0 and 5.0)

to or lower than 8000 mg kg^{-1} or for C equal to or lower than 15000 mg kg^{-1} using a S:B ratio of 4.0 or 3.3. For the other treatments, leached Pb was in each case below the limit of 10 mg kg^{-1} for non-hazardous waste landfill disposal. Results confirm that the main factor controlling the Pb concentration in the leachate is the final pH, and they are in agreement with those obtained by Jing *et al.* (2004). This finding on Pb leachability highlights the possibility of successfully treating also heavy Pb soil contamination using minimal binder percentage (i.e. S:B = 5.0:1) for silty-clayey soils or using higher binder percentages (depending on the objectives of the treatment) for sandy soils.

2.1.4. Durability

For S/S treated soils long-term performance assessment durability tests were performed and results are shown in Figures 4 (a) and 4 (b).

Results were observed to have an average margin of error of ±9%. For the F/T test a slight increase of the weight loss was observed with increasing C for C ≤ 15000 mg kg^{-1}. A significant average variation (about 15%) in terms of mass loss was observed between both tested soils. C > 15000 mg kg^{-1} significantly worsened the performance of the S/S matrices, shown by an increase of the mass loss during the durability cycles. Specifically, for the samples where C was 25000 mg kg^{-1} a weight loss of 21, 24 and 49% was achieved respectively for the soil A (S:B 3.3:1), soil B (S:B 3.3:1) and soil B (4.0:1). For the other experimental conditions, all the tested samples were disintegrated after 2 cycles, for soils A (S:B 5.0:1) and B (S:B 5.0:1) and after 8 cycles for soil A (S:B 4.0:1). These results clearly show that, despite the overall best durability performance of the sandy soils, at high Pb contamination level, they could be more vulnerable than fine texture soils such as soil B. This confirms that in sandy soil, the large retardation of the hydration reactions observed, due to the high Pb level, may significantly worsen the characteristics of the S/S treated soils and consequently their performance.

For the W/D test a small increase in the weight loss was observed with an increasing of C for C equal to or lower than 8000 mg kg^{-1}. For all the samples no difference was observed between C = 15000 and 25000 mg kg^{-1}, whereas an increase of about 10% was observed for soil B in comparison with soil A. The observed decrease in weight loss with an increasing Pb level in soil may be attributed to the water saturation phenomena regarding the samples during their immersion in water for 5 hours at each cycle of W/D. The presence of a higher amount of water in the sample matrix could play an important role in the improvement of the hydration kinetics that gave the best performance in terms of W/D durability to the treated soils.

Based on the durability acceptance criteria proposed by WTC 1991 and reported by Stegemann and Cote (1990) (weight loss < 30%), a minimal S:B ratio is sufficient to

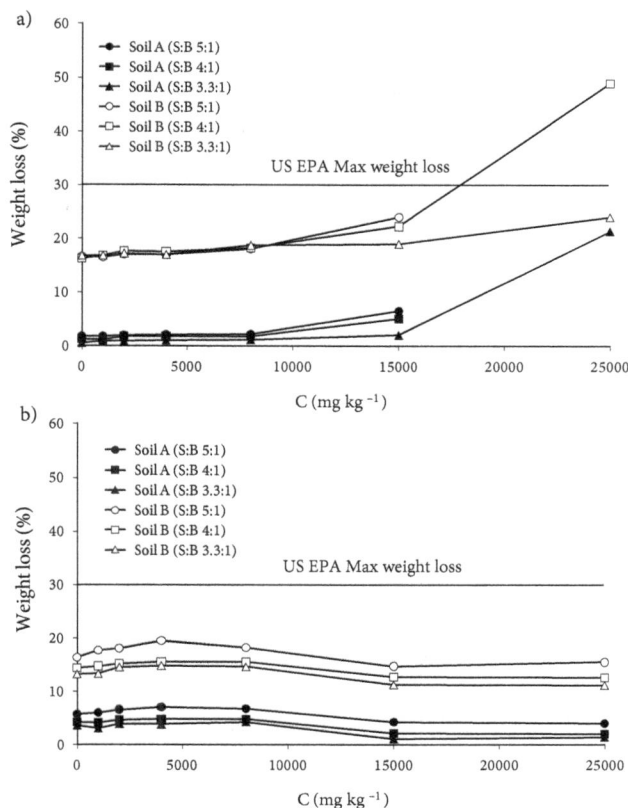

Fig. 4. Effect of Pb soil concentration on the weight loss of S/S treated soils during freeze-thaw (a) and wet-dry (b) durability test for soil A and B (S:B ratio of 3.3, 4.0 and 5.0)

successfully treat a Pb contamination level equal to or lower than 15000 mg kg^{-1}, whereas a S:B ratio of 3.3 is needed to treat contamination levels up to 25000 mg kg^{-1}.

2.2. Assessment of S/S management options

Based on the results obtained by the experimental phase of this work, a performance threshold approach was used to assess potential management options for the S/S treatments. Four management options, namely (1) in situ S/S treatment; (2) hazardous waste landfill disposal; (3) non-hazardous waste landfill disposal; and (4) inert waste landfill disposal or reuse of S/S products for construction materials were chosen for the two investigated soils (sandy and silty-clay soils). Performance thresholds were taken from the performance acceptance criteria reported in Table 4 and management options were represented as minimal percentage of binder to use (S:B ratio) versus the Pb contamination level of the soil (C) for which the selected performance thresholds are all satisfied. The ranges of the values shown on the x and y axes represent the limits of the experimental conditions investigated. The Pb soil concentration limit which can be treated and the related limiting factor and maximum S:B ratio required for sandy and silt-clay soils is reported in Table 5.

Table 5. Pb soil concentration limit and limiting factor for sandy and silt-clay soils as a function of the scenario adopted

Scenario – S:B ratio	Sand soil		Silt-clay soil	
	Pb concentration limit (mg kg⁻¹) (S:B ratio required)	Limiting factor	Pb concentration limit (mg kg⁻¹) (S:B ratio required)	Limiting factor
1	8000 (3.3)	Setting time	4000 (4.0)	UCS
2	–	–	–	–
3	15000 (4.0)	UCS	4000 (5.0)	UCS
4	15000 (4.0)	UCS-Leachability	4000 (5.0)	UCS

2.2.1. In situ S/S treatment

The in situ S/S soil mixing treatment is carried out using mixing augers through which a grout is introduced and mixed with the soil resulting in stabilized/solidified overlapping soil-grout columns (Fleri, Whetstone 2007). Compared to the ex-situ treatments, they could be more hazardous due to the potential risk of a contamination propagation during the realization of the interventions; consequently, high performance acceptance criteria limits were adopted for the purpose.

A fundamental property for the S/S soil-grout columns is the final setting time, since a delay in the S/S mass curing could significantly enhance the leachability of the contaminants. A downward contaminant migration could take place during the in situ mixing processes due to the transitory increase of water amount in soil. For this reason the BS EN limit of 24 h was adopted. Moreover, to ensure a high mechanical resistance another US EPA restrictive limit of 3.45 MPa was applied for UCS. EU landfill acceptance criteria (2003/33/EC) limit (0.5 mg kg⁻¹) and WTC limit (30% weight loss) were adopted for leaching and durability tests, respectively. Figures 5 (a) and 5 (b) show the minimal percentage of binder to use (S:B ratio) as a function of C for which the selected limits are satisfied in the case of an in situ treatment, for a sandy and a silty clay soil, respectively. Results highlight that due to the strict limits adopted, the in situ treatment may be successfully adopted only for not heavy Pb contamination levels. Specifically, a maximum level of 8000 mg kg⁻¹ and 4000 mg kg⁻¹ is treatable for a sandy and a silty clay soil, respectively. Moreover, a S:B ratio of 4.0 is adoptable for both the soils up to 2000 mg kg⁻¹, whereas a S:B ratio of 3.3 is necessary for higher contaminations.

2.2.2. Hazardous waste landfill disposal

A maximum final setting time of 72 h and an UCS of 350 kPa at 28 days are suggested by USEPA guidelines for materials that are to be disposed to landfill (Perera *et al.*

2004) which takes into consideration events such as weight of overburden and land moving equipment. For leaching performance and durability, EU landfill acceptance criteria limit for hazardous waste disposal (50 mg kg⁻¹) and WTC limit (30% weight loss) were used. Results (Figs 5 (c) and 5 (d)) show that a maximum C of 15000 mg kg⁻¹ may be treated using a S:B of 4.0 and 5.0 for sandy and silty clay soil, respectively, whereas a maximum percentage of binder must be adopted for higher contamination levels. Specifically, for silty clay soil this limitation is due to the durability features, whereas for sandy soil it is due to all the adopted limits except for the leaching limit that is always satisfied. In addition, the worst results, in terms of setting time, were observed for sandy soil, resulting in an S:B values equal to or lower than 4.0 being required for its treatment.

2.2.3. Non-hazardous waste landfill disposal

Compared to the previous scenario, a higher UCS value of 3.5 MPa has been suggested by USEPA and WTC (Perera *et al.* 2004) for disposal to non-hazardous waste (sanitary) landfill because compaction of municipal waste might subject the S/S material to higher stresses because handling, placement and covering operations are not tailored for S/S material. Moreover, the EU landfill acceptance criteria limit for non-hazardous waste disposal (10 mg kg⁻¹) was used. The use of 3.5 MPa limit for UCS results in a strong limitation of the applicability field (Figs 5 (e) and 5 (f)) especially for silty clay soil for which a maximum C of 4000 mg kg⁻¹ may be treated using a S:B from 3.3 to 5.0. For sandy soil a higher C up to 15000 mg kg⁻¹ may be treated but using a S:B from 3.3 to 4.0.

2.2.4. Inert waste landfill disposal or reuse for construction

Compared to the sanitary landfill disposal scenario, a lower limit of 0.5 mg kg⁻¹ for leaching is required by the EU for disposal in inert waste landfill. For both soils, the use of this severe limit for leaching results in the same limitation as the previous scenario (Figs 5 (g) and 5 (h)).

Conclusions

A sandy and a silty-clay soil, spiked with lead at different rates, were S/S treated using a binder mixture of PC and FA (PC:FA 1:1) at different soil:binder ratios, and a testing protocol included setting time, UCS, leaching and durability test were performed to assess the effects of lead and binder level on physical properties of S/S matrices. Based on the experimental results and defining specific quality performance criteria, a performance threshold approach was used to assess potential management options for the S/S treatments. The following conclusions have been drawn according to the results presented above:

1. Soil texture, the percentage of binders used and lead concentration in soil significantly influence

the performance of the S/S treated soils in terms of curing, compressive and weather cycling strength;

2. The observed influence of lead content on studied parameters may be useful in predicting setting

time, UCS, leaching and durability of soils treated by PC and FA at any contamination level;

3. Lead soil concentration higher than 15000 mg kg⁻¹ heavily reduces the applicability of the S/S techniques requiring a large amount of binder to satisfy the

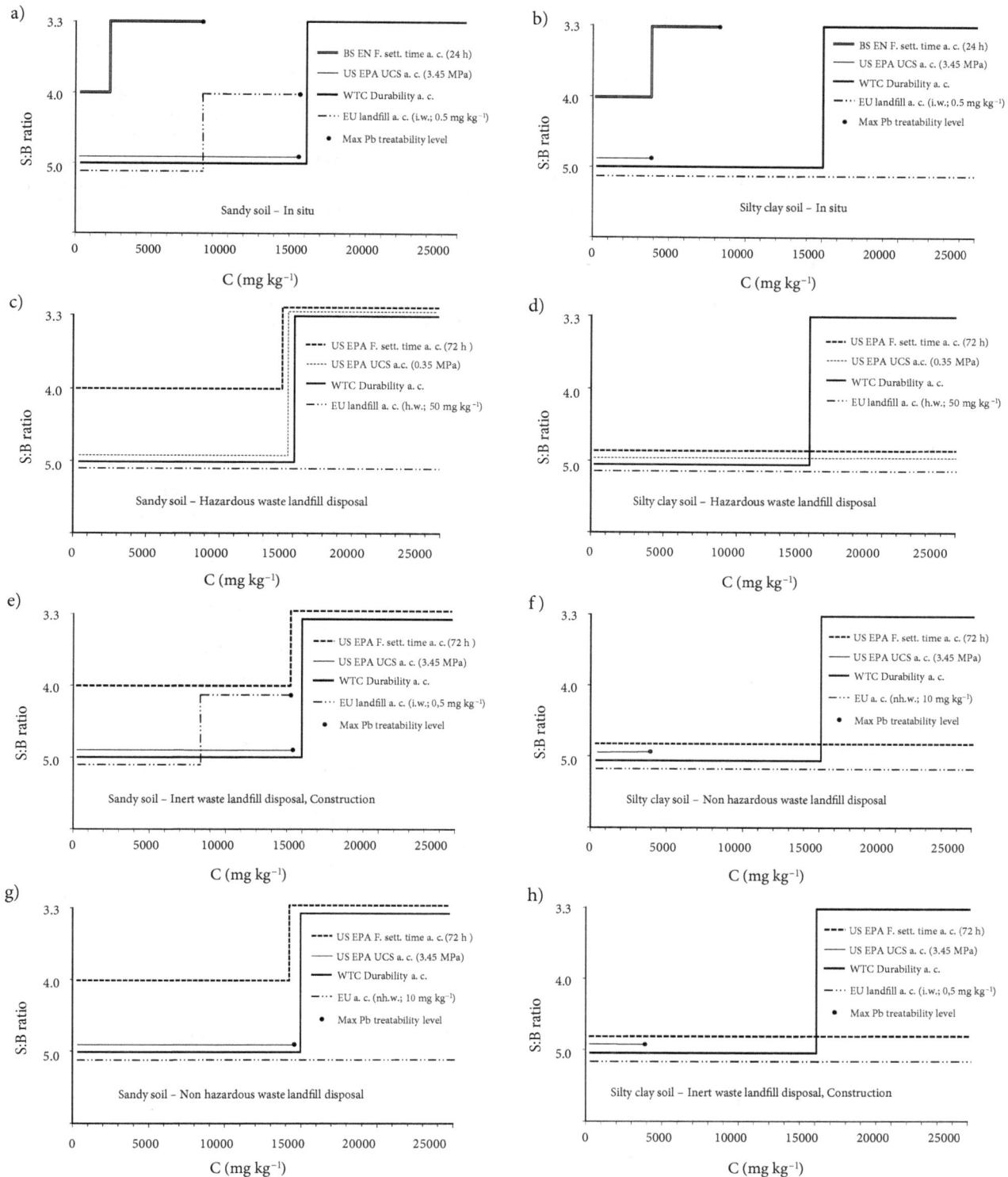

Fig. 5. Percentage of binder to use (S:B ratio) versus C for which the quality performance criteria thresholds are satisfied for sandy and silty clay soils and different scenarios

selected performance criteria; this could make the treatment very expensive. Specifically, despite the best compressive strength observed at the lowest lead concentration values, soils performance was shown to be limited by setting time and UCS features due to the lengthening of the hydration reaction;

4. In terms of leachability, lead soil concentration does not represent a limitation, indeed also heavy contamination may be successfully treatable also using minimal binder percentage for any type of soil, except for the in situ and inert waste landfill disposal scenarios, for which, due to the severe limits adopted, a higher soil:binder ratio of 4.0 is required;

5. A strong limitation of the applicability field was observed for an in situ treatment, S/S products reuse or disposal to a landfill, especially for silty clay soils for which only lead concentration lower than 4000 mg kg^{-1} may be treated in the case of landfill disposal or reuse, while a maximum contamination of 8000 mg kg^{-1} must be treated using a soil:binder ratio of 3.3:1 in the case of sandy soil in situ treatment;

6. The obtained results are of practical interest and may be used for preliminary cost assessment of full scale remediation activities of lead polluted soils.

References

Al-Ansary, M. S.; Al-Tabbaa, A. 2007. Stabilisation/solidification of synthetic petroleum drill cuttings, *Journal of Hazardous Materials* 141(2): 410–421. http://dx.doi.org/10.1016/j.jhazmat.2006.05.079

Antemir, A.; Hills, C. D.; Carey, P. J.; Magnié, M. C.; Polettini, A. 2010. Investigation of 4-year-old stabilised/solidified and accelerated carbonated contaminated soil, *Journal of Hazardous Materials* 181(1–3): 543–555. http://dx.doi.org/10.1016/j.jhazmat.2010.05.048

Bech, J.; Duran, P.; Roca, N.; Poma, W.; Sánchez, I.; Roca-Pérez, L.; Boluda, R.; Barceló, J.; Poschenrieder, C. 2012. Accumulation of Pb and Zn in *Bidens triplinervia* and *Senecio sp.* spontaneous species from mine spoils in Peru and their potential use in phytoremediation, *Journal of Geochemical Exploration* 123: 109–113. http://dx.doi.org/10.1016/j.gexplo.2012.06.021

Chen, Q. Y.; Tyrer, M.; Hills, C. D.; Yang, X. M.; Carey, P. 2009. Immobilisation of heavy metal in cement-based solidification/stabilisation: a review, *Waste Management* 29(1): 390–403. http://dx.doi.org/10.1016/j.wasman.2008.01.019

Cheng, Y.; Guo, Z.; Liu, X.; Yin, H.; Qiu, G.; Pan, F.; Liu, H. 2009. The bioleaching feasibility for Pb/Zn smelting slag and community characteristics of indigenous moderate-thermophilic bacteria, *Bioresource Technology* 100(10): 2737–2740. http://dx.doi.org/10.1016/j.biortech.2008.12.038

Dermatas, D.; Meng, X. 2003. Utilization of fly ash for stabilization/solidification of heavy metal contaminated soils, *Engineering Geology* 70(3–4): 377–394. http://dx.doi.org/10.1016/S0013-7952(03)00105-4

European Communities Council Decision 2003/33/EC. Establishing criteria and procedures for the acceptance of waste at landfill.

Falciglia, P. P.; Cannata, S.; Romano, S.; Vagliasindi, F. G. A. 2012. Assessment of mechanical resistance, g-radiation shielding and leachate g-radiation of stabilised/solidified radionuclides polluted soils: preliminary results, *Chemical Engineering Transactions* 28: 127–132.

Fleri, M. A.; Whetstone, G. T. 2007. In situ stabilisation/solidification: project lifecycle, *Journal of Hazardous Materials* 141(2): 441–456. http://dx.doi.org/10.1016/j.jhazmat.2006.05.096

Gervais, C.; Ouki, S. K. 2002. Performance study of cementitious systems containing zeolite and silica fume: effects of four metal nitrates on the setting time, strength and leaching characteristics, *Journal of Hazardous Materials* 93(2): 187–200. http://dx.doi.org/10.1016/S0304-3894(02)00005-5

Harbottle, M. J.; Al-Tabbaa, A.; Evans, C. W. 2007. A comparison of the technical sustainability of in situ stabilisation/solidification with disposal to landfill, *Journal of Hazardous Materials* 141(2): 430–440. http://dx.doi.org/10.1016/j.jhazmat.2006.05.084

Janoš, P.; Vávrová, J.; Herzogová, L.; Pilařová, V. 2010. Effects of inorganic and organic amendments on the mobility (leachability) of heavy metals in contaminated soil: a sequential extraction study, *Geoderma* 159(3–4): 335–341. http://dx.doi.org/10.1016/j.geoderma.2010.08.009

Jiang, M.; Wang, Q.; Jin, X.; Chen, Z. 2009. Removal of Pb(II) from aqueous solution using modified and unmodified kaolinite clay, *Journal of Hazardous Materials* 170(1): 332–339. http://dx.doi.org/10.1016/j.jhazmat.2009.04.092

Jing, C.; Meng, X.; Korfiatis, G. P. 2004. Lead leachability in stabilized/solidified soil samples evaluated with different leaching tests, *Journal of Hazardous Materials* 114(1–3): 101–110. http://dx.doi.org/10.1016/j.jhazmat.2004.07.017

Kundu, S.; Gupta, A. K. 2008. Immobilization and leaching characteristics of arsenic from cement and/or lime solidified/stabilized spent adsorbent containing arsenic, *Journal of Hazardous Materials* 153(1–2): 434–443. http://dx.doi.org/10.1016/j.jhazmat.2007.08.073

Lee, D. 2007. Formation of leadhillite and calcium lead silicate hydrate (C-Pb-S-H) in the solidification/stabilization of lead contaminants, *Chemosphere* 66(9): 1727–1733. http://dx.doi.org/10.1016/j.chemosphere.2006.07.012

Leonard, S. A.; Stegemann, J. A. 2010. Stabilization/solidification of petroleum drill cuttings: leaching studies, *Journal of Hazardous Materials* 174(1–3): 484–491. http://dx.doi.org/10.1016/j.jhazmat.2009.09.078

Malviya, R.; Chaudhary, R. 2006. Factors affecting hazardous waste solidification/stabilization: a review, *Journal of Hazardous Materials* 137(1): 267–276. http://dx.doi.org/10.1016/j.jhazmat.2006.01.065

Moon, D. H.; Dermatas, D. 2007. Arsenic and lead release from fly ash stabilized/solidified soils under modified semi-dynamic leaching conditions, *Journal of Hazardous Materials* 141(2): 388–394. http://dx.doi.org/10.1016/j.jhazmat.2006.05.085

Ouki, S. K.; Hills, C. D. 2002. Microstructure of Portland cement pastes containing metal nitrate salts, *Waste Management* 22(2): 147–151. http://dx.doi.org/10.1016/S0956-053X(01)00063-0

Perera, A. S. R.; Al-Tabbaa, A.; Reid, J. M.; Stegemann, J. A. 2004. *State of practice reports*, UK stabilization/solidification treatment and remediation – testing and performance criteria, in Al-Tabbaa, A.; Stegemann, J. A. (Eds.). *Stabilization/Solidifi-*

cation Treatment and Remediation, Advances in S/S for Waste and Contaminated Land. London: A. A. Balkema Publishers.

Polettini, A.; Pomi, R.; Sirini, P.; Testa, F. 2001. Properties of Portland cement – stabilised MSWI fly ashes, *Journal of Hazardous Materials* 88(1): 123–138. http://dx.doi.org/10.1016/S0304-3894(01)00292-8

Qian, G. R.; Shi, J.; Cao, Y. L.; Xu, Y. F.; Chui, P. C. 2008. Properties of MSW fly ash–calcium sulfoaluminate cement matrix and stabilization/solidification on heavy metals, *Journal of Hazardous Materials* 152(1): 196–203. http://dx.doi.org/10.1016/j.jhazmat.2007.06.118

Ryu, B. G.; Park, G. Y.; Yang, J. W.; Baek, K. 2011. Electrolyte conditioning for electrokinetic remediation of As, Cu, and Pb-contaminated soil, *Separation and Purification Technology* 79(2): 170–176. http://dx.doi.org/10.1016/j.seppur.2011.02.025

Stegemann, J. A.; Cote, P. L.; 1990. Summary of an investigation of test methods for solidified waste evaluation, *Waste Management* 10(1): 41–52. http://dx.doi.org/10.1016/0956-053X(90)90068-V

Svensson, M.; Allard, B. 2008. Leaching of mercury-containing cement monoliths aged for one year, *Waste Management* 28(3): 597–603. http://dx.doi.org/10.1016/j.wasman.2007.02.031

Tang, X.; Yang, J. 2012. Long-term stability and risk assessment of lead in mill waste treated by soluble phosphate, *Science of the Total Environment* 438: 299–303. http://dx.doi.org/10.1016/j.scitotenv.2012.08.054

Torres, L. G.; Lopez, R. B.; Beltran, M. 2012. Removal of As, Cd, Cu, Ni, Pb, and Zn from a highly contaminated industrial soil using surfactant enhanced soil washing, *Physics and Chemistry of the Earth* 37–39: 30–36.

Yin, C. Y.; Mahmud, H. B.; Shaaban, M. G. 2006. Stabilization/solidification of lead-contaminated soil using cement and rice husk ash, *Journal of Hazardous Materials* 137(3): 1758–1764. http://dx.doi.org/10.1016/j.jhazmat.2006.05.013

Pietro Paolo FALCIGLIA. A graduate of University of Catania, Italy, received a Master of Science from Superior School of Catania in 2004, and a PhD in Civil and Environmental Engineering from University of Salerno (Italy) in 2008. Currently he is a Research Associate and Contract Professor of Remediation and Reclamation of Polluted Sites at the Civil and Environmental Engineering Department of the University of Catania. Publications: author/co-author of over 40 scientific papers. Research interests include waste management and treatment, remediation of polluted sites, wastewater treatment.

Abir AL-TABBAA. She graduated from Bristol University in 1983 in Civil Engineering and obtained an MPhil and PhD degrees in Soil Mechanics in 1984 and 1987 respectively from Cambridge University. In 1991 she took up a lectureship at the University of Birmingham and in 1997 returned to Cambridge University as a University Lecturer, now Reader in Geotechnical Engineering. She is a fellow of Sidney Sussex College. She is a Chartered Engineer and a Member of the Institution of Civil Engineers. She was a member of the British Geotechnical Association Executive Committee (2001–2004), a member of the Institution of Civil Engineers Geotechnical Engineering Advisory Panel (2003–2005), Ground Improvement Advisory Panel (2007–2012) and the UK Representative on the International Society of Soil Mechanics and Geotechnical Engineering Technical Committee TC211 on Ground Improvement. She was awarded the Institution of Civil Engineers Reed and Mallik Medal in 2003 for her paper on the five-year soil mixing treatment work at the West Drayton site near Heathrow Airport. Research interests include soil mix technology, field characterisation of soils, waste, waste management and contaminated land, groundwater, seepage and permeability, geomaterials and constitutive models, materials for extreme environments.

Federico G. A. VAGLIASINDI. A graduate of University of Catania, Italy, received a Master of Science from the Civil Engineering Department at Colorado State University, Fort Collins (CO) in 1991, and a Full Professor of Environmental and Sanitary Engineering at University of Catania, Italy, since March 2002. Has participated to research projects financed by the European Commission, the Italian National Research Council, MIUR (PRIN 1999, 2001, 2004, 2006; Project Aquatec PON 2002), AWWARF (American Water Works Association Research Foundation, USA), EPRI (Energy and Power Research Institute, USA). He has consulted several public administrations involved with the ongoing emergency in the solid waste management and contaminated site remediation in Sicily. He is a member of several professional association, has been a component of the Environmental and Sanitary Section of the Italian Great Risks Commission. Since April 2008, he has been a President of the Italian Association of Sanitary and Environmental Engineering Professors. His research interests include water and wastewater treatment and reuse, integrated solid waste management, contaminated site remediation.

15

MAPPING OF HEAVY METAL CONTAMINATION IN ALLUVIAL SOILS OF THE MIDDLE NILE DELTA OF EGYPT

Mohamed S. SHOKR[a,b], Ahmed A. EL BAROUDY[a], Michael A. FULLEN[b], Talaat R. EL-BESHBESHY[a], Ramadan R. ALI[c], Abd ELHALIM[a], Antonio J. T. GUERRA[d] and Maria C. O. JORGE[d]

[a]*Soils and Water Department, Faculty of Agriculture, Tanta University, Tanta, Egypt*
[b]*The University of Wolverhampton, Wolverhampton WV1 1LY, UK*
[c]*Soils and Water Use Department, National Research Centre, Giza, Egypt*
[d]*Department of Geography, Federal University of Rio de Janeiro, Brazil*

Abstract. Areas contaminated by heavy metals were identified in the El-Gharbia Governorate (District) of Egypt. Identification used remote sensing and Geographical Information Systems (GIS) as the main research tools. Digital Elevation Models (DEM), Landsat 8 and contour maps were used to map physiographic units. Nine soil profiles were sampled in different physiographic units in the study area. Geochemical analysis of the 33 soil samples was conducted using X-ray fluorescence spectrometry (XRF). Vanadium (V), nickel (Ni), chromium (Cr), copper (Cu) and zinc (Zn) concentrations were measured. V, Ni and Cr concentrations exceeded recommended safety values in all horizons of the soil profiles, while Cu had a variable distribution. Zn concentrations slightly exceeded recommended concentration limits. Concentrations were mapped in each physiographic unit using the inverse distance weighted (IDW) function of Arc-GIS 10.1 software. Pollution levels were closely associated with industry and urban areas.

Keywords: soil contamination, x-ray fluorescence spectrometry, remote sensing, geographical information systems, Middle Nile Delta, Egypt.

Introduction

Sustainable agriculture is mainly related to environmental, agronomic, ethical and socio-economic issues (Abd Elgawad *et al.* 2007). One aspect of sustainability is accumulation of heavy metals in soils, which may cause serious problems, if certain levels are exceeded. In recent years, much concern has been articulated over problems of soil contamination with heavy metals. These metals can accumulate in plants and animals and then in humans through the food chain (Govil *et al.* 2001; Lu, Bai 2010; Romic, M., Romic, D. 2003). Thus, heavy metals may damage human health and the environment (Jankaite, Vasarevičius 2005).

The Nile Delta (area ~20,000 km²) represents only 2.3% of the area of Egypt, but it has ~46% of the total cultivated area (55,040 km²) and accommodates ~45% of Egypt's population (Fanos 2002), with densities ≤1600 inhabitants per km² (Zeydan 2005). On the Nile Delta ~63% land is agricultural, due to suitable soil properties and the presence of irrigation systems (Dawoud 2004). The River Nile divides into two branches, the Rosetta and Damietta, and the Delta region is located between them (Dumont 2009). The Nile Delta (area 404,686 ha) depends on drainage water for irrigation (Abu 2011).

There are three major layers in the middle Delta aquifer (Atwia *et al.* 2006). The uppermost layer is composed of clay deposits, the second layer is formed from sandy clay deposits and the third layer is composed of saturated sand and gravel. Thus, the thin clay layers and presence of sandy clay lenses facilitate percolation of sewage water to the aquifer. Many activities, including agricultural development and industrial activities and inadequate rural sanitation, have impacts on eutrophication and contamination status, ecological value and environmental conditions in the Nile Delta (Zeydan 2005).

Heavy metal contamination of soil may present risks and hazards to humans and the ecosystems through: direct consumption or contact with contaminated soil, the food chain (soil-plant-human or soil-plant-animal-human), drinking of contaminated ground-water, decreased food quality (safety and marketability) via

Corresponding author: Mohamed Shokr
E-mail: m.s012@yahoo.com

phytotoxicity, reduction in land usability for agricultural production causing food insecurity, and land tenure problems (McLaughlin *et al.* 2000; Ling *et al.* 2007). Huge amounts of fertilizers are regularly added to soils in intensive farming systems to provide sufficient nitrogen (N), phosphorus (P) and potassium (K) for crop growth. The compounds used to supply these elements contain trace amounts of heavy metals which, after continued fertilizer application, may significantly increase soil metal contents (Raven *et al.* 1998). Integration of remote sensing information within a GIS database can quickly provide detailed soil survey information at low cost. GIS databases can also help derive Digital Elevation Models (DEM), which can help derive landscape attributes utilized in landform characterization (Brough 1986; Dobos *et al.* 2000).

It is critical to analyse the distribution and concentration of metals. This will enable identification of contamination levels and assess associated impacts, on both the environment and human health. Soils are a vital sink for these metals, because of their high metal retention capacities. The assessment and mapping of soil heavy metals can assist the development of strategies to promote sustainable use of soil resources, decrease soil degradation and expand crop production. Remediation of soils polluted by heavy metals is a major global ecological issue. Remote sensing is one of the most important methods used for soil survey, mapping and environmental investigations (Lillesand, Kiefer 2003). Geostatistical interpolation is used to survey and interpret the spatial distribution of pollutants in soil (He, Jia 2004; Woo *et al.* 2009). The inverse distance weighted (IDW) function is helpful when the purpose is to investigate overall pollution patterns (Zheng 2006). The Middle Nile Delta is affected by different pollution sources, because of the increasing number and types of industries, urban expansion, increased traffic volumes, use of drain-water and waste deposits (Abu Khatita 2011). The latter may well present a long-term danger. Usually waste deposits just settle within the normal Nile sediments and no special effort is made to construct barriers, which hinder the migration of water from these deposits into ground-water. High concentrations of vanadium (V) can damage human health, while the inhalation of airborne V-compounds can affect eyes, throat and lungs, produce weakness, ringing in the ears, nausea, vomiting, headaches and damage nerve systems (Lagerkvist, Oskarsson 2007). In Egypt, measured chromium (Cr) contents in soils range between 11.6–179 ppm, and depend on soil types and land management (Abdel-Sabour *et al.* 2002). Cr toxicity depends on its oxidation status. While Cr^{3+} is considered relatively harmless, Cr^{6+} is highly toxic. Cr uptake can cause diarrhoea, bleeding in the stomach and intestines, liver and kidney damage and cramp. Nickel (Ni) compounds are relatively non-toxic for plants and animals, but there is an increased risk of respiratory tract cancer, due to exposure to nickel sulphide and oxides (Sundermann, Oskarsson 1991).

Copper (Cu) is an essential element for all life-forms. In plants, Cu is required in small amounts (5–15 ppm) (Bowen 1979). The amount of Cu in soils may affect crop growth and yields. The application of Cu salts to Cu-deficient soils increases crop yields, because it compensates for Cu deficiency in plants (Baker, Senft 1995). Coal fly-ash contains 48 $\mu g/g$ of Cu (Wong, M. H.; Wong, J. W. C. 1986). In Ohio (USA), measured Cu concentrations in indoor dust were twice that of outdoor dust (Tong 1998). Cu toxicity in humans is relatively rare, because they can tolerate levels ≤12 mg/day (WHO 1996). However, Cu deficiency in humans causes anaemia, bone and cardiovascular disorders, mental and nervous system deterioration and defective keratinization of hair.

Zinc (Zn) is the fourth most used metal in the world, after iron (Fe), aluminium (Al) and Cu (Bradl, Xenidis 2005). Zn uptake can lead to health disorders, including pancreatic diseases. Inhalation of Zn-oxide (particle size 0.2–1 μm) during Zn-processing causes metal fume fever, which is characterized by a sore throat, cough, fever, vomiting and pneumonitis (Ohnesorge, Wilhelm 1991).

The main aim of this research is to identify land contaminated by heavy metals in the El-Gharbia Governorate (District) of Egypt. This was undertaken using remote sensing, Geographical Information Systems (GIS) and X-ray fluorescence (XRF) spectrometry (Fig. 1).

1. Materials and methods

1.1. Study area

The study area occupies the Middle part of the Nile Delta of Egypt. It is bounded by 30°45′20″–31°10′50″E and 30°35′10″–31°10′05″N, and covers an area of 1927.4 km² (Fig. 2). Based on the US Soil Taxonomy (USDA 2010) the soil temperature regime of the study area is Thermic and the soil moisture regime is Torric. The mean annual

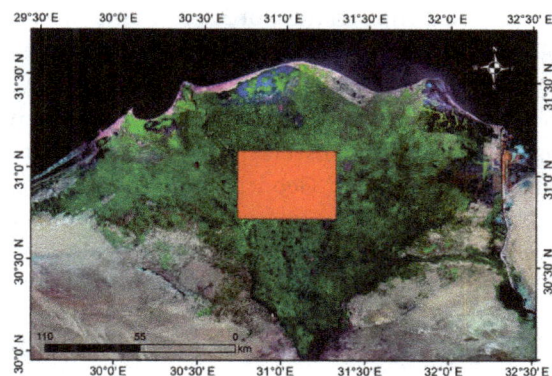

Fig. 1. Location of the study area in the middle Nile Delta of Egypt. ▮ – study area

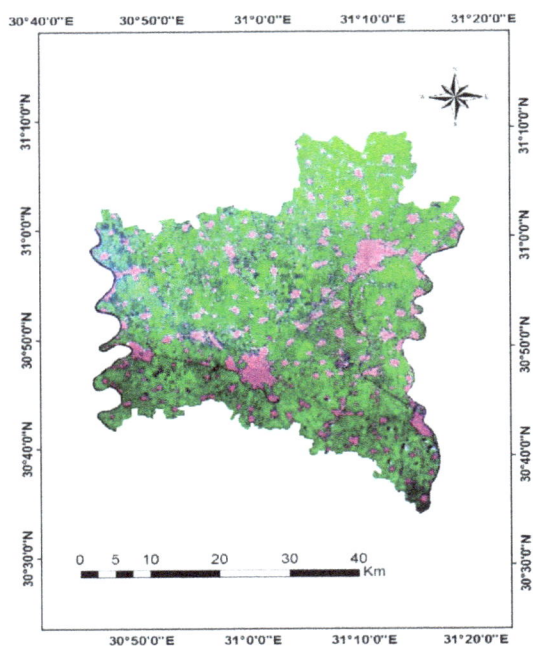

Fig. 2. Landsat 8 mosaic of the study area

temperature reaches its maximum in June, July and August and often exceeds 30 °C. The mean minimum temperature (11.2 °C) usually occurs in January, February or March at Tanta Meteorological Station (Climatologically Normal for Egypt 2011). Precipitation is unequally distributed through the rainy season. Annual rainfall is very low and mostly falls in winter; with a mean 3.8 mm/year. Rain mainly falls in the cold season (November–March) and the minimum amount is in June and September. The area belongs to the late Pleistocene era, which is evidenced by the deposits of the Neonile, which are composed of medium and fine silts (Said 1993).

1.2. Digital image processing and physiographic mapping

Digital image processing was completed for two Landsat 8 satellite images (path 177/row 38 and path 177/row 39), with a spatial resolution of 30 m, acquired in May 2014. The images were pre-processed, including radiometric correction (used to modify digital values of pixels to remove noise). Images were geometrically rectified using the Universal Transverse Mercator (UTM) co-ordinates, with the World Geodetic System datum (WGS 1984) and then maps were constructed. Images were atmospherically corrected using the FLAASH module (ITT 2009). Data were calibrated to radiance using the inputs of image type, acquisition date and time. Images were subject to linear stretching by 2%, smooth-filtered, and their histograms were matched, adopting the procedures of Lillesand and Kiefer (2007) and mosaicked using ENVI 5.1 software. The extraction of landform

units used high spatial resolution images, so the spatial resolution of satellite image was enhanced using the data merge function of Envi5.1 software. Merging is performed by using multispectral bands (~30 m) as low spatial resolution, and band 8 (panchromatic band) with ~15 m resolution. Landforms were extracted using contour maps (scale 1:25,000) and enhanced satellite images. Both enhanced satellite images were processed with DEM in ERDAS Imagine 8.7, to extract the landform information (Dobos et al. 2002). The initial landform maps were ground-truthed using field observations.

1.3. Spatial distribution of heavy metals

Spatial interpolation is widely used when data are collected at distinct locations (e.g. soil profiles) for producing continuous information (Ali, Moghanm 2013). Inverse distance weighted (IDW) is an interpolation method which uses measured values surrounding the prediction location. The measured values closest to the prediction location have more influence on the predicted value than those farther away, thus giving greater weight to points closest to the prediction location, and the weights decrease as a function of distance (Shepard 1968). Geostatistical relationships among the known points (IDW) of Arc-GIS 10.1 software were used to interpolate heavy metal concentrations in the study area. The spatial interpolation method (IDW) was used with 12 neighbouring samples for estimation of each grid point. A power of two was used to weight the nearest points.

1.4. Assessment of contamination risk

The Geoaccumulation Index (Igeo) was originally used to evaluate bottom sediment contamination. However, it has been successfully used to evaluate soil contamination (Gowd et al. 2010). The Igeo Index means the assessment of contamination depends on comparing heavy metal concentrations in soils to background values. The calculation of the Geoaccumulation Index uses the equation:

$$\mathrm{Igeo} = \log 2 \frac{C_n}{1.5 B_n}, \qquad (1)$$

where C_n = the measured concentration of the element in soil; B_n = the geochemical background concentration of the heavy metal.

The Geoaccumulation Index (Igeo) is shown in Table 1.

1.5. Soil analysis

Soil samples were collected from nine profiles in El-Gharbia Governorate. The selected profiles represent the different soil units. Pedological descriptions of profiles were conducted using the procedures of FAO (2006) (Table 2). About 1 kg was collected from each horizon of

Table 1. The Geoaccumulation Index (Igeo) for assessing contamination levels in soil (Rahman *et al.* 2012)

Igeo Class	Igeo value	Contamination level
0	Igeo ≤ 0	Uncontaminated
1	0 < Igeo < 1	Uncontaminated/moderately contaminated
2	1< Igeo < 2	Moderately contaminated
3	2 < Igeo < 3	Moderately/strongly contaminated
4	3 < Igeo< 4	Strongly contaminated
5	4 < Igeo< 5	Strongly/extremely contaminated
6	5 < Igeo	Extremely contaminated

each profile. Soil samples were air-dried and large stones and organic debris were removed before sieving. Samples were gently ground, homogenized, sieved through a 2.0 mm sieve and then crushed to a fine (<125 μm) powder. Oven-dry samples were ignited at 375 °C for 16 hours (overnight), adopting the procedures of Ball (1964). Sub-samples of 8.5 g of soil powder were added to 1.5 g of wax (Lico waxc micropowder PM, Hoechst wax)) and then compressed under 12 tonnes pressure by a semi-automatic hydraulic press to make a pellet. The geochemical composition of soil pellets were analysed using an XRF spectrometer model Epsilon3 XLE. XRF analyses were performed at the University of Wolverhampton, UK.

Table 2. Pedological descriptions of soil profiles

Profile Number	Depth (cm)	Colour	Texture	Structure	Soil consistency
1	0–50	5YR 3/4 5YR5/4	Loam	Sub-angular blocky	Sticky, plastic
	50–85	7.5YR 5/6 7.5YR 6/6	Sandy loam	Sub-angular blocky	Slightly sticky, slightly plastic
	85–120	10YR 5/8 10YR 6/8	Loam	Sub-angular blocky	Sticky, plastic
	120–150	10YR 5/8 10YR 7/8	Sandy loam	Sub-angular blocky	Slightly sticky, slightly plastic
2	0–45	5YR 4/6 5YR 5/6	Sandy Loam	Sub-angular blocky	Slightly sticky, slightly plastic
	45–85	10YR 5/8 10YR 6/8	Sandy Clay Loam	Sub-angular blocky	Sticky, plastic
	85–110	10YR 5/8 10YR 7/8	Sandy Loam	Sub-angular blocky	Slightly sticky, slightly plastic
3	0–75	10 YR 5/6 10YR 6/8	Sandy Loam	Sub-angular blocky	Slightly sticky, slightly plastic
	75–100	10YR 5/6 10YR 7/8	Loam	Sub-angular blocky	Sticky, plastic
	100–150	10YR 5/8 10YR 7/8	Silt loam	Sub-angular blocky	Sticky, plastic
4	0–60	5YR 4/8 5YR 4/6	Loam	Sub-angular blocky	Sticky, plastic
	60–100	10YR 5/8	Sandy Loam	Sub-angular blocky	Slightly sticky, slightly plastic
	100–120	10YR 7/6	Loam	Sub-angular blocky	Sticky, plastic
	120–150	5YR 4/8 5YR 4/6	Sandy Loam	Sub-angular blocky	Slightly sticky, slightly plastic
5	0–45	10YR 5/8 10YR 7/8	Sandy Loam	Sub-angular blocky	Slightly sticky, slightly plastic
	45–65	7.5YR 5/6 7.5YR 5/8	Loam	Sub-angular blocky	Sticky, plastic
	65–110	10YR 5/8 10YR 6/8	Silt loam	Sub-angular blocky	Sticky, plastic
	110–150	10YR 5/8 10YR 6/8	Sandy Loam	Sub-angular blocky	Slightly sticky, slightly plastic
6	0–35	10YR 5/6 10YR 6/8	Sandy Loam	Sub-angular blocky	Slightly sticky, slightly plastic

End of Table 2

Profile Number	Depth (cm)	Colour	Texture	Structure	Soil consistency
6	35–65	10YR 6/8 10YR 7/8	Loam	Sub-angular blocky	Sticky, plastic
	65–100	10YR 5/8 10YR 7/8	Loam	Sub-angular blocky	Sticky, plastic
	100–150	5YR 5/6 5YR 4/8	Sandy Loam	Sub-angular blocky	Slightly sticky, slightly plastic
7	0–55	5YR 4/8 5YR 6/8	Sandy Loam	Sub-angular blocky	Slightly sticky, slightly plastic
	55–110	5YR 4/8 5YR 6/8	Loam	Sub-angular blocky	Sticky, plastic
	110–150	7.5YR 5/8 7.5YR 6/4	Sandy Loam	Sub-angular blocky	Slightly sticky, slightly plastic
8	0–30	7.5YR 4/4 7.5YR 5/8	Silt loam	Sub-angular blocky	Sticky, plastic
	30–60	7.5YR 4/6 7.5YR 6/8	Silt loam	Sub-angular blocky	Sticky, plastic
	60–100	7.5YR 5/8 7.5YR 6/8	Silt loam	Sub-angular blocky	Sticky, plastic
	100–150	7.5YR 5/8 7.5YR 5/8	Silt loam	Sub-angular blocky	Sticky, plastic
9	0–45	7.5YR 5/8 7.5YR 6/8	Sandy Loam	Sub-angular blocky	Slightly sticky, slightly plastic
	45–105	10YR 5/8 10YR 6/8	Loam	Sub-angular blocky	Sticky, plastic
	105– 130	10YR 5/8 10YR 6/8	Sandy Loam	Sub-angular blocky	Slightly sticky, slightly plastic
	130–150	7.5YR 6/8 7.5YR 5/8	Loam	Sub-angular blocky	Sticky, plastic

Table 3. Physiographic units on the soil map

Physiographic unit	Landforms	Mapping unit	Soil profile	Profile elevation (masl)	Area (km²)	Area (%)
Flood plain	High terraces	T1	9	12	232.21	12.05
	Moderately high terraces	T2	4	8	431.99	22.41
	Low Terraces	T3	1	0	417.8	21.68
	High Decantation Basin	D1	3	10	39.53	2.05
	Low Decantation Basin	D2	5	6	236.29	12.26
	High overflow Basin	OB1	6	7	244.461	12.68
	Low Overflow basin	OB2	7	5	206.45	10.71
	Levees	L	8	9	103.82	5.39
	Swales	S	2	8	14.89	0.77
	Total	–	–	–	1927.441	100.00

*masl = metres above sea level.

1: River terraces: these soils represent the late Pleistocene deltaic plain and occur at the edge of decantation basins (these are basins in which sedimentation, particularly of silt and clay, occur during floods). The soils are formed on terraces at various heights above the valley floor.

2: Basins: these are artificially enclosed areas of a river or harbour, designed so that water levels are unaffected by tides.

3: River levees: these are a type of dam that runs along the banks of rivers or canals. Levees reinforce the banks and help prevent flooding. By confining the flow, levees can also increase water velocity.

4: Swales: these are low tracts of land, usually consisting of moist and marshy lands. The term can refer to both natural and artificial landscape features. Artificial swales are often designed to manage water runoff, filter pollutants and increase rainwater infiltration.

2. Results and discussion

2.1. Physiographic map of the study area

The satellite images show that the study area is a flood-plain and includes high terraces (12.04% of area), moderately high terraces (22.41%), low terraces (21.67%), high decantation basins (2.05%), low decantation basins (12.26%), high overflow basins (12.68%), low overflow basins (10.71%), river levees (5.38%) and swales (0.77%). The main physiographic soil units of the study area are reported in Table 3 and Figure 3.

2.2. Heavy metal contamination

XRF analyses of the soil samples identified the presence of SiO_2, Al_2O_3, P_2O_5, K_2O, CaO, MgO, Na_2O and Fe_2O_3 (major) and Cr, Cu, Zn, Ni, Br, Rb, Sr, Y, Zr, Nb, Sn, Te, Ba, Eu, Yb, Re, Ga, Ir, Mo, As and Pb (minor). Concentrations of the heavy metals Cr, Cu, Ni, V and Zn for each profile are reported in Table 4. For the metals Te, Mo, As and Pb, results are not reported, because their concentrations were below detection limits. Spatial interpolation maps (Figs 4–7) of heavy metal concentrations were prepared using the IDW function (inverse distance weighted) interpolation method in Arc GIS 10.1.

Vanadium

The concentrations and the interpolation map for V in the soil samples are given in Table 4 and Figure 4. V concentrations ranged from 194.0–744.4 mg/kg with a weighted mean ranging from 206.79–450.58 mg/kg (Table 5). The highest measured concentration of V was in the upper horizon of Profile 2, which represents a swales unit and is located 270 m north of Mansuriyyat Al-Farastaq village, ~6.5 km south-west from the centre of the town of Kfr Elzayat (population in 2015 was 448,965). The Igeo Index showed that all soil samples were in the uncontaminated/moderately contaminated categories, except for first horizon of Profile 2 in the swales mapping unit, which is classified as moderately/strongly contaminated (Table 6). The high deposition of V might be caused by the numerous local factories. V concentrations are higher than the permissible limits (90 mg/kg), recommended by Bowen (1979) in all soil profile horizons (Table 5). The spatial interpolation shows an increasing trend from north-east to south-west. The highest weighted mean (weighting concentration by representative area) (450.58 mg/kg) was found in 0.77% of the study area. From the interpolation map of V in the study area (Fig. 4) we can conclude that the order of concentration ascending in the mapping units is: low decantation basin (D2), high overflow basin (OB1), low overflow basin (OB2), moderately high terraces (T2), high decantation basin (D1), levees (L), high terraces (T1), low terraces (T3) and swales (S).

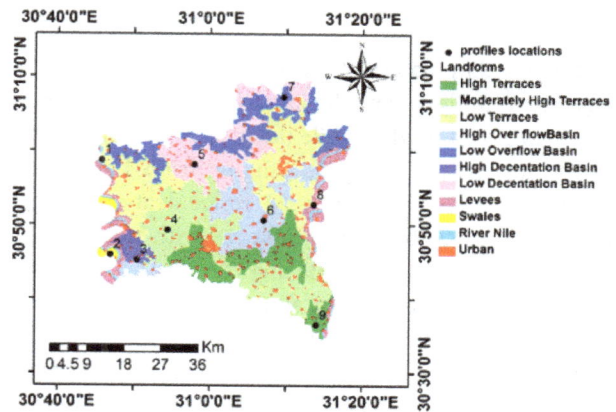

Fig. 3. The main landforms of the study area and profile locations

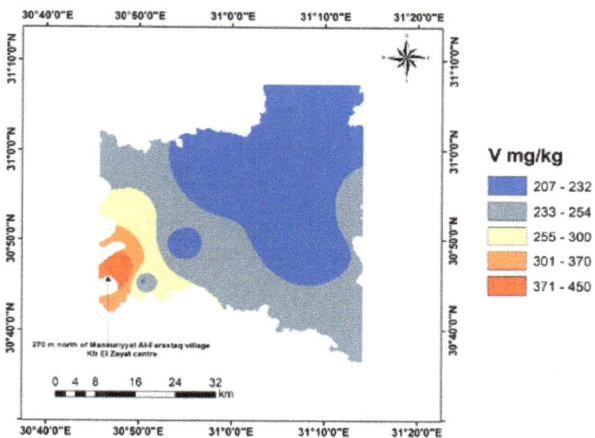

Fig. 4. Spatial interpolation of the weighted mean of vanadium

Chromium

Anthropogenic sources of Cr include alloys, chrome plating, pigments, chemical catalysts, dyes, tanning, wood impregnation and refractory bricks (Reimann, de Caritat 1998). The highest concentration of Cr (519 mg/kg) was in the top-soil of Profile 2, which may be due to the many local factories. The lowest (140.3 mg/kg) was in the top-soil of Profile 6, which represents a high over-flow basin (Table 4). The mean weight of Cr concentrations ranged from 152.84–314.73 mg/kg (Table 5). All concentrations exceeded the recommended values given by Bowen (1979) and, according to the Igeo Index, most soil samples are in the uncontaminated/moderately contaminated category (Table 6). Cr concentrations increased from east to west and south of the study area (Fig. 5). The highest Cr concentrations tended to be in the swales unit and the lowest in the high overflow basin unit.

Nickel

Baghdady, Sippola (1984) reported that the mean total Ni content in Egyptian alluvial soils is 64.4 mg/kg, ranging from 20–74 mg/kg, while the mean NH_4OAC–EDTA extractable Ni is 1.9 mg/kg, ranging from 1.0–2.2 mg/kg. However, in this study, Ni concentrations ranged from 60.60–267.30 mg/kg (Table 4), with mean weight ranging from 69.60–148.39 mg/kg (Table 5). Ni concentrations were higher in alluvial soils than previous studies and exceeded the permissible limit (50 mg/kg) (Bowen 1979). According to the Igeo Index, all samples are in the uncontaminated and moderately contaminated categories (Table 6). Table 4 reports Ni concentrations in soil profiles and Figure 6 shows the spatial trends, which increased from north to south and west. The highest concentrations were in swales, which occupy 14.89 km² of the study area. The interpolation of Ni shows high spatial variability, with the lowest values in the high decantation basin units.

Copper

Table 4 and Figure 7 report Cu concentrations and the spatial interpolation of weighted mean Cu concentrations, respectively. Cu contents in horizons ranged from

Table 4. XRF analysis of soils collected from the study area

Profile no	Mapping unit	Depth (cm)	Metal concentrations (mg/kg)				
			V	Cr	Ni	Cu	Zn
1	T3	0–50	227.1	179.7	68.30	78.60	94.30
		50–85	265.9	158.1	77.80	60.90	85.40
		85–120	250.4	167.3	73.50	81.40	93.30
		120–150	221.8	163.3	61.60	60.10	80.80
2	S	0–45	744.4	519.0	267.30	288.90	377.60
		45–85	244.9	161.4	63.50	50.90	75.50
		85–110	250.8	192.4	70.20	71.40	86.80
3	D1	0–75	221.0	170.7	63.80	94.50	90.20
		75–100	241.4	179.6	63.00	73.00	86.90
		100–150	238.9	166.6	81.60	76.60	85.00
4	T2	0–60	203.1	159.2	72.30	118.70	308.00
		60–100	258.3	166.3	70.90	0.00	88.70
		100–120	222.8	166.2	65.50	75.20	84.90
		120–150	206.1	179.8	70.70	72.30	91.60
5	D2	0–45	194.0	149.1	74.40	95.20	103.10
		45–65	197.5	159.6	69.20	93.30	98.40
		65–110	225.3	143.0	73.20	95.50	95.00
		110–150	205.0	150.5	72.20	94.60	98.80
6	OB1	0–35	210.1	140.3	60.60	131.80	124.50
		35–65	216.0	152.7	76.70	76.40	109.40
		65–100	220.9	164.3	84.50	97.50	103.20
		100–150	219.5	153.7	85.00	0.00	100.60
7	OB2	0–55	228.2	180.8	76.50	93.50	94.70
		55–110	230.3	170.7	73.30	89.10	84.70
		110–150	196.5	168.7	64.50	74.10	84.30
8	L	0–30	218.7	164.1	68.90	0.00	93.20
		30–60	241.3	151.3	73.90	0.00	97.10
		60–100	247.3	155.4	82.10	0.00	89.50
		100–150	231.9	154.9	75.40	85.80	93.80
9	T1	0–45	250.7	156.8	78.20	0.00	112.10
		45–105	232.8	158.1	80.90	74.30	103.30
		105– 130	223.7	168.8	76.40	74.50	89.50
		130–150	231.1	168.4	69.70	74.10	94.00

Table 5. Heavy metal concentrations in soil samples and concentration limits recommended by Bowen (1979)

Profile No	Mapping unit	Mean weight of metals concentrations (mg/kg)				
		V	Cr	Ni	Cu	Zn
1	T3	240.53	168.48	70.39	71.42	89.29
2	S	450.58	314.73	148.39	152.92	201.65
3	D1	230.36	170.81	69.60	84.94	87.91
4	T2	221.04	166.14	70.70	71.96	176.49
5	D2	206.79	149.06	72.76	94.87	98.89
6	OB1	216.93	152.84	77.53	68.79	108.54
7	OB2	220.51	173.87	72.12	86.71	88.26
8	L	235.24	156.15	75.58	28.60	93.19
9	T1	236.42	160.86	83.28	52.01	102.40
Concentration limits (mg/kg)		90	70	50	25	90

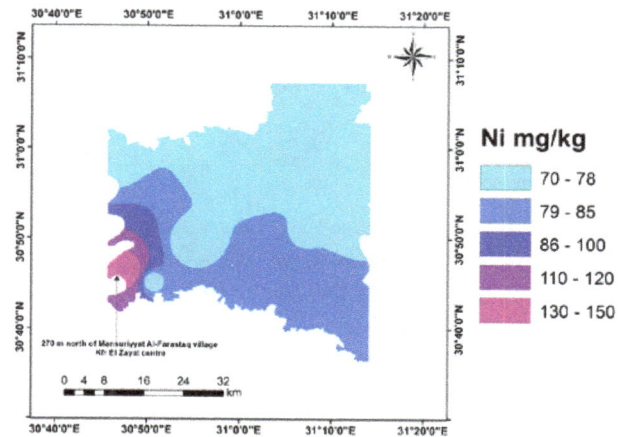

Fig. 6. Spatial interpolation of the weighted mean of nickel

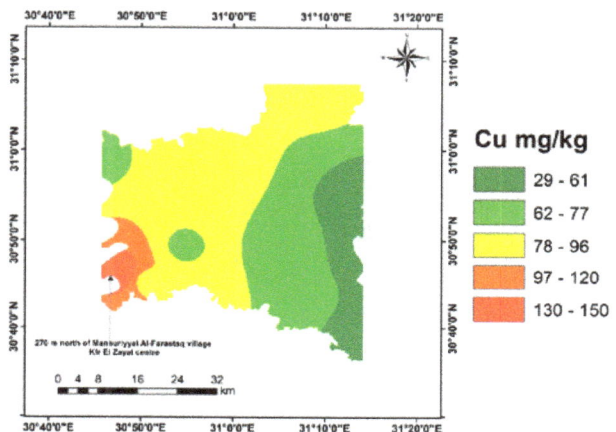

Fig. 7. Spatial interpolation of the weighted mean of copper

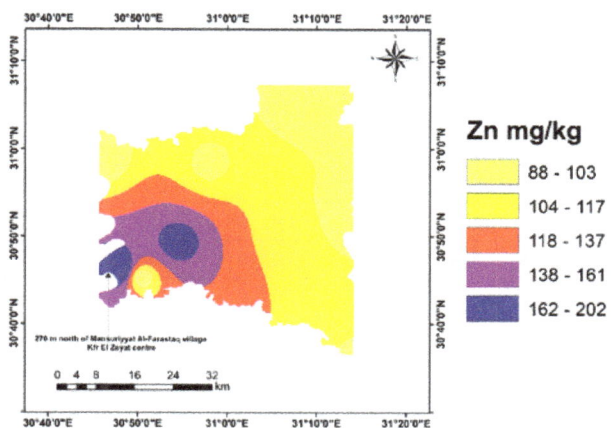

Fig. 8. Spatial interpolation of the weighted mean of zinc

0–288.9 mg/kg and mean weight ranged from 28.90–152.92 mg/kg (Table 5). All concentrations exceeded the permissible limit of 25 mg/kg (Bowen 1979), except for the second horizon of Profile 4 and the first, second and the third horizons of Profile 8 (Table 4). In addition, in Profile 9 the deepest horizon exceeded the limit, whereas concentrations in the upper layer were 0. This is probably due to percolation and illuviation of Cu associated with irrigation water. These profiles represent moderately high terraces, levees and high terraces, respectively. The Igeo Index showed that soil samples were in three contamination categories (uncontaminated/moderately contaminated, moderately contaminated and moderately/strongly contaminated) (Table 6). Two important sources of Cu in the Nile Delta are: (i) applications of Cu-based liquid fungicides, and (ii) use of $CuSO_4$ as an algicide in treating and controlling problematic macro-algal blooms in the Nile, especially during summer (Abdel-Moati, El-Sammak 1997). The lowest Cu concentrations were in the river levees units and the highest values were in the swale units.

Zinc

Zn concentrations slightly exceed the permissible concentration limit of 90 mg/kg (Bowen 1979) (Table 5). Exceptions include the upper layers of Profiles 2 and 4, where concentrations greatly exceeded permissible limits (377.6 and 308.0 mg/kg, respectively). The highest concentrations were in the upper horizon, but in Profile 8 the highest concentration was in the subsurface (Table 4). This could be caused by infiltration of irrigation water through the profile. The mean weight of Zn ranged between 88.26–201.65 mg/kg. According to the Igeo Index, the Zn concentrations of all soil samples fell into the uncontaminated

category, except for the first horizons of Profiles 2 and 4, which were moderately contaminated (Table 6). The spatial interpolation of Zn is presented in Figure 8. The highest concentration was in the south-west of the study area, which is located 270 m north of Mansuriyyat Al-Farastaq village. This could be due to atmospheric deposition, originating from local industrial plants. The highest Zn concentrations were in the swales top-soil and moderately high terrace units.

2.3. Major soil oxides

Soil samples were analysed for heavy metals and major oxides (Table 7). Results for SiO_2, Al_2O_3, P_2O_5, K_2O, CaO, MgO, Na_2O and Fe_2O_3 were compared with average concentrations of major oxides in soil (Bohn *et al.* 2001)

(Table 5). SiO_2 concentrations varied from 51.40–56.55% and were all less than the representative average value of 72.64% (Bohn *et al.* 2001). Al_2O_3 concentrations varied from 16.28–24.42% and the mean concentration of profiles ranged from 18.81–22.9%. Al_2O_3 concentrations in all samples exceeded the representative mean value of 13.22%. K_2O concentrations ranged from 1.05–1.62% with weighted mean values ranging from 1.14–1.35%, near the 1.2% representative mean. CaO concentrations ranged between 2.58–6.69% and the mean weighted value

Table 6. Igeo Index concentrations and associated contamination levels

Profile no	Depth, cm	V	C level	Cr	C Level	Ni	C Level	Cu	C level	Zn	C level
1	0–50	0.75	UN/M	0.77	UN/M	ND	UN	1.06	M	ND	UN
	50–85	0.97	UN/M	0.59	UN/M	0.05	UN/M	0.69	UN/M	ND	UN
	85–120	0.89	UN/M	0.67	UN/M	ND	UN	1.11	M	ND	UN
	120–150	0.71	UN/M	0.63	UN/M	ND	UN	0.68	UN/M	ND	UN
2	0–45	2.46	M/S	2.30	M/S	1.83	M	2.94	M/S	1.48	M
	45–85	0.85	UN/M	0.62	UN/M	ND	UN	0.44	UN/M	ND	UN
	85–110	0.89	UN/M	0.87	UN/M	ND	UN	0.92	UN/M	ND	UN
3	0–75	0.71	UN/M	0.70	UN/M	ND	UN	1.33	M	ND	UN
	75–100	0.83	UN/M	0.77	UN/M	ND	UN	0.96	UN/M	ND	UN
	100–150	0.82	UN/M	0.66	UN/M	0.12	UN/M	1.03	M	ND	UN
4	0–60	0.58	UN/M	0.60	UN/M	ND	UN	1.66	M	1.18	M
	60–100	0.93	UN/M	0.66	UN/M	ND	UN	ND	–	ND	UN
	100–120	0.72	UN/M	0.66	UN/M	ND	UN	1.00	M	ND	UN
	120–150	0.61	UN/M	0.77	UN/M	ND	UN	0.94	UN/M	ND	UN
5	0–45	0.52	UN/M	0.50	UN/M	ND	UN/M	1.34	M	ND	UN
	45–65	0.54	UN/M	0.60	UN/M	ND	UN	1.31	M	ND	UN
	65–110	0.73	UN/M	0.44	UN/M	ND	UN	1.34	M	ND	UN
	110–150	0.60	UN/M	0.51	UN/M	ND	UN	1.33	M	ND	UN
6	0–35	0.63	UN/M	0.41	UN/M	ND	UN	1.81	M	ND	UN
	35–65	0.67	UN/M	0.54	UN/M	0.03	UN/M	1.02	M	ND	UN
	65–100	0.71	UN/M	0.64	UN/M	0.17	UN/M	1.37	M	ND	UN
	100–150	0.70	UN/M	0.54	UN/M	0.18	UN/M	ND	–	ND	UN
7	0–55	0.75	UN/M	0.78	UN/M	0.02	UN/M	1.31	M	ND	UN
	55–110	0.77	UN/M	0.70	UN/M	ND	UN	1.24	M	ND	UN
	110–150	0.54	UN/M	0.68	UN/M	ND	UN	0.98	UN/M	ND	UN
8	0–30	0.69	UN/M	0.64	UN/M	ND	UN	ND	–	ND	UN
	30–60	0.83	UN/M	0.52	UN/M	ND	UN	ND	–	ND	UN
	60–100	0.87	UN/M	0.56	UN/M	0.13	UN/M	ND	–	ND	UN
	100–150	0.78	UN/M	0.56	UN/M	0.007	UN/M	1.19	M	ND	UN
9	0–45	0.89	UN/M	0.57	UN/M	0.06	UN/M	ND	–	ND	UN
	45–105	0.78	UN/M	0.59	UN/M	0.1	UN/M	0.98	UN/M	ND	UN
	105– 130	0.72	UN/M	0.68	UN/M	0.02	UN/M	0.99	UN/M	ND	UN
	130–150	0.77	UN/M	0.68	UN/M	ND	UN	0.98	UN/M	ND	UN

C Level = Contamination level; UN = Uncontaminated; UN/M = Uncontaminated/moderately contaminated; M = Moderately contaminated; M/S = Moderately/strongly contaminated; ND = Not detected

ranged between 3.70–5.95%. CaO and Na$_2$O concentrations exceeded the representative means of 1.44% and 0.99%, respectively. Fe$_2$O$_3$ concentration ranged between 9.73–12.23%, whereas the representative mean is 5.77%. P$_2$O$_5$ concentrations ranged from 0.15–0.49%. The weighted mean concentrations of P$_2$O$_5$ samples exceed the 0.18% representative mean (Table 8). Thus, these deltaic soils are predominantly siliceous, with slight enrichment of the alumina component.

2.4. Relationships between trace and major elements

V, Cr, Ni, Cu and Zn concentrations are significantly correlated (Table 9). There are no significant correlations between major elements and heavy metal concentrations, except for V, Ni and Zn. V and Ni have significant positive medium and strong correlations with Fe, respectively. This may indicate the sorption of these elements by Fe hydroxides. Ni has a strong positive association with Al, whereas a strong significant association was found between Zn and

Table 7. Summary of major oxide concentrations in soil samples of the study area

Profile No.	Mapping unit	Depth, cm	Major oxide concentrations (%)							
			SiO$_2$	Al$_2$O$_3$	P$_2$O$_5$	K$_2$O	CaO	MgO	Na$_2$O	Fe$_2$O$_3$
1	T3	0–50	52.67	20.10	0.31	1.29	3.66	4.18	1.44	10.93
		50–85	56.46	23.92	0.25	1.29	3.18	4.53	1.44	12.03
		85–120	55.12	21.48	0.23	1.16	3.88	4.43	1.77	11.54
		120–150	54.42	19.28	0.21	1.19	4.17	3.98	2.02	10.38
2	S	0–45	ND	ND	ND	ND	ND	ND	ND	ND
		45–85	52.26	18.57	0.24	1.21	4.98	4.66	1.55	11.16
		85–110	55.35	21.50	0.20	1.10	4.50	4.62	1.89	11.28
3	D1	0–75	54.33	20.74	0.39	1.45	4.53	4.83	1.48	10.51
		75–100	52.83	18.36	0.20	1.10	5.03	4.30	1.72	11.20
		100–150	54.03	20.96	0.18	1.08	4.63	4.59	1.66	11.03
4	T2	0–60	55.87	19.11	0.43	1.60	3.70	5.11	1.49	10.61
		60–100	56.55	21.37	0.20	1.31	4.26	5.26	1.47	10.77
		100–120	53.60	18.24	ND	1.13	6.69	4.96	1.70	10.78
		120–150	52.15	16.48	0.20	1.07	6.34	4.55	1.84	10.28
5	D2	0–45	52.70	18.80	0.46	1.30	5.36	5.16	1.42	10.54
		45–65	54.20	19.19	0.20	1.24	4.38	5.43	1.43	10.96
		65–110	54.88	19.60	0.17	1.20	3.56	5.57	1.50	11.03
		110–150	53.93	19.20	0.28	1.25	4.43	5.39	1.45	10.84
6	OB1	0–35	54.64	19.19	0.49	1.62	5.22	4.99	1.48	10.82
		35–65	55.46	22.14	0.29	1.35	4.03	4.99	1.39	11.44
		65–100	55.71	23.59	0.20	1.25	3.47	5.00	1.36	11.74
		100–150	56.04	23.64	0.19	1.18	3.40	4.99	1.33	11.75
7	OB2	0–55	56.43	19.43	0.24	1.39	4.74	5.88	1.60	10.72
		55–110	55.29	19.4	0.15	1.22	3.06	4.99	2.44	10.59
		110–150	53.4	17.14	0.17	1.28	4.31	4.62	2.80	9.72
8	L	0–30	52.38	19.40	0.32	1.22	4.23	4.35	1.59	10.38
		30–60	54.42	19.47	0.28	1.21	4.35	4.90	1.62	11.11
		60–100	55.01	23.20	0.21	1.09	3.45	4.80	1.50	11.67
		100–150	56.06	24.42	0.17	1.09	2.58	4.79	1.40	12.23
9	T1	0–45	51.40	17.93	0.31	1.47	4.67	4.93	1.52	11.06
		45–105	52.32	19.01	0.20	1.14	4.32	4.78	1.64	11.43
		105–130	52.36	21.47	0.16	1.05	3.12	4.52	1.70	11.73
		130–150	52.09	21.08	0.15	1.05	3.17	4.44	1.71	11.68

* ND = Not detected.

K_2O and a medium correlation with P_2O_5. Such associations indicate strong affinity for these elements for Fe, Al and K oxides. Al_2O_3 and Fe_2O_3 display a strong positive significant correlation. In addition, P_2O_5 is significantly correlated with K_2O. There are strong significant negative correlations between Al_2O_3 and CaO, between CaO and Fe_2O_3 and a medium significant negative correlation between Na_2O and Fe_2O_3.

Conclusions

This study shows that concentrations of V, Ni and Cr exceeded recommended limits in the soils of the Middle Nile Delta. Cu concentrations were very variable. Zn concentrations slightly exceed the recommended limit. V, Cr, Ni, Cu and Zn concentrations are significantly correlated. There are no significant correlations between major elements and heavy metal concentrations, except for V, Ni

and Zn. V and Ni have significant positive medium and strong correlations with Fe, respectively. The Igeo Index of V showed that all soil samples were in the uncontaminated/moderately contaminated categories, except for the first horizon of Profile 2, which is classified as moderately/strongly contaminated, while the Igeo Index for Cr showed most soil samples are in the uncontaminated/moderately contaminated category. The Igeo Index for Ni reveals that all samples are in the uncontaminated and moderately contaminated categories. For Cu, soil samples were in three contamination categories (uncontaminated/moderately contaminated, moderately contaminated and moderately/strongly contaminated). All Zn concentrations were in the uncontaminated category, except for the first horizons of Profiles 2 and 4, which were moderately contaminated. The highest heavy metal concentrations dominate the south-west of El-Gharbia Governorate and is mainly attributed to human activities, especially pollution

Table 8. Mean weight of major oxide concentrations in soil samples and representative average limits (Bohn *et al.* 2001)

Profile No.	Mapping unit	Mean weight of major element concentrations (%)							
		SiO_2	Al_2O_3	P_2O_5	K_2O	CaO	MgO	Na_2O	Fe_2O_3
1	T3	54.47	21.14	0.25	1.23	3. 70	4.27	1.63	11.21
2	S	ND	ND	ND	ND	ND	ND	ND	ND
3	D1	53.97	20.40	0.28	1.27	4.64	4.66	1.58	10.79
4	T2	55. 06	19.07	ND	1.35	4.77	5.01	1.57	10.60
5	D2	53.88	19.19	0.28	1.24	4.44	5.38	1.45	10.82
6	OB1	55.51	22.29	0.28	1.33	5.95	4.99	1.38	11.47
7	OB2	55.20	18.81	0.18	1.29	4.01	5.21	2.23	10.40
8	L	54.71	22.09	0.23	1.14	3.49	4.72	1.50	11.48
9	T1	52.01	19.36	0.22	1.21	4.07	4.73	1.62	11.40
Conc. limits (%)		70.29	>13.22	0.18	1.20	1.44	0.99	0.99	5.77

*ND = Not detected

Table 9. Correlation coefficients between trace and major elements in soils of the study area

	V	Cr	Ni	Cu	Zn	SiO_2	Al_2O_3	P_2O_5	K_2O	CaO	MgO	Na_2O
Cr	0.97***											
Ni	0.97***	0.96***										
Cu	0.64***	0.71***	0.68***									
Zn	0.73***	0.75***	0.77***	0.68***								
SiO_2	0.19	0.11	0.28	0.18	0.19							
Al_2O_3	0.4*	−0.13	0.56***	−0.17	−0.09	0.61***						
P_2O_5	−0.33	−0.35	−0.26	0.29	0.49**	−0.06	−0.22					
K_2O	−0.25	−0.32	−0.19	0.32	0.56***	0.23	−0.18	0.78***				
CaO	−0.23	−0.15	−0.42**	0.14	−0.09	−0.38	−0.69***	0.37	0.1			
MgO	−0.25	0.35	0.27	0.24	0.21	0.41*	−0.03	0.14	0.39*	0.08		
Na_2O	−0.1	0.40*	−0.33	−0.05	−0.20	−0.16	−0.43	−0.37	−0.21	0.06	−0.30	
Fe_2O_3	0.51**	−0.17	0.56***	−0.18	−0.11	0.26	0.79***	−0.31	−0.35	−0.55***	−0.07	−0.51***

Note: *P < 0.05, **P < 0.01, ***P < 0.001, n = 33 soil samples from the Kfr Elzayat urban area (population in 1995 was 448,965). In terms of the distribution of heavy metals in the different physiographic units, the swales unit contained the highest values, as this is in Kfr El Zayat, which has many factories. We recommend that heavy metal contamination be studied within entire soil profiles and not just topsoils, because these metals affect soil and crop quality and can cause ground-water pollution. Protection against this hazard is vital for sustainable land management. Precise measures and efficient methods to improve soil and water quality must be conducted, in order to prevent soil and water pollution and to avoid the need for costly remediation in the future.

References

Abd Elgawad, M.; Hamdi, A. A.; Mahmoud, M. S.; Samir, I. G. 2007. Status of some heavy metals in Fayoum district soils, in *3rd Conference for Sustainable Agricultural Development*, 12–14 November 2007, Fayoum, Egypt, 507–526.

Abdel-Moati, M. A. R.; El-Sammak, A. A. 1997. Man-made impact on the geochemistry of the Nile Delta Lakes. A study of metal concentrations in sediments, Water, *Air and Soil Pollution* 97(3): 413–429. http://dx.doi.org/10.1007/BF02407476

Abdel-Sabour, M. F.; Abdou, F. M.; Elwan, I. M.; Al-Salama, Y. J. 2002. Effect of soil contamination due to wastewater irrigation on Cr fractions in some soils of Egypt, in *Proceedings of 6th Radiation Physics Conference*, Assuit, Egypt, 27–30.

Abu Khatita, A. M. 2011. *Assessment of soil and sediment contamination in the Middle Nile Delta area (Egypt)- Geo-Environmental study using combined sedimentological, geophysical and geochemical methods*. Alexander-Universität Erlangen-Nürnberg zur Erlangung, Germany.

Abu, Z. M. 2011. *Egyptian policies for using low quality water for irrigation*. Water Resource Center, Egypt, Cairo.

Ali, A. A.; Moghanm, F. S. 2013. Variation of soil properties over the landforms around Idku Lake, Egypt, *The Egyptian Journal of Remote Sensing and Space Sciences* 16(1): 91–101. http://dx.doi.org/10.1016/j.ejrs.2013.04.001

Atwia, M. G.; Abou Heleika, M. M.; El-Shishtawy, A. M.; Sharp, J. M. 2006. Hydrostratigraphy of the Central Nile Delta, Egypt, using geoelectric measurements, *EGS-AGU-EUG Joint Assembly*, Nice, France 5: 6–11.

Baghdady, N. H.; Sippola, J. 1984. Extractability of polluting elements Cd, Cr, Ni and Pb of soil with three methods, *Acta Agriculturae Scandinavica* 34(3): 345–348. http://dx.doi.org/10.1080/00015128409435402

Baker, D. E.; Senft, J. P. 1995. Copper, in B. J. Alloway (Ed.). *Heavy metals in soils*. London: Blackie. http://dx.doi.org/10.1007/978-94-011-1344-1_8

Ball, D. F. 1964. Loss-on-ignition as estimate of organic matter and organic carbon in non-calcareous soils, *Journal of Soil Science* 15(1): 84–92. http://dx.doi.org/10.1111/j.1365-2389.1964.tb00247.x

Bohn, L. H.; McNeal, L. B.; O'Connor, A. G. 2001. *Soil chemistry*. 2nd ed. New York: John Wiley.

Bowen, H. J. M. 1979. *Environmental chemistry of the elements*. London: Academic Press.

Bradl, H.; Xenidis, A. 2005. Chemical remediation techniques, in H. B. Bradl (Ed.). *Heavy metals in the environment: origin, interaction and remediation*. Amsterdam: Elsevier. http://dx.doi.org/10.1016/S1573-4285(05)80022-5

Brough, P. A. 1986. *Principles of geographical information systems for land resources assessment*. Oxford: Oxford University Press.

Climatologically normal for Egypt. 2011. The normal for El-Gharbia Governorate Station (1960–2011), Ministry of Civil Aviation: Meteorological.

Dawoud, M. A. 2004. Design of national groundwater quality monitoring network in Egypt, *Environmental Monitoring and Assessment* 96: 99–118. http://dx.doi.org/10.1023/B:EMAS.0000031718.98107.eb

Dobos, E.; Micheli, E.; Baumgardner, M. F.; Biehl, L.; Helt, T. 2000. Use of combined digital elevation models and satellite radiometric data for regional soil mapping, *Geoderma* 97: 367–391. http://dx.doi.org/10.1016/S0016-7061(00)00046-X

Dobos, E.; Norman, B.; Bruce, W.; Luca, M.; Chris, J.; Erika, M. 2002. The use of DEM and satellite images for regional scale soil database, in *17th World Congress of Soil Science (WCSS)*, 14–21 August 2002, Bangkok, Thailand.

Dumont, H. J. 2009. A description of the Nile Basin, and a synopsis of its history, ecology, biogeography, hydrology, and natural resources, in H. J. Dumont (Ed.). *The Nile: origin, environments, limnology and human use*. New York: Springer. http://dx.doi.org/10.1007/978-1-4020-9726-3_1

Fanos, A. M. 2002. *Report on the modified Mega Delta Workshop*.

FAO. 2006. *Guidelines for soil description*. 4th ed. FAO, Rome. ISBN 92-5-105521-1.

Govil, P. K.; Reddy, G. L. N.; Krishne, A. K. 2001. Contamination of soil due to heavy metals in Patanchera Industrial Development Area: Andra Pradesh, India, *Environmental Geology* 41: 461–469. http://dx.doi.org/10.1007/s002540100415

Gowd, S. S.; Reddy, M. R.; Govil, P. K. 2010. Assessment of heavy metal contamination in soils at Jajmau (Kanpur) and Unnao industrial areas of the Ganga Plain, Uttar Pradesh, India, *Journal of Hazardous Materials* 174: 113–121. http://dx.doi.org/10.1016/j.jhazmat.2009.09.024

He, J. Y.; Jia, X. 2004. Arc GIS geostatistical analyst application in assessment of MTBE contamination, in *ESRI User Conference*, Fremont, USA.

ITT. 2009. ITT *Corporation ENVI 4.7 software*. New York: White Plains.

Jankaite, A.; Vasarevičius, S. 2005. Remediation technologies for soils contaminated with heavy metals, *Journal of Environmental Engineering and Landscape Management* 13(2): 109–113.

Lagerkvist, J-S. B.; Oskarsson, A. 2007. Vanadium, in G. Nordberg, B. Fowler, M. Nordberg and L. Friberg, L. (Eds.). *Handbook on the toxicology of metals*. 3rd ed. San Diego: Academic Press.

Lillesand, T. M.; Kiefer, R. W. 2003. *Remote sensing and image interpretation*. 2nd ed. New York: John Wiley.

Lillesand, T. M.; Kiefer, R. W. 2007. *Remote sensing and image interpretation*. 5th ed. New York: John Wiley.

Ling, W.; Shen, Q.; Gao, Y.; Gu, X.; Yang, Z. 2007. Use of bentonite to control the release of copper from contaminated soils, *Australian Journal of Soil Research* 45(8): 618–623. http://dx.doi.org/10.1071/SR07079

Lu, S. G.; Bai, S. Q. 2010. Contamination and potential mobility assessment of heavy metals in urban soils of Hangzhou, China, *Environmental Earth Sciences* 60(7): 1481–1490. http://dx.doi.org/10.1007/s12665-009-0283-2

McLaughlin, M. J.; Zarcinas, B. A.; Stevens, D. P.; Cook, N. 2000. Soil testing for heavy metals. *Communications in Soil Science and Plant Analysis* 31(11–14): 1661–1700. http://dx.doi.org/10.1080/00103620009370531

Ohnesorge, F. K.; Wilhelm, M. 1991. Zinc, in E. Merian (Ed.). *Metals and their compounds in the environment, occurrence, analysis and biological relevance.* Verlagsgesellschaft, Weinheim, Basel: VCH.

Rahman, S.; Khanam, D.; Adyel T.; Islam, M. Sh.; Mohammad Ahsan, A.; Akbor, M. A. 2012. Assessment of heavy metal contamination of agricultural soil around Dhaka Export Processing Zone (DEPZ), Bangladesh: Implications of seasonal variations and indices, *Applied Sciences* 2(3): 584–601. http://dx.doi.org/10.3390/app2030584

Raven, P. H.; Berg, L. R.; Johnson, G. B. 1998. *Environment.* 2nd ed. New York: Saunders College Publishing.

Reimann, C.; de Caritat, P. 1998. *Chemical elements in the environment.* Berlin: Springer-Verlag. http://dx.doi.org/10.1007/978-3-642-72016-1

Romic, M.; Romic, D. 2003. Heavy metal distribution in agricultural top-soils in urban areas, *Environmental Geology* 43: 795–805.

Said, R. 1993. *The River Nile: geology, hydrology and utilization.* Oxford: Pergamon Press.

Shepard, D. 1968. *A two-dimensional interpolation function for irregularly-spaced data.* New York: ACM. http://dx.doi.org/10.1145/800186.810616

Sundermann, F. W. (Jr.); Oskarsson, A. 1991. Nickel, in E. Merian (Ed.). *Metals and their compounds in the environment; occurrence, analysis and biological relevance.* Weinteim, Basel: VCH.

Tong, S. T. Y. 1998. Indoor and outdoor household dust contamination in Cincinnati, Ohio, USA, *Environmental Geochemistry and Health* 20(3): 123–133. http://dx.doi.org/10.1023/A:1006561832381

USDA. 2010. *Keys to soil taxonomy.* 11th ed. United States Department of Agriculture, Natural Resources Conservation Service (NRCS).

WHO. 1996. *Trace elements in human nutrition and health.* World Health Organization: Geneva.

Wong, M. H.; Wong, J. W. C. 1986. Effects of fly ash on soil microbial activity, *Environmental Pollution* A(40): 127–144. http://dx.doi.org/10.1016/0143-1471(86)90080-2

Woo, K. S.; Jo, J. H.; Basu, P. K.; Ahn, J. S. 2009. Stress intensity factor by p-adaptive refinement based on ordinary Kriging interpolation, *Finite Elements in Analysis and Design* 45(3): 227–234. http://dx.doi.org/10.1016/j.finel.2008.10.002

Zeydan, B. A. 2005. The Nile Delta in a global vision, in *9th International Water Technology Conference, IWTC9 2005,* Sharm El-Sheikh, Egypt, 31–40.

Zheng, C. 2006. Using multivariate analyses and GIS to identify pollutants and their spatial patterns in urban soils in Galway, Ireland, *Environmental Pollution* 142(3): 501–511. http://dx.doi.org/10.1016/j.envpol.2005.10.028

Mohamed S. SHOKR. Assistant Lecturer, received the degrees of B.Sc. and M.Sc from Tanta University, Egypt. Currently, he is Assistant Lecturer in the Soils and Water Department, Tanta University and has a scholarship from the Egyptian Government to complete part of his Ph.D. studies in the University of Wolverhampton, UK. His research activities are concerned with pedology, soil sustainability, remote sensing and GIS.

Ahmed A. EL BAROUDY. Received the degrees of B.Sc., M.Sc. and Ph.D. from The University of Tanta, Egypt. Currently, he is Assoc. Prof. of Soil in the Soil and Water Department of the University of Tanta. His research activities are mainly concerned with pedology, soil degradation, desertification, land evaluation, remote sensing and GIS modelling. He has published widely in Soil Science, having authored 21 refereed and conference papers.

Michael A. FULLEN. Professor Fullen received the degrees of B.Sc. and M.Sc. from The University of Hull (UK), a Ph.D. from the UK Council for National Academic Awards (CNAA) and a D.Sc. from The University of Wolverhampton. Currently, he is a Professor of Soil Technology at the University of Wolverhampton, UK. His research activities are mainly concerned with soil erosion, soil conservation, desertification and desert reclamation and his fieldwork is mainly based in Europe and Asia. He has published widely in Soil Science (as of March 2016, he has authored one book, 221 refereed papers, 217 conference papers and 26 consultancy reports). He is a referee for 48 journals and a member of the Editorial Board of 24 journals.

Talaat R. EL-BESHBESHY. Received the degrees of B.Sc., M.Sc. and Ph.D. from The University of Minia, Egypt. Currently, he is Professor of Soil Science and Plant Nutrition in the Soil and Water Department of the University of Tanta. His research activities are mainly concerned with plant nutrition, hydroponics, soil fertility and soil chemistry. He has published widely in Soil Science, having authored 25 refereed and conference papers.

Ramadan R. ALI. Professor of Soil Science at the National Research Centre, Cairo, since 1999. He was appointed as a Research Assistant in the Soils and Water Use Department. In 1993, he was awarded the degree of M.Sc. and was then appointed as Assistant Researcher. He obtained a PhD in 2013 and was then appointed as a Researcher. During 2003–2015 he published 56 scientific papers. He became Assistant Professor in 2008 and then Professor in 2014. He also participated in 24 research projects, published three book chapters and supervised several M.Sc. and Ph.D. degrees.

Abd ELHALIM. Dr, received the degrees of B.Sc. and M.Sc. from The University of Tanta and a PhD from The University of Tanta through a scientific association with the University of Rostock, Germany. Currently, he is Assoc. Prof. of Soil Physics and Water Relations at the University of Tanta. His research activities are mainly concerned with soil physics, soil erosion, soil conservation, irrigation water management, evapotranspiration models and desert reclamation. He has published widely in Soil Science (he has authored 12 refereed papers and three conference papers).

Antonio J. T. GUERRA. Professor Guerra received the degrees of B.Sc. and M.Sc. from the Federal University of Rio de Janeiro (Brazil) and a PhD from King's College London (UK). He undertook his first Post-doctoral Fellowship at Oxford University (UK) and his second at the University of Wolverhampton (UK). Currently, he is Professor of Physical Geography at the Department of Geography, Federal University of Rio de Janeiro, where he co-ordinates LAGESOLOS (the Laboratory of Environmental Geomorphology and Soil Degradation). His research activities are mainly concerned with soil erosion, land degradation, soil rehabilitation and his fieldwork is mainly based in Rio de Janeiro and São Paulo States (Brazil). He has published nearly 100 papers in refereed papers in both Brazilian and international journals. He has co-edited 17 books on geomorphology, soil erosion, environmental geomorphology, tourism, and soil rehabilitation in Brazil.

Maria C. O. JORGE. Received the degree of B.Sc. from the Federal University of Paraná (Brazil) and M.Sc. from the State University of Rio Claro (Brazil). Currently she is completing her PhD at the Federal University of Rio de Janeiro and she has just finished the sandwich component of her PhD at the University of Wolverhampton (2015). So far she has published 14 refereed papers in both Brazilian and international journals. She has also co-edited two books on land degradation and soil rehabilitation in Brazil. Her research activities are concerned with geotourism, geoconservation and geodiversity and her field work is conducted in Ubatuba Municipality, São Paulo State, Brazil.

16

ACCUMULATION OF HEAVY METALS IN DIFFERENT BODY TISSUES OF GIBEL CARP *CARASSIUS GIBELIO* SEPARATELY EXPOSED TO A MODEL MIXTURE (Cu, Zn, Ni, Cr, Pb, Cd) AND NICKEL

Gintaras SVECEVIČIUS[a], Raimondas Leopoldas IDZELIS[b], Eglė MOCKUTĖ[b]

[a] *Nature Research Centre, Akademijos g. 2, 08412 Vilnius, Lithuania*
[b] *Vilnius Gediminas Technical University, Saulėtekio al. 11, 10223 Vilnius, Lithuania*

Abstract. Heavy metals (HMs) are common persistent pollutants of aquatic ecosystems, which have a property to migrate and accumulate in water organisms. Little information has been compiled on HM accumulation and the interactions between them in fish exposed to their mixtures at environmentally-relevant concentrations. The aim of the present study was to determine accumulation patterns of Cu, Zn, Ni, Cr, Pb and Cd in the muscle, gills and liver of Gibel carp after 14-day exposure to HM model mixture (HMMM) and to Ni separately, at concentrations corresponding to Lithuanian inland water standards (Cu-0.01; Zn-0.1; Ni-0.01; Cr-0.01; Pb-0.005; Cd-0.005 mg/l, respectively). Laboratory tests were conducted on adult Gibel carp under semi-static conditions. The amounts of HMs in the water and body tissues were determined using atomic absorption spectrophotometry (AAS). Heavy metal accumulation order in body tissues of Gibel carp was as follows: muscle > gills > liver. The highest amounts found were of Zn (15.2 mg kg), while the lowest of Cd (0.012 mg/kg). In the muscle and gills, HMs were accumulated in the following order: Zn > Cu > Ni > Cr > Pb > Cd, while in the liver: Zn > Cu > Pb > Ni > Cr > Cd. Meanwhile, the Ni concentration in HMMM-exposed fish liver and gills were significantly higher than in Ni-exposed fish. Data obtained showed that HM accumulation in Gibel carp body tissues was metal and tissue specific, i.e. different tissues showed a different capacity for accumulating HMs. Metals in mixture promoted Ni accumulation in test fish gills and liver under different experimental conditions due to the synergistic effects among them. Although the HM content in Gibel carp different body tissues increased in all cases (average of 49 to 224%) they did not exceed Maximum-Permissible-Amounts (MPA) indicated in the Lithuanian Hygiene Standard.

Keywords: water pollution, fish, Gibel carp, heavy metals, accumulation, Bioconcentration factor (BCF), Maximum-Permissible-Concentration (MPC), Maximum-Permissible-Amounts (MPA).

Introduction

Heavy metals (HMs) are a serious threat to the aquatic environment because of their toxicity, accumulation and magnification in organisms, causing severe damage to the organ-systems leading to innumerable health hazards (Roy 2010).

Copper, zinc, nickel, chromium, cadmium and lead are widely used in various anthropogenic activities. These are common persistent water pollutants, and are assigned to priority hazardous substances (Directive 2000/60/EC of the European Parliament and the Council 2000; US EPA 2009). All these metals migrate in ecosystems, accumulate in aquatic animals, presumably through binding to metallothioneins or other metal-binding proteins (Valavanidis *et al.* 2006).

Fish are unique among the vertebrates, a consequence of having two routes of metal acquisition, from the water and from the diet (the gill and intestinal epithelium uptake processes) (Bury *et al.* 2003). However, the result of the bioaccumulation process is strongly associated with the metal depuration rate and metabolic activity in different fishes.

Numerous field and laboratory studies showed that HM accumulation in fish depends on a complexity of a lot biotic and abiotic factors such as: the species, its trophic level and feeding habits, age and size, interspecific differences in sensitivity to various metals, accumulative-metabolic activity, pollutant concentration in water, sediment and food-objects, physico-chemical characteristics of the water, chemical speciation of the metal and its bioavailability (Clearwater *et al.* 2002; Canli, Atli 2003; Green, Knutzen 2003; Papagiannis *et al.* 2004; Licata *et al.* 2005; Nemova 2005; Jezierska, Witeska 2006; Luczynska, Tonska 2006; Karthikeyan *et al.* 2007; Oymak *et al.* 2009; Polak-Juszczak 2009).

It should be noted that fish are not only an important source of animal protein in human nutrition, which quality on HM content should be continuously monitored in order to evaluate the risk to human health. Because of their property to accumulate persistent pollutants fish are an excellent bioindicators of ecotoxicological state of the aquatic environment (Chovanec *et al.* 2003; Nemova 2005; Has-Schön *et al.* 2006; Schmitt *et al.* 2007; Klavins *et al.* 2009).

Most studies of the effects of metals on fish are addressed to a particular metal. Meanwhile, in the natural environment, fish are exposed to different HM mixtures, which are usually more toxic than individual metal as their action is additive or more-than-additive (synergistic). It seems that interaction between different metals are related to their competitive uptake from the environment and to different distribution in fish tissues, which results from that certain metals affect the accumulation of other metals in fish (Jezierska, Sarnowski 2002).

Although the influence of the presence of particular metals in the medium on accumulation of others in fish has been proved experimentally (Van Hoof, Nauwelaers 1984; Ribeyre *et al.* 1995; Ghosh, Adhikari 2006; Ghosh *et al.* 2007; Palaniappan, Karthikeyan 2009) such investigations still remain rather scarce, however.

Furthermore in the accumulation tests, fish are usually exposed to relatively high non-realistic nearly lethal or even lethal HM concentrations, which are much higher than those environmentally–relevant in natural waters. The results from such studies could be hardly extrapolated to natural conditions (Van Hoof, Nauwelaers 1984; Karthikeyan *et al.* 2007; Palaniappan, Karthikeyan 2009).

Actually, in most natural waters, the concentrations of HMs are quite low and strictly depended on water chemistry and metal chemical speciation (the equilibrium between an ionic and metal-ligand complexation level) in the water and deposed in the bottom sediments (Van Briesen *et al.* 2010). Little information is compiled on accumulation in fish exposed to environmentally realistic, metal concentrations simulating those, which could occur in an ordinary environment. Therefore, the modelling of the situation which could actually take place in natural waters is required.

Recently, the implementation of the European Parliament and Council Directive 2000/60/EC establishing a Community action in the field of water policy, and taking into account the European Commission recommendation 2006/283/EC, new Lithuanian inland water standards – Maximum-Permissible-Concentrations (MPC) for representative polluting substances have been accepted. The question arises whether these limits are actually safe and can completely protect aquatic life from harmful toxic effects.

The aim of the present study was to determine experimentally the patterns of accumulation of HMs (Cu, Zn, Ni, Cr, Pb, Cd) in different body tissues of Gibel carp (muscle, gills, liver), after 14-day exposure to their model mixture (HMMM) and to Ni separately at concentrations corresponding to MPC. Nickel was chosen because according to the results of monitoring studies of HM accumulation in fish and bottom sediments in Lithuania (Projekto... 2004), this metal often exceeds the levels recommended for human consumption in fish from natural water bodies. Actually, this study is a continuation of recently-started series of the works on experimental investigation of HM accumulation in freshwater fish of Lithuania (Idzelis *et al.* 2008, 2010).

1. Material and methods

The tests were conducted on Gibel carp *Carassius gibelio* (Bloch 1782) adults. The fish were collected in Arnionys fish hatchery ponds (Švenčionys District, Lithuania). Average total length of test fish was 104±5 mm, and the total weight was 21±3 g (mean±SEM, $n = 30$, respectively). The fish were acclimated to laboratory conditions for one week prior to testing. They were kept in flow-through 1000-litre holding tanks supplied with aerated deep-well water (minimum flow rate 1 litre per 1 gram of their wet body mass per day), under natural illumination and were fed commercial fish feed (SKRETTING T-2P Supra) daily in the morning; the total amount was no less than 1% of their wet body mass per day. During the tests, the fish were fed in the same manner.

Reagent grade HM salts ("REACHIM" Company, Russia) were used as the toxicants. Stock solution was prepared by dissolving necessary amount of the salt in distilled water, the final concentration being recalculated according to the amount of HM ion.

Deep-well water was used as the dilution water. Average hardness of the water was approximately 284 (271–296) mg/l as $CaCO_3$, alkalinity was 200 (190–210) mg/l as $CaCO_3$, pH ranged from 7.9 to 8.1, temperature was maintained at 14 to 16 °C, dissolved oxygen concentration was maintained at 8 to 10 mg/l, and dissolved organic carbon (DOC) was less than the detection limit (<0.3 mg/l).

The tests were conducted under semi-static conditions on three groups of 10 individuals using glass tanks of 30-litre total volume (20×30×50 cm in size) filled to a level of 2/3 with continuously aerated dilution water. Test fish were exposed for the 14-day period to a HM model mixture (HMMM) and separately to Ni (Table 1). Test solutions and clean water were renewed every day, and test fish were transferred into freshly prepared solutions after they were fed.

After the testing was completed, fish (of control and metal-exposed groups) were scarified and stored frozen. Later, they were used in the removal of needed tissues. In total, 30 test fish individuals were investigated.

The following body tissues were selected for the determination of the amounts of HMs in fish: muscle without scales, gills (whole organ) and liver (whole organ).

After the morphometric analysis of fish has been performed, muscle, gills and liver were separated from bone tissue, and samples were taken (averagely 3–5 g of each for one fish).

The samples were dried in an oven and digested in aqua regia (concentrated HNO_3 and HCl at a ratio of 1:3 v/v) for 50 minutes at a temperature ~180 °C using microwave digestion system ETHOS (Milestone, USA). After cooling solutions were filtered through a 0.45 μm glass filter and diluted with deionised water up to 50 ml. Metal concentrations were measured by atomic absorption spectrophotometry on AAS Buck Scientific 210VGP with flame or graphite furnace techniques in accordance with standardized procedures (LST ISO 11047:2004) final concentration being expressed as mg/kg of wet weight.

The amount of oxygen in the tanks as well as temperature and pH was measured routinely with a hand-held multi-meter (WTW Multi 340i/SET, Germany). Nominal HM concentrations were checked during blank tests with an atomic absorption spectrophotometer (SHIMADZU AA-6800, Japan) by graphite furnace technique using proprietary software. Each sample was analysed three times. Mean measured concentrations were within 5–10% of the target.

It is well known that metal accumulation in fish is not static, but a dynamic process and bioconcentration is occurring when uptake from the water is greater than excretion. Bioconcentration factor (BCF) is used to quantify the bioconcentration potential in aquatic biota and is defined as the ratio of the concentration of the chemicals in the biota to that of water at equilibrium (Ivanciuc *et al.* 2006). Actually, the BCF is a measure of the extent of chemical partitioning between an organism and the surrounding environment. The values of BCF greater than 1,000 are considered high, less than 250 low, and those between are classified as moderate.

In our case BCF equals the milligrams of metal per kilogram fresh weight of the sample divided by the milligrams of the same metal per litre in the water.

The data were analysed statistically through STATISTICA 6.0 (StatSoft Inc., Tulsa, Oklahoma, USA) software. All the samples underwent distribution normality test. Then the significance of the differences between results was established using Student's *t*-test at $p \leq 0.05$.

Table 1. Heavy metal (HM) concentrations in a model mixture

Heavy metal (HM)	Source	Maximum-Permissible-Concentration (MPC) (mg/l)
Cu	$CuSO_4 \cdot 5H_2O$	0.01
Zn	$ZnSO_4 \cdot 7H_2O$	0.1
Ni	$NiSO_4 \cdot 7H_2O$	0.01
Cr	$K_2Cr_2O_7$	0.01
Cd	$Cd(CH_3COO)_2 \cdot 2H_2O$	0.005
Pb	$Pb(NO_3)_2$	0.005

2. Results

Obtained data showed (Figs 1–3) that in most cases, the amount of HMs in body tissues of HMMM-exposed fish was significantly higher as compared to control fish.

An exception was Cu in the muscle and liver, and Cd in the liver. The total amount of HMs in control fish liver was 0.51 mg/kg, in gills – 2.20 mg/kg and in muscle – 6.10 mg/kg, while in HMMM-exposed fish, liver was 1.91, gills – 3.0 and in muscle – 16.10 mg/kg, respectively.

Heavy metal accumulation order in Gibel carp muscle and gills was as follows: Zn > Cu > Ni > Cr > Pb > Cd, while in liver – Zn > Cu > Pb > Ni > Cr > Cd.

The amount of Ni in HMMM-exposed fish gills increased by 118, Cd – 64, Zn – 37, Cu – 34, Cr – 22 and Pb – 18%, respectively. The amount of Zn in HMMM-exposed fish liver increased by 1,398, Cr – 63, Ni – 47, Pb – 21, Cu – 5.6 and Cd – 0.8%, respectively.

Meanwhile, in muscle, the amounts of Zn increased by 183, Pb – 178, Cr – 92, Cd – 79, Ni – 30 and Cu – 0.3%, respectively, as compared with control fish.

The amount of Ni in test fish exposed to Ni separately increased in gills by 50, in muscle – 41 and in liver 31%, respectively, as compared with control fish.

On the average, the total amount of HMs in HMMM-exposed fish liver was by 224±196, in muscle – 86±27 and in gills – 49±13% higher as compared with control fish (mean±SEM, $n = 6$, respectively).

Quantitatively maximum levels in the Gibel carp body tissues found were of Zn (Fig. 1), while the minimum of cadmium (Fig. 2). Copper was in the second place. Zinc was accumulated mostly in the muscle, at least in the liver.

Copper bioconcentration factor (BCF) in Gibel carp body tissues was as follows: in muscle – 30, gills – 29, and liver – 22.

Zinc BCF in Gibel carp body tissues was as follows: in muscle – 152, gills – 24, and liver – 14.

Nickel BCF in Gibel carp body tissues were as follows: in muscle – 28, gills – 14, and liver – 12.

Meanwhile, in the tests with Ni separately, its BCF in Gibel carp body tissues was as follows: in muscle – 30, gills – 10, and liver – 11.

Chromium BCF in Gibel carp body tissues was as follows: in muscle – 15, gills – 4, and liver – 10.

Lead BCF in Gibel carp body tissues was as follows: in muscle – 32, gills – 31, and liver – 13.

Cadmium BCF in Gibel carp body tissues was as follows: in muscle – 7, gills – 4, and liver – 3.

After the average values of factors (BCF_{av}) were calculated, it has been established that the highest HM

accumulation intensity was in Gibel carp muscle ($BCF_{av.}$ = 42.0±18.7), then in gills (BCF_{av} = 16.6±4.3) and the lowest – in liver (BCF_{av} = 12.1±2.2) (mean±SEM, $n = 7$, respectively).

However, no statistically significant differences between all these values were found ($p > 0.1$, one-way ANOVA). This indicates that HMs were accumulated in Gibel carp body tissues with averagely equal intensity. This means that the accumulation of HMs in a given tissue passed in a balanced countervailing way: an increase in one metal concentration was compensated by the

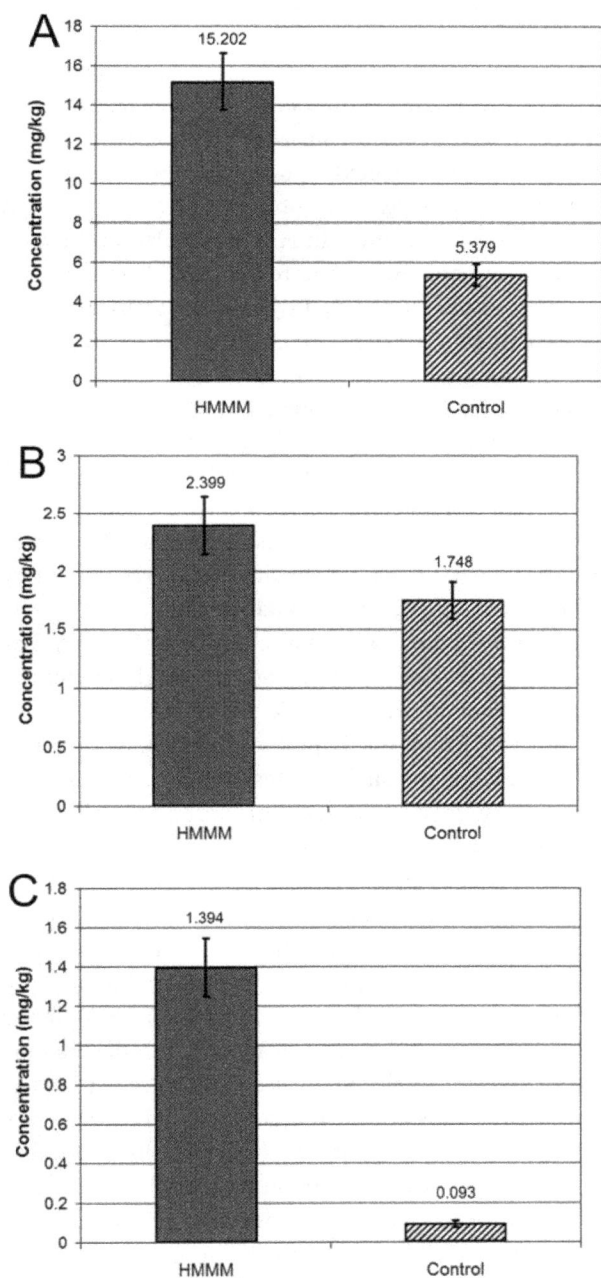

Fig. 1. Zinc concentration in body tissues of mixture-exposed Gibel carp (HMMM): muscle (A), gills (B) and liver (C) (Mean±SD). All test values are significantly different from control ($p < 0.05$)

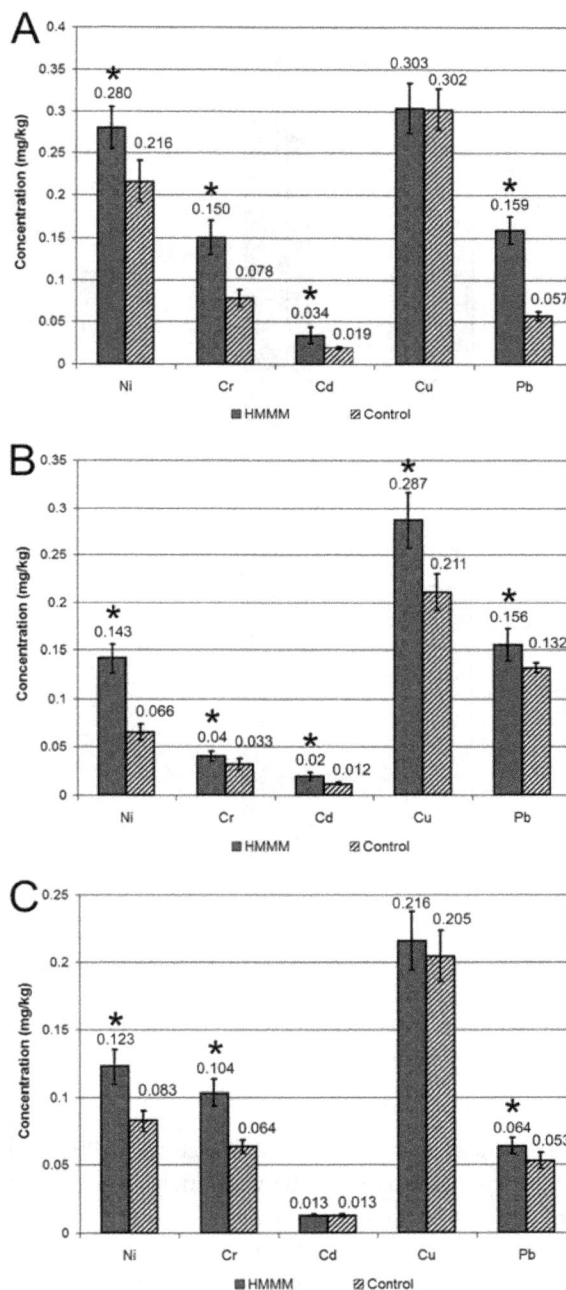

Fig. 2. Heavy metal concentration in body tissues of mixture-exposed Gibel carp (HMMM): muscle (A), gills (B) and liver (C) (Mean±SD). Asterisks (*) denote values significantly different from control ($p \leq 0.05$)

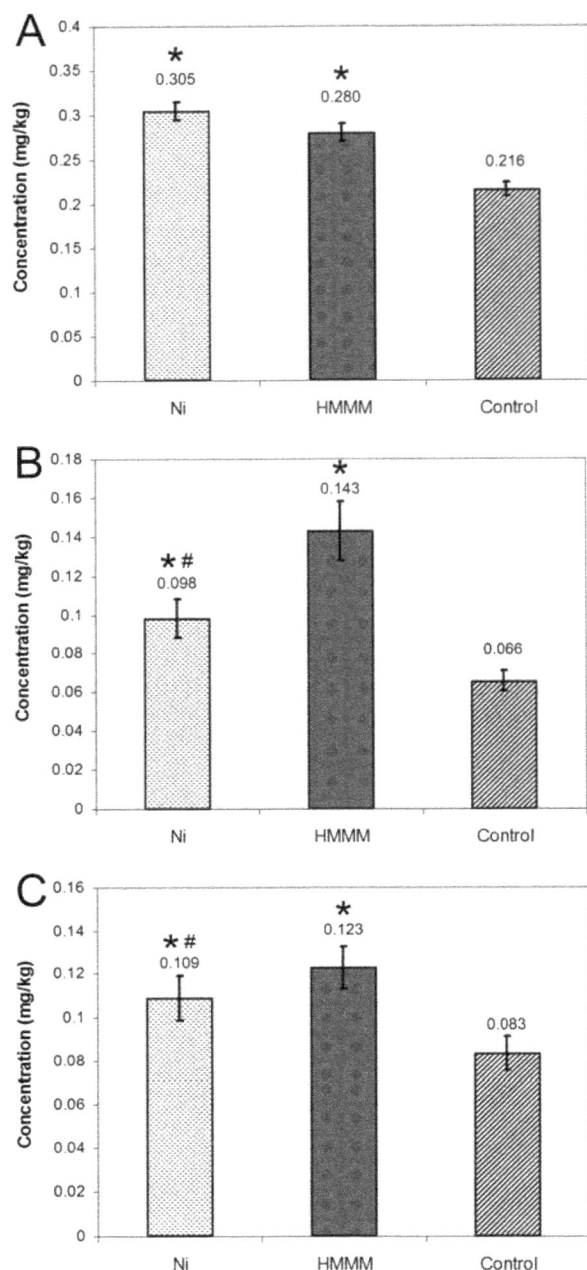

Fig. 3. Comparison of Ni concentration in Ni-exposed (Ni) and mixture-exposed (HMMM) Gibel carp body tissues: muscle (A), gills (B) and liver (C) (Mean±SD). Asterisks (*) denote values significantly different from control, and grades (#) denote significant differences between Ni and mixture-exposed fish at $p \leq 0.05$, respectively

Accumulation of Ni was significantly influenced by the presence of HMMM in the water. In the tests with Ni, no significant difference between its contents in HMMM-exposed and Ni-exposed fish muscle was found (Fig. 3). Meanwhile, Ni concentrations in HMMM-exposed fish liver and gills were significantly higher than those in Ni-exposed fish (by 11 and 31%, respectively). This indicates that the presence of HM mixture in the water promotes Ni accumulation in test fish gills and liver under different experimental conditions. Presumably, this occurs due to the synergistic effects among the metals.

3. Discussion

Data obtained showed that HM accumulation in Gibel carp was metal and tissue specific, i.e. a different tissue showed different capacity for accumulating HMs. In general, all tissues contained high concentrations of Zn and Cu, but a much lower concentrations of Ni, Cr, Pb, and Cd. Such great differences in HM accumulation could be explained, apparently, by their reliance to different categories of HM as described by Roy (2010): essential metals, non-essential metals, etc.

The essential trace elements are those minerals which though needed in very low quantities are vital to the proper functioning of the various biological systems. A heavy metal is termed essential if its deficiency results in the impairment of a function. Nickel and Cr are non-essential HMs. These elements are harmless below their "threshold level" but when they exceed the threshold level, they become toxic and impair various vital functions (Roy 2010). Meanwhile, Cd and Pb are non-essential toxic heavy metals. These metals have no well known biological functions in the animal body and are toxic even in minimal quantities. Their toxicity is due in part to their competition with essential metals for binding sites and also their interference with sulfhydryl groups, which are essential for the normal functioning of enzymes and structural proteins (Da Silva et al. 2005).

Zinc and Cu are essential metals, their role in fish organism is important as they take part in metabolic activities, and their concentrations are under homeostatic control (Bury et al. 2003; Papagiannis et al. 2004). Many authors report that Cu accumulation in fish occurs only under exposure to elevated Cu concentrations (Grosell, Wood 2002). This conclusion could be successfully applied to our results because no differences in Cu content in the liver and muscle was found between the control and HMMM-exposed fish. The significant Cu increase was found in the gills which are the target organ of Cu toxicity and accumulation in fish (Kamunde et al. 2002a, b). The gills are obviously very susceptible to waterborne metals, and often show various metal induced lesions (Jezierska, Sarnowski 2002). The gills of teleost fish play an important

less accumulation of the other and vice versa. This indicates an obvious interaction between metal accumulation relationships in tissues.

Similar results were obtained in a previous study (Idzelis et al. 2010) for rainbow trout and stone loach. Heavy metals were accumulated in different body tissues of both fish species with approximately the same total intensity.

role in ion regulation, gas exchange, acid-base balance, and nitrogenous waste excretion, thus playing a key role of the interface of the fish and their environment. The gills are assumed to be major sites of metal uptake due to the large surface area and direct contact with the aquatic environment (Grosell, Wood 2002). The total metal level in the gills depends upon the absorption of metals not only in the gill tissue but also due to their complexation with the increased amount of mucus onto the surface of the gills, and a number of recent studies confirm it (Oymak *et al.* 2009).

However, the above-mentioned statements could hardly be applied to Zn which content increased at the most of any in all Gibel carp tissues. Such intense increase in Zn concentration a bit surprises as many authors state that diet is normally the primary source of Zn, and waterborne Zn can become significant only in the case when dietary supply is low (Bury *et al.* 2003). Furthermore, the highest content of Zn was found in muscle, the tissue which is generally recognized as the non-metal-accumulating site in freshwater fish (Has-Schön *et al.* 2006). The highest increase in Zn content was found in Gibel carp liver (15-fold). The liver has the ability to accumulate large amounts of HMs from the water and also plays an important role in their storage, redistribution, detoxification transformation and metabolisation, mainly through the induction of metal-binding proteins such as metallothioneins (Bury *et al.* 2003; Licata *et al.* 2005). Furthermore, Zn was fond to be a promoter of the synthesis of more metallothioneins (Sigel *et al.* 2009). The level of metallotioneins has been proposed to be an important biomarker of HM exposure in fish (Valavanidis *et al.* 2006; Schmitt *et al.* 2007; Roy *et al.* 2011).

Generally, liver accumulates high concentrations of metals, irrespectively of the uptake route. The liver is considered a good monitor of water pollution with metals since their concentrations accumulated in this organ are often proportional to those present in the environment. Metal levels in the liver rapidly increase during exposure, and remain high for a long time of depuration, when other organs are already cleared (Jezierska, Witeska 2006).

The data obtained in this study obviously contradict some findings by other authors. For example, Ribeyre *et al.* (1995) investigated the actions and interactions between five elements (Cu, Ag, Se, Zn, and Hg) toward their accumulation by zebrafish (*Brachydanio rerio*) from the direct route at the whole organism level at environmentally–relevant concentrations and found that the accumulation of Zn was not significantly affected by the concentrations of Zn in the water or by any of the other elements.

It is unlikely that Zn uptake was from the diet since both control and test fish were fed in the same manner with well-balanced proprietary fish feed. In any case, active uptake of Zn and other metals into fish tissues from the water is evident. It remains just to state that such

remarkable increase in Zn concentration in all Gibel carp tissues could be explained at most of the high exposure concentration (0.1 Zn mg/l) among the metals and synergistic action between them. These differences in the results need further experimental revising.

Several studies have been devoted to the analysis of Ni accumulation in fish separately and at the presence of other metals in the water. Thus, Van Hoof and Nauwelaers (1984) exposed roach (*Rutilus rutilus*) for 24 hours to acutely lethal Ni concentrations of 50 and 100 mg/l separately and in binary mixtures with Cr and Cu at 80 and 1 mg/l, respectively and found that in the separate exposures, the highest Ni concentration was found in the gills and the lowest in the muscle. While, simultaneous exposures to Ni+Cu and Ni+Cr resulted in a significant Ni levels increase in the gills and opercula. Sreedevi *et al.* (1992) exposed common carp (*Cyprinus carpio*) for 4–12 days to sublethal and lethal Ni concentrations (8 and 40 mg/l, respectively) and found that the accumulation intensity was time and concentration-depended. The accumulation of Ni was in the order: gill > kidney > liver > brain > muscle. Palaniappan and Karthikeyan (2009) studied the accumulation and elimination of Cr separately and in the binary mixture with Ni in *Cirrhinus mrigala* at sublethal concentrations of 2.9 mg/l for seven days and found that the combined exposure induced higher nickel accumulation in fish by 10–28%. This, in turn, determined an increase of 7–12% Cr in the presence of Ni. The accumulation of both nickel and chromium was in the order: kidney > liver > gill > muscle. Similarly, Vinodhini and Narayanan (2008) exposed common carp (*Cyprinus carpio*) for 1–62 days to a mixture of metals (Cr, Ni, Cd and Pb) at sublethal concentrations corresponding to their 0.01 of 48-hour LC50 and found that the accumulation order varied greatly depending on the fish organ, the proportion of Cd and Pb being the highest in the tissues. Karthikeyan *et al.* (2009) reported that the accumulation of Ni in *Cirrhinus mrigala* exposed to sublethal concentrations ranging from 1.05 to 4.6 mg Ni/l was significantly influenced by water pH and hardness. It increased when pH increased and decreased when hardness increased. However, the accumulation order remained unaffected: kidney > liver > gills > muscle.

Summarizing all these results it should be concluded that our data confirm these findings only partly. It is evident that the presence of the metal mixture in the water stimulated Ni accumulation in the gills and liver of Gibel carp, but not in the muscle. However, the accumulation orders in HMMM-exposed and Ni-exposed Gibel carp were quite different as compared with those above-mentioned: muscle > gills > liver and muscle > gills = liver, respectively. Therefore, muscle is the primary tissue of Ni accumulation as in the case with Zn. So far, this phenomenon is difficult to explain. Perhaps, it depends not only on Gibel carp species-specific accumulative-metabolic activity properties but also on the exposure concentration. The reason for this as well could be low concentrations of exposure, which were several orders of magnitude lower than those in above-mentioned experiments. Perhaps, HM accumulation from the direct route at low concentrations follows quite different patterns as compared to those nearly-lethal of even lethal ambient concentrations, and this pattern becomes evident only at low concentrations. All of these queries still need to have the answers.

It is remarkable that here the muscle was found to be a target organ for HM accumulation in Gibel carp. Overall, this study has raised a number of questions, the answers to which must be obtained from further, more exhaustive experimental investigations.

In this study, Maximum-Permissible-Amounts (MPA) in fish indicated in the Lithuanian hygiene standard HN 54:2001: Zn-40, Cu-10, Ni-0.5, Cr-0.3, Pb-0.2 and Cd-0.05 mg/kg of raw mass, respectively (Projekto... 2004) in all cases were not exceeded, i.e. they were below the recommended levels for human consumption. However, in previous studies under the same experimental conditions, it has been established that Ni, Pb and Cd concentrations exceeded above-mentioned limits in the body tissues of roach, perch, stone loach and rainbow trout (Idzelis *et al.* 2008, 2010). In natural environments, HMs can migrate into fish organism not only from surrounding medium but also through food-objects (trophic chain) thus, further increasing their amount in the tissues. Furthermore, HMs even at low concentrations can increase their accumulation in fish due to the actions and interactions between the metals. Therefore, the obtained results require continuous control of HMs in fish, although their concentrations in ambient water do not exceed MPC.

These experimental results coincide with data of environmental monitoring where it has been established that HM concentrations in fishes from natural water bodies rarely exceed MPA (Projekto... 2004). An exception was determined for Ni and Pb, which often exceed these limits.

Heavy metal accumulation order in Gibel carp organ tissues was following: muscle > gills > liver. Meanwhile, under the same experimental conditions in the tissues of roach and perch, they were accumulated as follows: liver > muscle > gills, in stone loach: gills> liver > muscle, and quite contrary in rainbow trout: muscle > liver > gills (Idzelis *et al.* 2008, 2010). Without discussing possible mechanisms it is evident that HM accumulates from the water into the tissues of ecologically different fish species (roach and Gibel carp refer to freshwater limnophilous benthophagous species, stone loach – freshwater rheophylous benthophagous, perch – freshwater limnophilous euryphagous and rainbow trout – anadromous rheophylous-limnophilous euryphagous) quantitatively diversely and specifically.

Although, at present, a lot of data on HM accumulation in fish from natural water-bodies and experimentally are compiled the interspecific differences of this process still remains investigated insufficiently. Evidently, that this depends on the HM chemical origin, the presence of other HMs in the water, fish species-specific ecological, behavioural and their organism biochemical-physiological properties. Further, more exhaustive experimental investigation into HM accumulation patterns in various fish species is evident.

Conclusions

1. Heavy metal (HM) concentration increased significantly in most cases in body tissues of Gibel carp (muscle, gills and liver) after 14-day exposure to HM model mixture (HMMM) and to Ni separately, at concentrations corresponding to Maximum-Permissible-Concentrations (MPC) accepted for the inland waters of Lithuania: Cu-0.01, Zn-0.1, Ni-0.01, Cr-0.01, Pb-0.005 and Cd-0.005 mg/l, respectively.

2. Heavy metal accumulation order in body tissues of Gibel carp was following: muscle > gills > liver. The highest amounts found were of zinc (15.2 mg/kg), while the lowest of cadmium (0.012 mg/kg).

3. In the muscle and gills, HMs were accumulated in such decreasing order: Zn > Cu > Ni > Cr > Pb > Cd, while in the liver: Zn > Cu > Pb > Ni > Cr > Cd. Heavy metals were accumulated in Gibel carp body tissues with approximately the same total intensity: no significant differences between average values of bioconcentration factors (BCF_{av}) were found ($p > 0.1$).

4. Accumulation of Ni was significantly influenced by the presence of HMMM in the water. It promoted Ni accumulation in test fish gills and liver under different experimental conditions (by 11 and 31%, respectively). Presumably, this occurs due to the synergistic effects among the metals.

5. Although the concentrations of HMs in Gibel carp body tissues increased, they were below the recommended levels (Maximum-Permissible-Amounts) (MPA) for human consumption.

6. Data obtained here, and in previous studies under the same experimental conditions confirmed that HMs accumulation in different fish species is highly specific and their amounts sometimes exceed MPA (especially Ni, Pb and Cd). These results coincide with the data of environmental monitoring of HMs in fishes from natural water bodies and, apparently, depend on the HM chemical origin, fish ecological, behavioural and their organism biochemical-physiological process properties.

7. The obtained results require continuous control of HMs in fish although their concentrations in ambient water do not exceed MPC.

References

Bury, N. R.; Walker, P. A.; Glover, C. N. 2003. Nutritive metal uptake in teleost fish, *The Journal of Experimental Biology* 206(Pt 1): 11–23. http://dx.doi.org/10.1242/jeb.00068

Canli, M.; Atli, G. 2003. The relationships between heavy metal (Cd, Cr, Cu, Fe, Pb, Zn) levels and the size of six Mediterranean fish species, *Environmental Pollution* 121(1): 129–136. http://dx.doi.org/10.1016/S0269-7491(02)00194-X

Chovanec, A.; Hofer, R.; Schiemer, F. 2003. Fishes as bioindicators, in B. A. Markert; A. M. Breure; H. G. Zechmeister (Eds.). *Bioindicators and biomonitors, principles, concepts and applications.* Amsterdam: Elsevier, 639–676. http://dx.doi.org/10.1016/S0927-5215(03)80148-0

Clearwater, S. J.; Farag, A. M.; Meyer, J. S. 2002. Bioavailability and toxicity of dietborne copper and zinc to fish, *Comparative Biochemistry and Physiology. Part C: Toxicological Pharmacology* 132(3): 269–313.

Da Silva, A. L. O.; Barrocas, P. R. G.; Jacob, S. C.; Mreira, J. C. 2005. Dietary intake and health effects of selected toxic elements, *Brazilian Journal of Plants Physiology* 17(1): 79–93. http://dx.doi.org/10.1590/S1677-04202005000100007

Directive 2000/60/EC of the European Parliament and of the Council establishing a framework for Community action in the field of water policy. 2000, *CELEX-EUR Official Journal of the European Communities* L 327: 1–72.

Ghosh, L.; Adhikari, S. 2006. Accumulation of heavy metals in freshwater fish – an assessment of toxic interactions with calcium, *American Journal of Food Technology* 1(2): 139–148. http://dx.doi.org/10.3923/ajft.2006.139.148

Ghosh, L.; Adhikari, S.; Ayyappan, S. 2007. Assessment of toxic interactions of heavy metals and their effects on accumulation in tissues of freshwater fish, *Research Journal of Environmental Toxicology* 1(1): 37–44. http://dx.doi.org/10.3923/rjet.2007.37.44

Green, N. W.; Knutzen J. 2003. Organohalogens and metals in marine fish and mussels and some relationships to biological variables at reference localities in Norway, *Marine Pollution Bulletin* 46(3): 362–377. http://dx.doi.org/10.1016/S0025-326X(02)00515-5

Grosell, M.; Wood, C. M. 2002. Copper uptake across rainbow trout gills: mechanisms of apical entry, *The Journal of Experimental Biology* 205(8): 1179–1188.

Has-Schön, E.; Bogut, I.; Strelec, I. 2006. Heavy metal profile in five fish species included in human diet, domiciled in the end flow of river Neretva (Croatia), *Archives of Environmental Contamination and Toxicology* 50(4): 545–551. http://dx.doi.org/10.1007/s00244-005-0047-2

Idzelis, R. L.; Kesminas, V.; Svecevičius, G.; Misius, V. 2008. Accumulation of heavy metals (Cu, Zn, Ni, Cr, Pb,Cd) in tissues of perch (Perca fluviatilis L.) and roach Rutilus rutilus (L.) under experimental conditions, *Journal of Environmental Enginnering and Landscape Management* 16(4): 205–212. http://dx.doi.org/10.3846/1648-6897.2008.16.205-212

Idzelis, R. L.; Kesminas, V.; Svecevičius, G.; Venclovas, A. 2010. Experimental investigation of heavy metal accumulation in tissues of stone loach Noemacheilus barbatulus (L.) and rainbow trout Oncorhynchus mykiss (Walbaum) exposed to a model mixture (Cu, Zn, Ni, Cr, Pb, Cd), *Journal of Environmental Engineering and Landscape Management* 18(2): 111–117. http://dx.doi.org/10.3846/1648-6897.2010.13

Ivanciuc, T.; Ivanciuc, O.; Klein, D. J. 2006. Modeling the bioconcentration factors and bioaccumulation factors of polychlorinated biphenyls with posetic quantitative super-structure/activity relationships (QSSAR), *Molecular Diversity* 10(2): 133–145. http://dx.doi.org/10.1007/s11030-005-9003-3

Jezierska, B.; Sarnowski, P. 2002. The effect of mercury, copper and cadmium during single and combined exposure on oxygen consumption of Oncorhynchus mykiss (WAL.) and Cuprinus carpio (L.) larvae, *Archives of Polish Fisheries* 10(1): 15–22.

Jezierska, B.; Witeska, M. 2006. The metal uptake and accumulation in fish living in polluted waters, in I. Twardowska et al. (Eds.). *Soil and water pollution monitoring, protection and remediation.* Springer, 3–23.

Kamunde, C.; Grosell, M.; Higgs, D.; Wood, C. M. 2002a. Copper metabolism in actively growing rainbow trout (Oncorhynchus mykiss): interactions between dietary and waterborne copper uptake, *The Journal of Experimental Biology* 205: 279–290.

Kamunde, C.; Clayton, C.; Wood, C. W. 2002b. Waterborne vs. dietary copper uptake in rainbow trout and the effects of previous waterborne copper exposure, *The American Journal of Physiology-Regulatory, Integrative and Comparative Physiology* 283(1): 69–78.

Karthikeyan, S.; Palaniappan, P. L. R. M.; Sabhanayakam, S. 2007. Influence of pH and water hardness upon nickel accumulation in edible fish Cirrhinus mrigala, *Journal of Environmental Biology* 28(2): 489–492.

Klavins, M.; Potapovics, O.; Rodinov, V. 2009. Heavy metals in fish from lakes in Latvia: concentrations and trends of changes, *Bulletin of Environmental Contamination and Toxicology* 82(1): 96–100. http://dx.doi.org/10.1007/s00128-008-9510-x

Licata, P.; Trombetta, D.; Cristani, M.; Naccari, C.; Martino, D.; Calo, M.; Naccari, F. 2005. Heavy metals in liver and muscle of bluefin tuna (Thunnus thynnus) caght in the straits of Messina (Sicily, Italy), *Environmental Monitoring and Assessment* 107(1–3): 239–248. http://dx.doi.org/10.1007/s10661-005-2382-1

LST ISO 11047:2004. Dirvožemio kokybė. Kadmio, chromo, kobalto, vario, švino, mangano, nikelio ir cinko nustatymas ekstrahuojant dirvožemį karališkuoju vandeniu. Liepsnos ir elektroterminės atominės atominės atominės absorbcijos spektrometriniai metodai (tapatus ISO 11047:1998) [Soil quality. Determination of cadmium, chromium, cobalt, copper, lead, manganese, nickel and zinc in aqua regia extracts of soil. Flame and electrothermal atomic absorption spectrometric methods (idt ISO 11047:1998)]. Lietuvos standartizacijos departamentas. 18 p.

Luczynska, J.; Tonska, E. 2006. The effect of fish size on the content of zinc, iron, copper, and manganese in the muscles of perch (Perca fluviatilis L.) and pike (Esox lucius L.), *Archives of Polish Fisheries* 14(1): 5–13.

Nemova, N. N. 2005. *Biochemical effects of accumulation of mercury in fish.* Moscow: Nauka. 164 p. (in Russian).

Oymak, S. A.; Karadede-Akin, H.; Dogan, N. 2009. Heavy metal in tissues of Tor grypus from Atatürk Dam Lake, Euphrates River-Turkey, *Journal of Biologia* 64(1): 151–155.

Palaniappan, P. L. R. M.; Karthikeyan, S. 2009. Bioaccumulation and depuration of chromium in the selected organs and whole body tissues of freshwater fish Cirrhinus mrigala individually and in binary solutions with nickel, *Journal of Environmental Sciences* 21(2): 229–236. http://dx.doi.org/10.1016/S1001-0742(08)62256-1

Papagiannis, I.; Kagalou, I.; Petridis, D.; Kalfakaou, V. 2004. Copper and zinc in four freshwater fish species from Lake Pamvotis (Greece), *Environmental International* 30(3): 357–362. http://dx.doi.org/10.1016/j.envint.2003.08.002

Polak-Juszczak, L. 2009. Temporal trends in the bioaccumulation of trace metals in herring, sprat, and cod from the southern Baltic Sea in the 1994–2003 period, *Chemosphere* 76(10): 1334–1339. http://dx.doi.org/10.1016/j.chemosphere.2009.06.030

Projekto „Sunkiųjų metalų kaupimasis žuvyse ir dugno nuosėdose" rezultatų apžvalga [online], [cited 20 May 2011]. 2004. Available from Internet: http://oldaaa.gamta.lt/VI/article.php3?article_id=1203 (in Lithuanian).

Ribeyre, F.; Amiard-Triquet, C.; Boudou, A.; Amiard, J. C. 1995. Experimental study of interactions between five trace elements – Cu, Ag, Se, Zn, and Hg – toward their bioaccumulation by fish (*Brachydanio rerio*) from the direct route, *Ecotoxicology and Environmental Safety* 32(1): 1–11. http://dx.doi.org/10.1006/eesa.1995.1078

Roy, S. P. 2010. Overview of heavy metals and aquatic environment with notes on their recovery, *Ecoscan: An International Quarterly Journal of Environmental Sciences* 4(2–3): 235–240.

Roy, S. U.; Chattopadhyay, B.; Datta, S.; Kumar, S. 2011. Metallothionein as a biomarker to assess the effects of pollution on Indian major carp species from wastewater-fed fishponds of East Calcutta wetlands (a Ramsar Site) Mukhopadhyay, *Environmental Research, Engineering and Management* 4(58): 10–17.

Schmitt, C. J.; Whyte, J. J.; Roberts, A. P.; Annis, M. L.; May, T. W.; Tillitt, D. E. 2007. Biomarkers of metal exposure in fish from lead-zinc mining areas of Southeastern Missouri, USA, *Ecotoxicology and Environmental Safety* 67(1): 31–47. http://dx.doi.org/10.1016/j.ecoenv.2006.12.011

Sigel, A.; Sigel, H.; Sigel, R. K. O. (Eds.). 2009. *Metallothioneins and related chelators. Metal ions in life sciences*, vol. 5. Cambridge: RSC Publishing. 514 p. http://dx.doi.org/10.1039/9781847559531

Sreedevi, P.; Suresh, A.; Sivaramakrishnan, B.; Prabhavathi, B. 1992. Bioaccumulation of nickel in the organs of the freshwater fish, *Cyprinus carpio*, and the freshwater mussel, *Lamellidens marginalis*, under lethal and sublethal nickel, *Chemosphere* 24(1): 29–36. http://dx.doi.org/10.1016/0045-6535(92)90564-8

US EPA. 2009. *National recommended water quality criteria* [online], [cited 24 August 2013]. Office of Water, Office of Science and Technology (4304T), Washington, DC. 25. Available from Internet: http://www.rsc.org/dose/

Valavanidis, A.; Vlahogianni, T.; Dassenakis, M.; Scoullos, M. 2006. Molecular biomarkers of oxidative stress in aquatic organisms in relation to toxic environmental pollutants, *Ecotoxicology and Environmental Safety* 64(2): 178–189. http://dx.doi.org/10.1016/j.ecoenv.2005.03.013

Van Briesen, J. M.; Small, M. J.; Weber, C.; Wilson, J. 2010. Modeling chemical speciation: thermodynamics, kinetics, and uncertainty, in G. Hanrahan (Ed.). *Modeling of pollutants in complex environmental systems*, vol. 2, 133–149.

Van Hoof, F.; Nauwelaers, J. P. 1984. Distribution of nickel in the roach *Rutilus rutilus* L. after exposure to lethal and sublethal concentrations, *Chemosphere* 13(9): 1053–1058. http://dx.doi.org/10.1016/0045-6535(84)90064-X

Vinodhini, R.; Narayanan, M. 2008. Bioaccumulation of heavy metals in organ of freshwater fish *Cyprinus carpio* (Common carp), *International Journal of Environmental Science and Technology* 5(2): 179–182. http://dx.doi.org/10.1007/BF03326011

Gintaras SVECEVIČIUS. Dr, Senior Research Worker, Division of Ecological Physiology and Toxicology, Laboratory of Ecology and Physiology of Hydrobionts, Institute of Ecology of the Nature Research Centre. Doctor of Natural Sciences, 1993; Laureate of Lithuanian Science Award, 2007. Publications: author/co-author of ~150 scientific papers. Research interests: ecotoxicology, ichthyotoxicology, fish behavioural toxicology.

Raimondas Leopoldas IDZELIS. Dr, Assoc. Prof., Department of Environmental Protection, Vilnius Gediminas Technical University (VGTU). Doctor of Natural Sciences, 1993. Publications: author/co-author of ~70 scientific papers, 2 study guides, co-author of 3 monographs. Research interests: landscape management, ecology, environmental protection, animal guide urbanization.

Eglė MOCKUTĖ. Master of Science, Vilnius Gediminas Technical University (VGTU), 2011. Publications: co-author of 1 research paper. Research interests: environmental protection, accumulation of heavy metals, pollution prevention.

17

EVALUATION OF NO$_X$ EMISSION AND DISPERSION FROM MARINE SHIPS IN KLAIPEDA SEA PORT

Eglė ABRUTYTĖ, Audronė ŽUKAUSKAITĖ, Rima MICKEVIČIENĖ,
Vytenis ZABUKAS, Tatjana PAULAUSKIENĖ

Faculty of Marine Engineering, Klaipėda University, Bijūnų g. 17, 91225 Klaipėda, Lithuania

Abstract. The aim of the presented research was to assess the NO$_x$ emission and dispersion from marine ships in Klaipeda sea port. NO$_x$ emissions from ships operating in Klaipeda sea port were calculated using the Lloyd's Register detailed ship movement method, after collecting the information about technical characteristics of each marine ship visiting the port and the time spent staying in the port. After calculating the emission, the modelling using AERMOD software was completed and the dispersion of pollutants over different seasons of the year was determined. When performing the evaluation of NO$_x$ emissions it was estimated, that most of these pollutants enter the atmosphere from stationary vessels moored to quays with active auxiliary motors; this accounts even for up to 72% of the total NO$_x$ emission from marine ships in Klaipeda port. It was calculated that a total of 945.6 tons of NO$_x$ compounds enter the air basin from ships which operate in Klaipeda port. It was determined that the seasonality and meteorological conditions are a significant factor affecting the dispersion of pollutants. During winter time, a higher dispersion of pollutants is typically found at the source of contamination, and in the summer pollutants are decomposed more quickly and their concentrations as formed above the port are 30% lower, however, 40–50% higher concentrations are formed over the Klaipeda city residential districts.

Keywords: Klaipeda port, atmospheric pollution, ships, nitrogen oxides.

Introduction

Concerns regarding the air quality in the port cities due to the pollutants emitted by ships are increasing all over the world; increasingly more attention is paid to evaluation and control of air pollution caused by ships. The impact on the city air basin made by growing number of ships, entering the port is assessed. Ways to reduce the negative environmental impact of sea navigation were searched. It is searched for ways to reduce the negative environmental impact of sea navigation (Corbett *et al.* 2007, 1999; Cooper 2003; Wang *et al.* 2007; Georgakaki *et al.* 2005; Saxea, Larsena 2004).

Growth in demand for transport services is inseparable from the country's overall economic growth. When the gross domestic product of the country grows by 2.5%, the demand for transport services increases by 2.7%. The official EU rules state, that the policy of the transport development must be an integral part of environmental policy. Transportation means in Lithuania emit approximately 500 thousand tons of pollutants each year (Baltrėnas *et al.* 2004). In port cities the economic activities and the associated atmospheric pollution is mostly concentrated in the coastal areas. Due to the growing global trade the flow of goods by maritime transport is also increasing. The diesel combustion process results in the highest atmospheric pollution; particulate matter and nitrogen oxides are one of the main components of pollution. The road transport is the main source of nitrogen oxides (NO$_x$), where about a half of amount of the nitrogen oxides is emitted in the Europe. Therefore, the highest concentrations of NO and NO$_2$ are formed in cities where the traffic is the most intense. Other important sources of pollution are: thermal power plants, industrial processes and shipping (Bailey, Solomon 2004; Isakson *et al.* 2001; Lonati *et al.* 2010; Luke *et al.* 2010; Schreier *et al.* 2006). In ships, the energy producing machinery that burns fuel oil is important regular source of air pollution emissions; it includes propulsion and auxiliary engines and steam boilers. The composition

Corresponding author: Audronė Žukauskaitė
E-mail: dekanatas.jtf@ku.lt

of fuel combustion products emitted by them is well-researched and can be readily evaluated according to chemical composition of the fuel, air and fuel ratio in the cylinder and combustion conditions (Smailys et al. 2003).

Various atmospheric pollution components exist in gases that are emitted by ships when combusting fuel. According to Eyring et al. (2009), the combustion of 1 kilogram of diesel fuel results in 3170 g of CO_2, 77 g of NO_x, 40 g of SO_2, 7.4 g of CO and 5.5 g of PM emitted into the atmosphere. Thus, the largest emissions into the atmosphere from ships include carbon dioxide (CO_2), nitrogen oxides (NO_x) and sulfur dioxide (SO_2). Due to the large emissions of these pollutants into the air basin, the environmental requirements are being elevated. This fact is receiving increasing attention during design of new vessels and chemical composition of fuel used in ships.

The age of ships visiting the port plays an important role in the emissions of NO_x, since engines are more sophisticated in newly built ships, which leads to significantly lower emissions. Accomplished studies have shown that the new ship engines manufactured since 2005, meet the requirements of the Technical Code for NO_x emissions better by 17% than the older ones. However, the engine life time is approximately 25 years, and the fleet composition typically remains stable for a long time, so it was estimated that over the period of 5 years the fleet is supplemented with only 4% of new vessels (Trozzi et al. 1995).

The amount of nitrogen and sulphur oxides (NO_x and SO_x), entering the atmosphere from ships, is increasing (Deniz, Durmusoglu 2008). It is forecasted, that by 2020 the emissions of NO_x will increase by two-thirds, and SO_x emissions – nearly twice (Eyring et al. 2009). Meanwhile, the emission limitations are especially tightened during implementation of international law regulations, such as MARPOL 73/78 convention Annex VI, regulating the prevention of air pollution from ships. NO_x emissions are limited by the special conditions provided in MARPOL 73/78 Annex VI, which depend on the ship engine speed. UN sets new standards for ship engines in order to reduce the amount of NO_x emitted into atmosphere. NO_x emission limits should be reduced by 20% per kWh for marine engines manufactured after 2011, i.e. from 7.7 to 14.4 g/kWh, depending on engine speed. And the NO_x emissions should be decreased by as much as 80% (from 2.0 to 3.4 g/kWh, depending on the ship engine speed) for ships built after the year 2016 (McCarthy 2009).

The Baltic Sea is an area extensively used for short sea shipping. At one time there are more than 2000 different ships sailing in the Baltic Sea. Methods for calculating emissions from ships are based on the number of vessels, distance travelled, power of ship engines and (or) fuel consumption (Endresen et al. 2003; Dalsoren et al. 2009; Miola, Ciuffo 2010; Cooper 2003). In some studies the emission calculations are performed based on the amount of fuel purchased in the country. However, when using this method, it is difficult to assess the amount of emissions within a defined territory, let alone to determine the contribution of ship pollutants to overall air basin in the port and port area.

Due to pollutants emitted into atmosphere, the urban air quality deteriorates, and acidity of soil and surface water increases, eutrophication takes place; it also contributes to the formation of the greenhouse effect and formation of ground level ozone. This has a negative effect firstly on human health, agricultural productivity, biodiversity and condition of forests (Oke 2004). Due to the air contamination, the number of people with asthma and other respiratory diseases and cardiovascular conditions increases; lung cancer morbidity rate and early mortality is influenced. It is therefore necessary to take measures to reduce this pollution (Bailey, Solomon 2004; Krozer et al. 2003).

One of the most important features of port city of Klaipeda is the fact that the city is narrow, on average not wider than 2.5–3 km, and is 11–12 km long, extending from north to south along the port; the nearest residential areas of the city are located not more than 250–300 meters from port quays. When dominant west winds blow, the entire city is trapped inside the dispersion trail of port air pollutants, which has a significant impact on the formation of sanitary condition of city air basin (Smailys et al. 2003). The influence of maritime transport is stronger in the cities which are located near lagoons than in other coastal areas, thus the air pollution monitoring and evaluation of contaminants formed via the main transportation channels becomes necessary (Premuda et al. 2011). The aim of these studies is to assess the NO_x emissions from marine ships and dispersion of pollutants in Klaipeda port, on the basis of analysis of arriving ships.

1. Research method

In order to calculate the emissions from ships operating in Klaipeda port the Lloyd's Register detailed ship movement method was selected, which is typically recommended when the detailed movement data of the vessels and their technical information is known. The calculations of NO_x emissions were performed considering that the ships use diesel, because black oil in the port area is forbidden. When performing the investigation, the data collected in 2009 about the ships which visited the Klaipeda port was used. Since the fleet composition varies insignificantly, and the age of ships does not exceed 24 years, this data may retain its representative character for a long time.

When calculating NO_x emissions from ships stationed at the quays and ships sailing inside the port, and

Table 1. Subdivision of Klaipeda port area

Part of port	Companies	Quays
I	SC "Klaipėdos Jūrų krovinių kompanija" KLASCO (northern part); SC "Klaipėdos nafta"; SC "Krovinių terminalas"	1–7
II	SC "Klaipėdos Jūrų krovinių kompanija" KLASCO (southern part); SC "Klaipėdos laivų remontas"; SC "Laivitė"; SC "Vakarų Baltijos laivų statykla"	8–64
III	Joint-stock stevedoring company "Klaipėdos Smeltė" (northern part); UAB "Klaipėdos keleivių ir krovinių terminalas"; SC Klaipėdos jūrų krovinių kompanija "Bega"	65–88
IV	SC "Senoji Baltija"; Jūrų keltų terminalas; Karinės jūrų pajėgos; Joint-stock stevedoring company "Klaipėdos Smeltė" (southern part); Mažųjų žvejybos laivų prieplauka; SC "Klaipėdos hidrotechnika"	89–126; 146–151; 121A–123A
V	SC "Vakarų laivų gamykla"; SC "Klaipėdos konteinerių terminalas"; SC "Malkų įlankos terminalas"	127–144; 131A–138A

also when evaluating the pollution dispersion, the Klaipeda port area was subdivided into 5 equal sections, each 2,4 km long. In this way it was attempted to ensure the precision of the research results. Table 1 shows what companies are located here and which quays are included into particular sections of the port.

In order to assess the total emission of NO_x, the ship sailing and stationary time (h) while staying in the port has to be matched against the emission rates (kg/h). The emission was evaluated separately for all five sections of the port. Ship sailing distance is considered to be equal to the distance from the port gates to the respective area of the port. In Klaipeda port aquatory the ship movement speed is limited to 6 knots. Consequently, after assessing the permissible ship speed inside the port and sailing distances, durations of sailing of each ship to the respective quay were estimated. Each ship stationary time was calculated in accordance to the data regarding arrival and departure time of ships (accuracy in minutes), provided by Klaipeda port direction.

Data on the main and auxiliary ship engine power ratings were collected from the database of Lloyd's Register, according to the IMO (International Maritime Organization) number of each ship, i.e. the ship identification number issued by International Maritime Organization. Emission rates are calculated separately for main and auxiliary engines.

During 2009, a total of 7529 vessels visited the Klaipeda port. Only sea ships were analysed, since they contain more powerful engines compared to the transport of internal waters, and they emit the largest quantities of pollutants. In total, the data about 1890 marine ships which visited the Klaipeda port during 2009 was collected. They entered the port 6837 times and their total stationary time while moored with active auxiliary engines accounted for 14668 days.

The AERMOD model was selected to evaluate the dispersion of NO_x pollutants in Klaipeda port. ISC-AERMOD View is a complete and powerful Windows air dispersion modelling system which seamlessly incorporates three models into one interface: ISCST3, AERMOD and ISC-PRIME. AERMOD is the next generation air dispersion model designed for short-range (up to 50 kilometers) dispersion of air pollutant emissions from stationary industrial sources. AERMOD is a steady-state plume model. The basis of the model is the straight-line, steady-state Gaussian plume equation (Petraitis 2010; Kowalski, Tarelko 2009).

NO_x emissions from marine ships operating in Klaipeda port were calculated using detailed ship movement method. Emissions from sailing and stationary ships were calculated separately for five sections of the port, therefore, when evaluating the pollution dispersion, each area of the port was considered as a separate area source of pollution. In order to assess the impact of seasonality on pollution dispersion, the modelling was carried out for months of April, August and December, and also the time duration from 2006 to 2009 was selected. In this way the dispersion of NO_x components was determined for selected months, according to prevailing four-year meteorological conditions; this ensures higher reliability of the results. During modelling of pollutant dispersion in Klaipeda port, NO_x concentrations at 1.5 meter height were determined. Average daily concentrations of pollutants from stationary and sailing ships in Klaipeda port were calculated; other nitrogen oxide emission sources and background pollution were not considered. Concentration areas matching the coordinate grid of Klaipeda city plan were formed with aim to assess the pollutant concentrations. Using the model, concentrations are calculated at the points of created grid; 150 meter calculation step was selected and values at 2911 points were modelled.

1.1. NO_x analysis method

The passive measurement method with diffusive samplers was used for analysis NO_x concentration. Diffusion samplers were located in the port area as shown in Figure 1. Eleven places have been chosen for the exposition. The diffusive samplers have been exposed for the period of seven days during different seasons of the year- during the spring and autumn seasons. In the selected locations the samples were hung at a height approximately 3.5 meters. NO_x concentration accumulated in samplers was determined (Smailys *et al.* 2009) using naphtylethylendiamino-dihydrochloride with spectrophotometer (JENWAY 6300, $\lambda = 540$ nm).

2. Results and their discussion

2.1. Analysis of ships visiting the Klaipeda port

Ships of medium and small tonnage visit Klaipeda port most often. Most part of tonnages of arriving ships fall into range between 1000 and 4999 t or is smaller than 500 t. Ships of large tonnage (larger than 50000 t) arrive in rare cases.

Ships of different tonnage are distributed in different areas of the port unevenly (Fig. 2). Port section II differentiates in particularly small tonnage ships; they comprise 71% of total number of small tonnage ships which visited Klaipeda port. There are a lot of tugboats and fishing boats in this area of the port with tonnage up to 500 t. Number of visiting vessels with tonnage from 1000 to 4999 t is similar in all areas of the port; none of the port sections distinguishes by number of ships with tonnage 500–900 t and these vessels are evenly distributed over entire port.

Even 53% of ships with tonnage 10000–49999 t belong to port section IV and 21% belong to section I. Even though the smallest number of arriving ships is characteristic to the port area I, it distinguishes by ships of largest tonnage. 22 visits of ships with tonnage 50000 t and larger were observed in Klaipeda port, and 16 of them were made in port section I; tankers and bulk carriers arrive here most often.

2.2. Analysis of sailing and stationary times of marine ships operating in Klaipeda port

Number of ship visits in separate areas of the port varied from 721 to 2456, while the calculated total stationary time of the ships staying at the quays in different areas of the port varied from 946.6 to 5801.3 days. Most ship visits (even 41% from total number) belong to port area II; respectively, the ship stationary time while staying at the quays is also longest in this section of the port. According

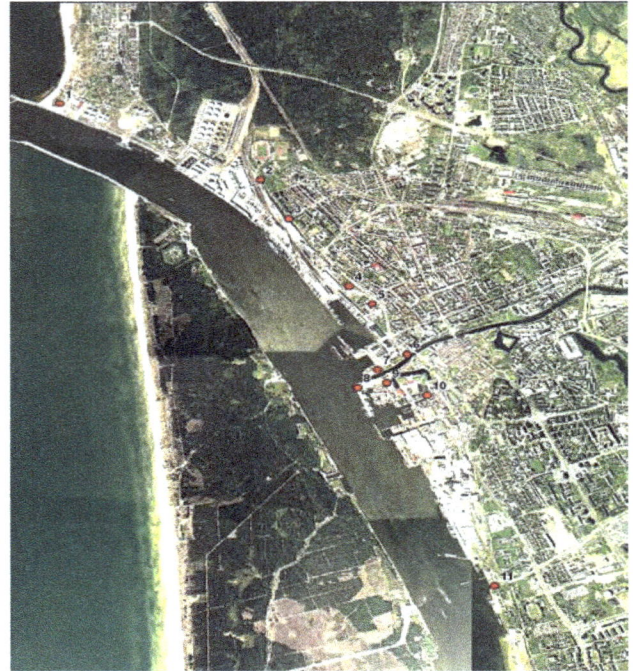

Fig. 1. Map of diffusive samplers' exposure on Klaipeda's streets

to number of visits and total stationary time duration, the port area which was attributed to section IV does not fall far behind. It has 23% of ship visits and 27% of stationary time, compared to the overall result of Klaipeda port. The smallest number of ship visits and the shortest vessel stationary time is characteristic to the port area I.

It can be noted that in separate areas of the port number of visits and stationary time at the quays are directly proportional. When the larger number of ship visits is present in respective area of the port, the total ship time spent at the quays with auxiliary engines active will be also longer (Fig. 3).

In Klaipeda port ship sailing time until respective quays is reached will be directly proportional to the distance to port gates, since in port aquatory the speed is limited to 6 knots for all vessels. Therefore, the ships entering the first area of the port, the distance of which to the port gates is approximately 2.4 km, will take 0.22 hours. The longest sailing time (1.08 hours) belongs to ships which sail to the port area V, since the distance from the quays located in this

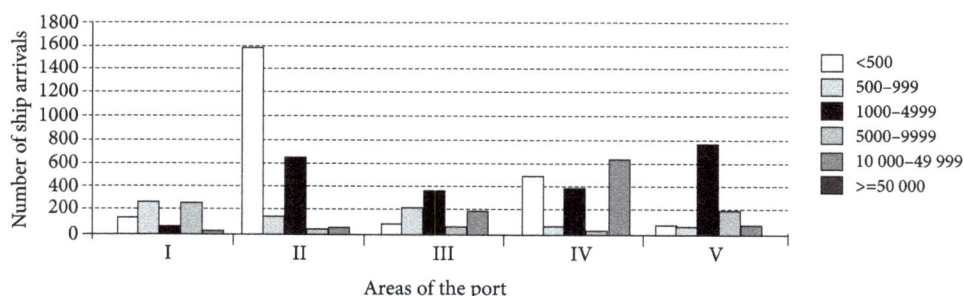

Fig. 2. Number of ship arrivals according to tonnage in separate areas of the port

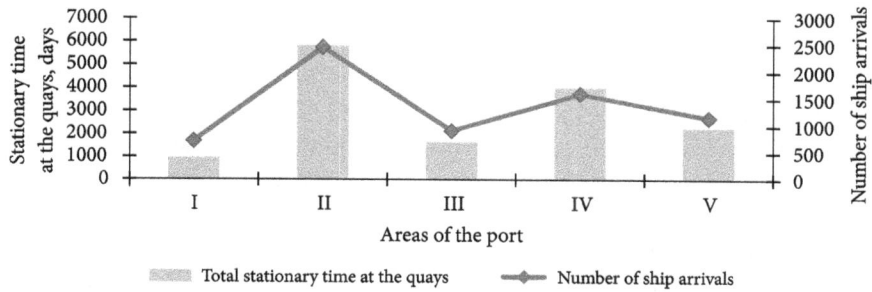

Fig. 3. Relation between number of ship arrivals and their total stationary time at the quays

2.3. NO$_x$ emissions from ships operating in Klaipeda port

The emission of NO$_x$ pollutants from ship sailing in Klaipeda port over one hour reaches from 0.95 kg to 358.8 kg, and from 0.1 kg to 28.5 kg over one hour from stationary ship with active auxiliary engines. Although sailing ships emit the amount of NO$_x$ pollution components over one hour about 10 times larger than those stationed at quays, the total NO$_x$ emission from stationary ships in Klaipeda port is higher (Table 2).

Relative distribution of overall NO$_x$ emission in different areas of the port indicates, that the largest amount of these pollutants enter the atmosphere from the port area IV, which is responsible for 27% of total NO$_x$ emission, and the smallest pollution is characteristic to the port section I, from which 12% of NO$_x$ components from the total sum of port emissions enter the air basin.

The emission depends not only on the ship stationary time duration. Data presented in Fig. 4 shows, that the NO$_x$ emission in the port area II is smaller by 65.3 t/year compared to the port section IV, and smaller by 53.7 t/year compared to the port section V, even though the ship stationary time while moored at the quays is the longest and the number of visits is the largest. Visits of ships with small power engines dominate in this area of the port.

Although the distance to the port gates in the port area II is 4.8 km, the total time of ships sailing into this area reaches 1056 hours due to the large number of arriving ships, and the NO$_x$ emission from ships sailing to the quays in this section is one of the smallest, 20.8 t/year, and exceeds only emissions in the port area I. Most of NO$_x$ is emitted in the port area IV, since it has both the largest number of arriving ships and also ships with large tonnage and powerful engines; furthermore, the distance from the quays in this section to the port gates is 9.6 km.

Thus, the calculated annual emission of NO$_x$ components in Klaipeda port reaches 945.62 tons. This data matches calculations accomplished by Smailys et al. (2003), where during the year 1995 approximately 981 tons of NO$_x$ components entered air basin from ships.

Table 2. Total NO$_x$ emission (t/year) from ships operating in Klaipeda port

Port area	Emission from ships sailing from port gates to quays and back, t	Emission from stationary ships, t	Total NO$_x$ emission in separate areas of the port, t	Relative distribution of overall NO$_x$ emission
I	11.31	60.14	113.77	12%
II	32.19	137.28	190.91	20%
III	41.72	86.45	140.08	15%
IV	119.37	202.59	256.23	27%
V	63.57	191.01	244.64	26%
Total:	268.16	677.46	945.62	

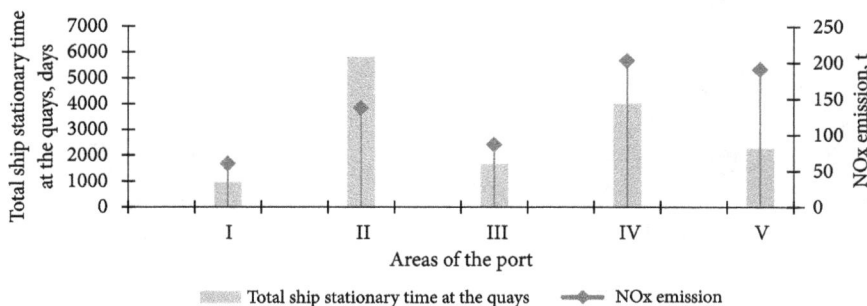

Fig. 4. Dependency between total ship stationary time at the wharves and NO$_x$ emission from the ships

In 2000 the European Commission study estimated the emissions from ships in European ports. In ports which are significantly larger than the port of Klaipeda the NO_x emissions were even 3 times higher; these ports include Rotterdam (Netherlands), where annual NO_x emission reaches up to 3800 tons, and Hamburg (Germany), were estimated annual port emission constitutes 2000 tons. In slightly smaller ports the determined pollution emissions are more similar to values obtained for Klaipeda port; for example in port of Gothenburg (Sweden) the annual NO_x emission reaches 1500 tons (Isakson *et al.* 2001; European Commission... 2002).

2.4. Results of NO_x pollution dispersion evaluation using the AERMOD model

After completing the calculations of NO_x emissions from ships, it was determined what amount of pollutants is emitted in each area of the port, and that a total of 945.62 t of NO_x components is emitted from marine ships in Klaipeda port into the air basin of Klaipeda city.

We can see from figures (Figs 5, 6, and 7), in which the average daily concentrations are presented, that the highest dispersion of pollutants is characteristic to the southern area of the port (port section V), where

concentration reaches up to 60 $\mu g/m^3$. Highest dispersion of pollutants and lowest concentrations were determined in April and August, when 20–30 $\mu g/m^3$ of NO_x components is found in Klaipeda city, and the 30–40 $\mu g/m^3$ concentration zone extends through entire port territory. Meanwhile, in December significantly higher pollution concentrations were determined both above the port and in adjacent territories. In April and August pollutants are mostly transferred into the Klaipeda city area, while in December the NO_x pollutants are more displaced to the western side, above the sea, especially in the southwestern part, where 40–60 $\mu g/m^3$ NO_x concentrations were determined; however the pollution concentrations in residential districts of Klaipeda city are considerably lower.

Similar studies (Matthias *et al.* 2010; Lonati *et al.* 2010) were conducted in the North Sea regions, where the dispersion of pollutants was modelled for summer and winter seasons. The obtained results coincide with the results of the modelling accomplished in this work. In both cases higher concentrations of nitrogen oxides were determined in winter than in summer. Results could be influenced by the fact that during the warm season of the year the chemical reactions taking place in the air are faster due to higher ambient temperatures, and pollutants are degraded more

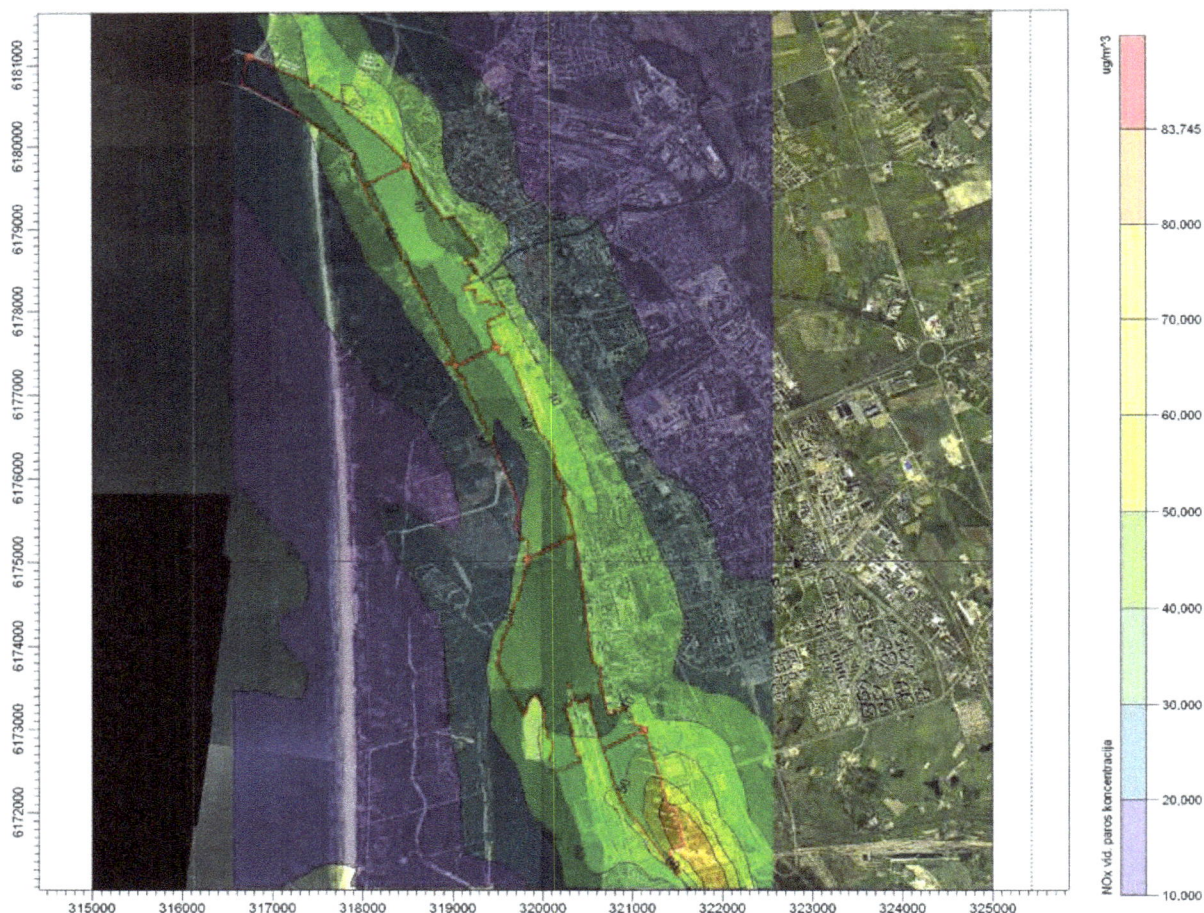

Fig. 5. Average daily concentrations of NO_x ($\mu g/m^3$) due to emissions from ships in April

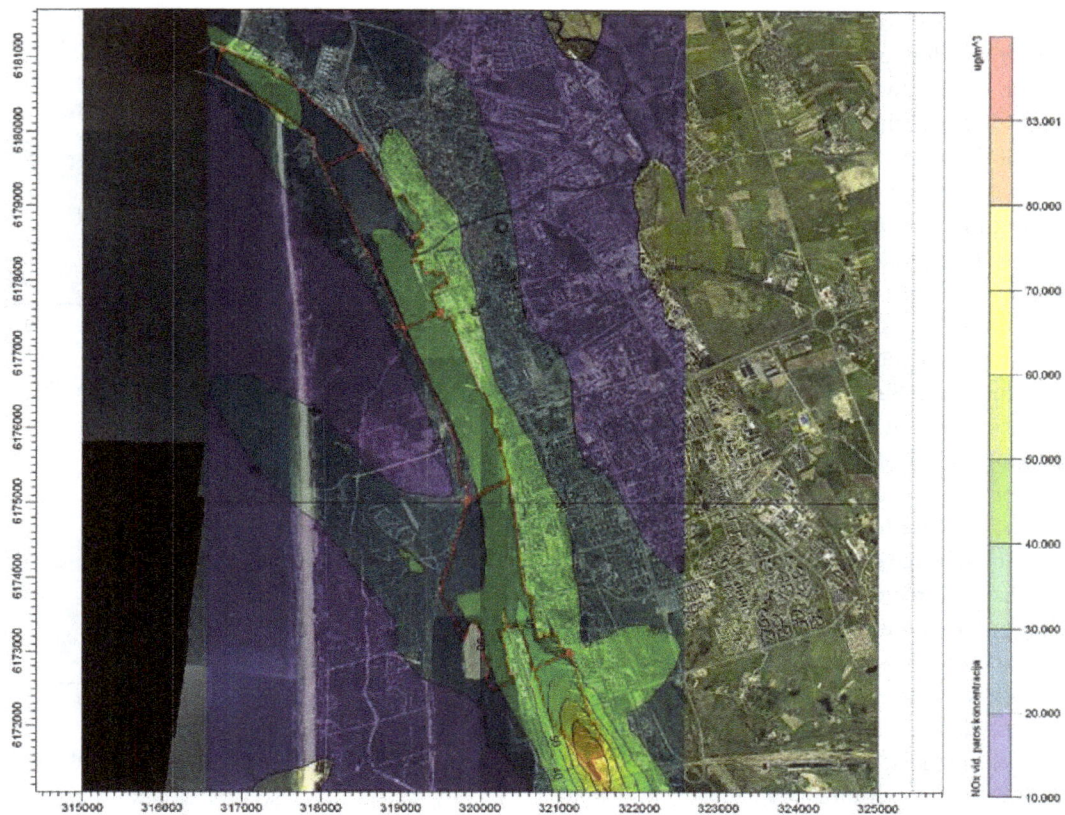

Fig. 6. Average daily concentrations of NO_x (µg/m³) due to emissions from ships in August

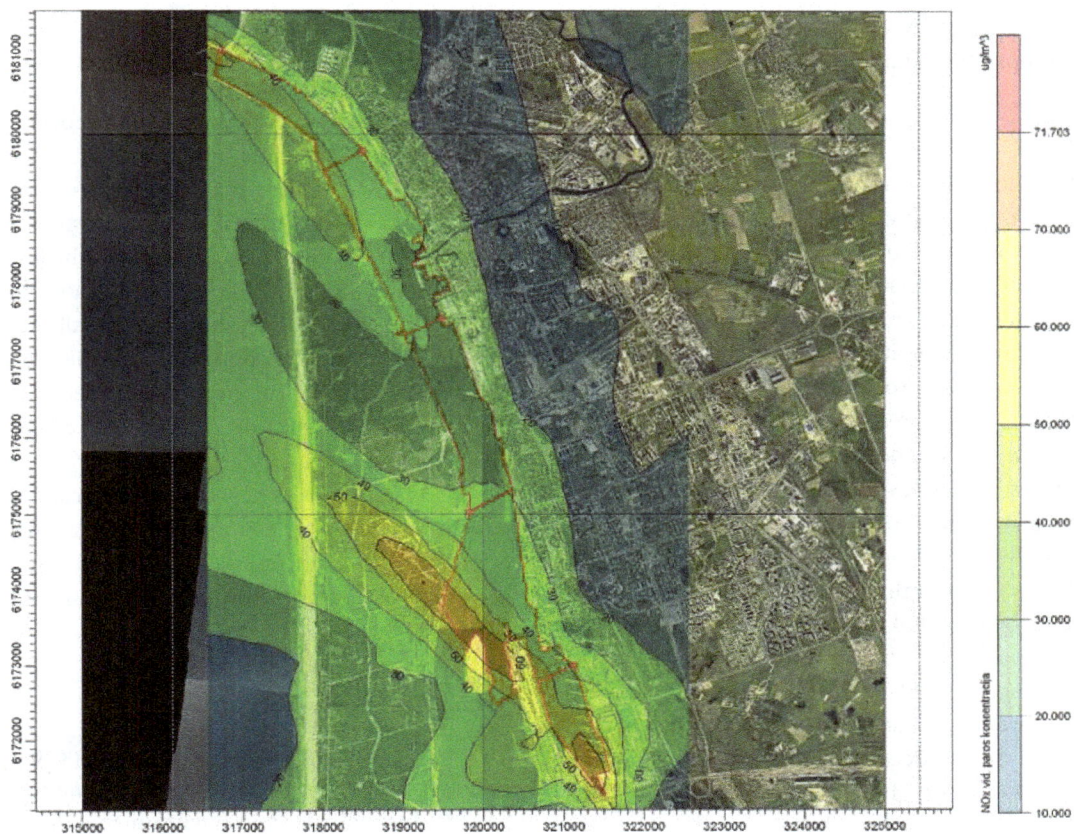

Fig. 7. Average daily concentrations of NO_x (µg/m³) due to emissions from ships in December

quickly; additionally, stronger vertical mixing of atmospheric layers is typical for summer, which also determines faster dispersion of pollutants and lower accumulation of pollutants in the near-ground atmosphere layer.

Analysis of diffusive samplers data was carried out and the change of average concentration of NO_x was determined during the months of April and August, at the same time as the ships' pollution dispersion modelling was being performed (Fig. 8).

In the Figure 8, it is shown the comparison of NO_x concentrations found by modelling and diffusive samplers in the selected locations. Similar tendencies of concentration ranges are observed in spring. The least NO_x concentrations are found in the 1st point in both cases by using diffusive samplers and by modelling. The highest concentrations of both cases are found in 8, 9, and 10th points. Direct comparison of received concentrations is impossible due to reason that research using samplers was provided in 2010 while summarized meteorological data in model was used. Also some inaccuracies maybe originate due to impact of buildings. Moreover, it is impossible to exclude an influence of auto transport if measure NO_x concentrations by using diffusive samplers. However average NO_x concentrations determined by both methods allow finding what city place in what time is more affected by ships entering to Klaipeda seaport. The similar conclusion was made by Smailys *et al.* (2013). It was noticed that in most cases the emissions from ships are estimated using theoretical calculations and include total amounts of pollutants and it is difficult to compare this amounts with experimental data.

Conclusions

1. After completing the analysis of ships operating in Klaipeda port, during which their types, tonnage, engine power and time while staying in port were assessed, it was determined, that these vessels accomplished 6837 arrivals and the total stationary time while moored to quays was 14668 days. It was estimated that the most part of NO_x enters the atmosphere from moored ships with active auxiliary engines; it forms 72% of the total NO_x emission from marine ships in Klaipeda port. It was calculated that a total of 945.6 tons/per year of NO_x components enter the air basin from ships operating in Klaipeda port.

2. After the modelling of pollutant dispersion was completed, it was determined that during winter time a higher dispersion of pollutants is typically observed near the pollution source, and during summer time pollutants are degraded more quickly and their concentrations less by up to 30% are formed above the port, but the concentrations above the residential districts of Klaipeda city become 40–50% higher. Higher input of NO_x pollutants into the air basin of Klaipeda city was observed during summer, when the dispersion of pollutants was wider, and

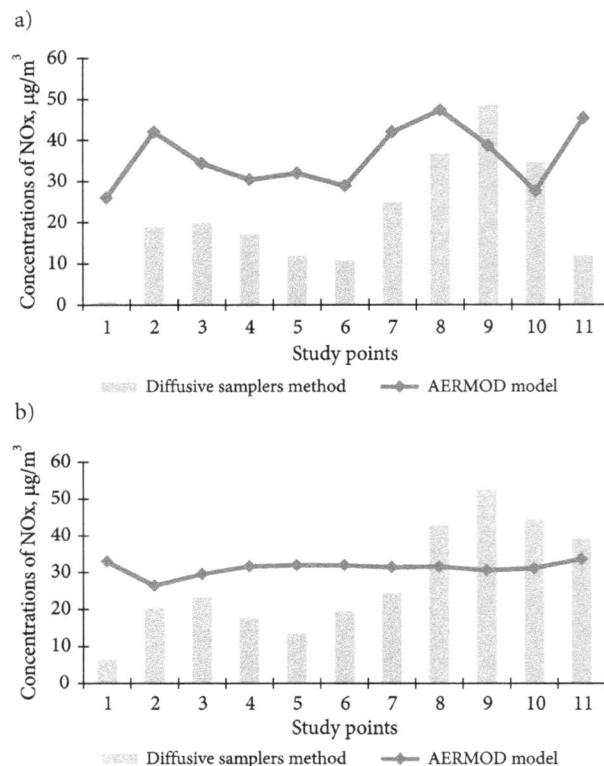

Fig. 8. Comparison of research of diffusive samplers conducted in: a) April; b) August and modelling results

during the cold time of a year pollutants were concentrated above the port and their mass extended towards the western direction, to the sea.

3. After evaluation of NO_x emission from marine ships and dispersion of pollutants in Klaipeda port, it was estimated, that pollutants from ships concentrate more in the southern part of the port during all seasons of a year. Highest NO_x pollution emissions from ships were determined in the southern part of the port, which contains the port sections IV and V. The section IV also distinguishes by the number of arriving ships and overall ship stationary time in the port; these numbers comprise 23% and 27% of the total result of the port, respectively; all this could have had some impact on formation of higher pollutant concentrations in this side of the port.

4. Average NO_x concentrations determined by both – passive measurement method with diffusive samplers and pollution dispersion modelling allow finding what city place in what time is more affected by ships entering to Klaipeda seaport.

Acknowledgments

Presented research was funded by a European Social Fund Agency grant for national project "Lithuanian Maritime Sectors' Technologies and Environmental Research Development" (Nb.VP1-3.1-ŠMM-08-K-01-019).

References

Baltrėnas, P.; Vaitiekūnas, P.; Mincevič, I. 2004. Investigation on the impact of transport exhaust emissions on the air, *Journal of Environmental Engineering and Landscape Management* 12(1): 3–11.

Bailey, D.; Solomon, G. 2004. Pollution prevention at ports: clearing the air, *Environmental Impact Assessment Review* 24(7–8): 749–774. http://dx.doi.org/10.1016/j.eiar.2004.06.005

Cooper, D. A. 2003. Exhaust emissions from ships at berth, *Atmospheric Environment* 37(27): 3817–3830. http://dx.doi.org/10.1016/S1352-2310(03)00446-1

Corbett, J.; Fischbeck, P. S.; Pandis, S. N.; Geophys, J. 1999. Global nitrogen and sulphur inventories for oceangoing ships, *Journal of Geophysical Research* 104(3): 3457–3470. http://dx.doi.org/10.1029/1998JD100040

Corbett, J.; Winebrake, J.; Green, E. H.; Kasibhatla, P.; Eyring, V.; Lauer, A. 2007. Mortality from ship emissions: a global assessment, *Environmental Science & Technology* 41(24): 3226–3232. http://dx.doi.org/10.1021/es071686z

Dalsoren, S. B.; Eide, M. S.; Endresen, O.; Mjelde, A.; Gravir, G.; Isaksen, I. S. A. 2009. Update on emissions and environmental impacts from the international fleet of ships: the contribution from major ship types and ports, *Atmospheric Chemistry and Physics* 9: 2171–2194. http://dx.doi.org/10.5194/acp-9-2171-2009

Deniz, C.; Durmusoglu, Y. 2008. Estimating shipping emissions in the region of the Sea of Marmara, Turkey, *Science of the Total Environment* 390(1): 255–261. http://dx.doi.org/10.1016/j.scitotenv.2007.09.033

Endresen, O.; Sorgard, E.; Sundet, J. K.; Dalsoren, S. B.; Isaksen, I. S. A.; Berglen, T. F.; Gravir, G. 2003. Emission from international sea transportation and environmental impact, *Journal of Geophysical Research: Atmospheres (1984–2012)* 108(D17). http://dx.doi.org/10.1029/2002JD002898

Eyring, V.; Ivar, S.; William, J.; Collins, D.; James, J.; Corbett, E.; Oyvind Endresen, F.; Roy, G.; Grainger, G.; Jana Moldanova, H.; Hans Schlager, A.; David, S. 2009. Transport impacts on atmosphere and climate: shipping, *Atmospheric Environment* 44(37): 4735–4771. http://dx.doi.org/10.1016/j.atmosenv.2009.04.059

European Commission. 2002. *Quantification of emissions from ships associated with ship movement between ports in the European Community.* Entec UK Limited. 21 p.

Georgakaki, A.; Coffey, R. A.; Lock, G.; Sorenson, S. C. 2005. Transport and Environment Database System (TRENDS): maritime air pollutant emission modelling, *Atmospheric Environmental* 39(13): 2357–2365. http://dx.doi.org/10.1016/j.atmosenv.2004.07.038

Isakson, J.; Persson, T. A.; Lindgren, E. S. 2001. Identication and assessment of ship emissions and their efects in the harbour of Goteborg, Sweden, *Atmospheric Environment* 35(21): 3659–3666. http://dx.doi.org/10.1016/S1352-2310(00)00528-8

Kowalski, J.; Tarelko, W. 2009. NO_x emission from a two-stroke ship engine. Part 1: modeling aspect, *Applied Thermal Engineering* 29(11–12): 2153–2159. http://dx.doi.org/10.1016/j.applthermaleng.2008.06.032

Krozer, J.; Mass, K.; Kothuis, B. 2003. Demonstration of environmentally sound and cost-effective shipping, *Journal of Cleaner Production* 11(7): 767–777. http://dx.doi.org/10.1016/S0959-6526(02)00148-8

Lonati, G.; Cernuschi, S.; Sidi, Sh. 2010. Air quality impact assessment of at-berth ship emissions: case-study for the project of a new freight port, *Science of the Total Environment* 409(1): 192–200. http://dx.doi.org/10.1016/j.scitotenv.2010.08.029

Luke, W.; Kelley, P.; Lefer, B. L.; Flynn, J.; Rappenglück, B.; Leuchner, M.; Dibb, J. E.; Ziemba, L. D.; Anderson, C. H.; Martin Buhr, M. 2010. Measurements of primary trace gases and NO_x composition in Houston, Texas, *Atmospheric Environment* 44(33): 4068–4080. http://dx.doi.org/10.1016/j.atmosenv.2009.08.014

Matthias, V.; Bewersdorff, I.; Aulinger, A.; Quante, M. 2010. The contribution of ship emissions to air pollution in the North Sea regions, *Environmental Pollution* 158(6): 2241–2250. http://dx.doi.org/10.1016/j.envpol.2010.02.013

McCarthy, J. E. 2009. Air pollution and greenhouse gas emissions from ships, *Congressional Research Service.* 18 p.

Miola, A.; Ciuffo, B. 2010. Estimating air emissions from ships: Meta-analysis of modeling approaches and available data sources, *Atmospheric Environment* 45(13): 2242–2251. http://dx.doi.org/10.1016/j.atmosenv.2011.01.046

Oke, S. A. 2004. On the environmental pollution problem: a review, *Journal of Environmental Engineering and Landscape Management* 12(3): 108–113. http://dx.doi.org/10.1080/16486897.2004.9636828

Petraitis, E. 2010. *Numerical simulation in environmental protection.* Vilnius: Technika. 162 p.

Premuda, M.; Masieri, S.; Bortoli, D.; Kostadinov, I.; Petritoli, A.; Giovanelli, G. 2011. Evaluation of vessel emissions in a lagoon area with ground based Multi axis DOAS measurements, *Atmospheric Environment* 45(29): 5212–5219. http://dx.doi.org/10.1016/j.atmosenv.2011.05.067

Saxea, H.; Larsena, T. 2004. Air pollution from ships in three Danish ports, *Atmospheric Environment* 38(24): 4057–4067. http://dx.doi.org/10.1016/j.atmosenv.2004.03.055

Smailys, V.; Strazdauskienė, R.; Gedgaudas, A. 2003. Air pollution from ships operating in Klaipėda seaport, *Sea and Environmental* 1(8): 22–44.

Smailys, V.; Strazdauskienė, R.; Bereisiene, K. 2009. Evaluation of possibility to identify port pollutants trace in Klaipeda city air pollution monitoring stations, *Enviromental Research, Engineering and Management* 4(50): 66–75.

Smailys, V.; Rapalis, P.; Strazdauskienė, R.; Ešmantaitė, V. 2013. Air pollution by NO_x from ships passing Klaipeda port channel, in *Proceedings of 17th international conference Transport Means*, 24–25 October, 2013, Kaunas, Lithuania, 97–100.

Schreier, M.; Kokhanovsky, A. A.; Eyring, V.; Bugliaro, L.; Mannstein, H.; Mayer, B.; Bovensmann, H.; Burrows, J. P. 2006. Impact of ship emissions on the microphysical, optical and radiative properties of marine stratus: a case study, *Atmospheric Chemistry and Physics* 6: 4925–4942. http://dx.doi.org/10.5194/acp-6-4925-2006

Trozzi, C.; Vaccaro, R.; Nicolo, L. 1995. Air pollutants emissions estimate from maritime traffic in the Italian harbours of Venice and Piombino, *Science of The Total Environment* 169(1): 257–263. http://dx.doi.org/10.1016/0048-9697(95)04656-L

Wang, C.; Corbett, J. J.; Firestone, J. 2007. Modeling energy use and emissions from north American shipping: application of the ship traffic, energy, and environmental model, *Environmental Science & Technology* 41(9): 3226–3232. http://dx.doi.org/10.1021/es060752e

Eglė ABRUTYTĖ. Bachelor of Science (Ecology and Environmental) (2005), Master of Science (Marine Environmental Engineering) (2012), Klaipėda University. Research interests: air pollution from ships, waste management.

Audronė ŽUKAUSKAITĖ. Doctor of Natural Sciences, Assoc. Prof. Dr, Department of Technological Processes, Klaipėda University. Publications: author of more than 50 research papers. Research interests: waste treatment technologies, biodegradation of oil products.

Rima MICKEVIČIENĖ. Dr Assoc. Prof. (since 2000), Department of Ship Engineering, Klaipėda University. Doctor of Technology Science (Transport Engineering, Diesel repair technologies of seagoing ships), Kaunas Technology University (KTU), 2000. Research interests: green shipping, green shipyards, shipbuilding and ship repair eco-technologies.

Vytenis ZABUKAS. Prof. Dr Habil, Department of Technological Processes, Klaipėda University. Doctor Habil of Material Engineering (Technologic Sciences). Publications: author of 110 research papers. Membership: a corresponding member of the International Academy of Ecology and Life Protection. Research interests: composite materials, technology of petroleum and environmental protections problems in petroleum plants and terminals.

Tatjana PAULAUSKIENĖ. Doctor of Technological Sciences (Vilnius Gediminas Technical University, 2008; air pollution by VOCs). Assoc. Prof., Senior Researcher, Department of Technological Processes Klaipėda University. Publications: author of 30 scientific publications. Research interests: air pollution and its reduction, environmental management.

18

EFFECTIVENESS RESEARCH ON A WAVY LAMELLAR PLATE-TYPE BIOFILTER WITH A CAPILLARY SYSTEM FOR THE HUMIDIFICATION OF THE PACKING MATERIAL APPLYING INTROINDUCED MICROORGANISMS

Kęstutis MAČAITIS[a], Antonas MISEVIČIUS[b], Algimantas PAŠKEVIČIUS[c],
Vita RAUDONIENĖ[d], Jūratė REPEČKIENĖ[e]

[a, b]*Environmental Institute, Vilnius Gediminas Technical University,
Saulėtekio al. 11, LT-10223 Vilnius, Lithuania*
[c, d, e]*Nature Research Centre, Vilnius University, Akademijos g. 2, LT-08412 Vilnius, Lithuania*

Abstract. To conduct research, a new generation plate-type air treatment biofilter for removing gaseous pollutants from air has been applied under laboratory conditions. A distinguishing feature of the packing material of the biofilter includes wavy lamellar polymer plates placed to each other and producing a capillary effect of humidification. While having such an arrangement, wavy lamellar plates also have rather wide spacing (6 mm), and therefore the employment of the structure of the plate-type packing material decreases the aerodynamic resistance of the device. A wavy porous plate is made of a polymer plate that ensures stiffness. Both sides of the wavy lamellar polymer plate have attached *steam exploded* birch fiber pellets under which, to increase plate capillarity, not-woven caulking material is put. This technological decision allows effectively enhancing the durability of the biopacking material. The work presents the results of research on the efficiency of the biodestruction process of acetone, xylene and ammonia. With reference to the conducted investigation, the high efficiency of air treatment and microbiological activity has been established. When pollutant gases (acetone, xylene and ammonia), under a velocity of 0.08 m s^{-1}, passed through the biopacking material, microbiological activity in the material reached on average 1×10^8 cfu/cm^2, and air treatment efficiency made 90.7%.

Keywords: biofilter, air treatment efficiency, capillary, biopacking material, microorganisms.

Introduction

The branches of industry like chemistry, varnishes and paints, food and oil refining terminals use a large amount of chemical materials that find different ways to be released into the atmosphere. Acetone, butanol, toluene, xylene, ammonia methane, etc. are among the most common volatile organic and inorganic compounds emitted to the atmosphere due to human factors and forming photochemical antioxidants, a high concentration of which is harmful to human health, plants and the environment in general (Baltrėnas *et al.* 2004; Jeong *et al.* 2006; Paulauskienė *et al.* 2011).

The emission levels of volatile organic compounds (acetone, xylene) to the atmosphere are significantly lower than those of combustion products, for example, CO_2, CO, SO_2 and NO_2. However, the impact of VOC on humans

and the natural environment is much stronger (Pielech-Przybylska *et al.* 2006; Paulauskienė *et al.* 2011; Yang *et al.* 2010). Also, the increasing amounts of VOC directly influence fluctuations in climate and are related to the depletion of the ozone layer (Delhomenie, Heitz 2005). Thus, the decontamination of these pollutants is an important task leading a reduction in a negative impact on the environment.

VOC (acetone, xylene) and ammonia are most frequently emitted from wastewater treatment plants in a number of industries such as foundries, chemical industry, electronics, paints, etc. (Wu *et al.* 2006; Jun, Wenfeng 2009).

At the moment, biological air treatment using certain cultures of microorganisms is one of the most promising air cleaning methods (Baltrėnas, Zagorskis 2009).

Corresponding author: Kęstutis Mačaitis
E-mail: kestutis.macaitis@vgtu.lt

Biofilters are employed for removing butanol, acetone, xylene, toluene and other volatile organic compounds from air. Equipment may be efficient when the concentration of the pollutant does not exceed 500 mg/m^3 (Baltrėnas, Zagorskis 2010).

The effectiveness of the biological air treatment process depends on the growth of microorganism cultures in the medium. At the initial moment of air treatment, under a continuous supply of pollutants to the biofilter and while activating microorganisms, they encounter surplus food and therefore grow (Domsh et al. 2007).

To initiate the biodestruction of pollutants in the biofilter and to stimulate the development of microorganisms, appropriate conditions are required. Physical factors such as humidity and temperature often affect the growth and reproduction of microorganisms (Baltrėnas et al. 2004). Water, in this case, is the most powerful medium where the metabolism reactions of materials take place; moreover, all chemical reactions occurring in live microorganisms necessitate water that makes approximately 75% and even more of the whole biomass (Zigmontienė, Žarnauskas 2011).

The fundamental element of the biological air treatment device is a filtering medium necessary as substrate for microorganisms, and at the same time, providing them with the needed nutrients. In practice, as the filtering media, the packing materials of natural origin, including compost, peat, wood chips, bark and activated sludge, are applied (Zigmontienė, Baltrėnas 2004).

The humidification systems arranged in biofilters have a strong impact on the efficiency of biological air treatment (Mohseni, Allen 2000; Shareefdeen et al. 2003). The optimal humidity of the packing material is 60–80%. At present, the applied biofilters use the humidified packing material employing the above placed humidification nozzles, water to which is supplied to the pump from a water tank. While applying the introduced humidification system, a large amount of electricity is used, anaerobic areas occur inside the filtering layer and a possibility of leaching biomass from the packing material arises, which causes a decrease in the efficiency of biofilters used for air treatment. In the events of electricity failure or the crash of the technological process, the packing material is not humidified and therefore may parch or cracks may appear.

The humidity of the packing material depends on the type of the material and the humidification system installed into the biofilter. In our researched case, the biofilter contains the implemented capillary system for the humidification of the packing material, i.e. the medium in the biofilter, due to narrow spacing (6 mm) between wavy lamellar plates, has risen with the help of not-woven caulking material and birch fiber pores.

One of the main requirements for biological air treatment equipment are the low aerodynamic resistance of the packing material. Aerodynamic resistance depends on a variety of factors like the porosity, form, fraction and humidity of the packing material. The aerodynamic resistance of the packing material also affects the treatment efficiency of the biofilter (Eldon et al. 2010). Lower aerodynamic resistance determines a better distribution of oxygen, which is involved in the metabolic processes of nutrients, in the packing material. Regular search for the methods that improve the aerodynamic processes occurring in the packing material without reducing the effectiveness of treatment is taking place (Baltrėnas, Zagorskis 2009).

Time for contact between the biopacking material and pollutant seems to be an important aspect of optimal and efficient air treatment. The lower is the air flow rate in the biofilter, the longer takes time for contact between the polluted air and the used packing material, which therefore increases the effectiveness of the biodestructive process.

Temperature is a crucial factor having an influence on the proliferation of microorganisms and on the intensiveness of biochemical reactions. Different groups of microorganisms have adapted for living under different temperatures. The microorganisms involved in the processes of pollutant biodestruction fall into a few categories, including psychrophilic, mesophilic and thermophilic. Research was conducted under a temperature of 25–30 °C in the medium and air in the biofilter. The maintained temperature was favourable for the growth and development of mesophilic microorganisms (Boswell 2010).

The conducted investigation has been aimed at using a capillary system for the humidification of the packing material made of porous wavy lamellar plates so that to establish the efficiency of the biofilter while emitting the air polluted with acetone, xylene and ammonia emissions and propagating introinduced microorganisms.

1. Research methods

1.1. Structure of the Biofilter

To conduct research, a bench of a laboratory biofilter has been employed (Fig. 1). The biofilter stands out for completely new design that allows reducing the aerodynamic resistance of the device, and the capillary system for humidifying the packing material improves the qualities of humidity retention in the packing material as well as decreases energy expenses.

The packing material of the biofilter uses wavy lamellar plates which, while moving in the contaminated air flow, increase contact between the employed pollutants and wavy lamellar partition walls containing microorganisms.

The structure of the biofilter consists of the biopacking material (cassette has been made of wavy lamellar

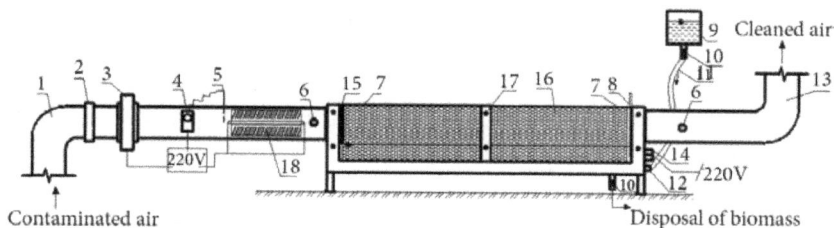

Fig. 1. Scheme for a wavy lamellar plate-type air treatment biofilter with a capillary system for the humidification of the packing material

plates), the system for maintaining the humidity and temperature of the packing material in the filter (air heater with a control thermostat and sensor; heating element of the bio-medium), ventilator (maintains a constant air flow rate in the packing material of the biofilter), air ducts for supplying and removing air from the device, air flow control valve (the adjustable air flow rate and, simultaneously, the supplied air rate), perforated card (evenly distributes air flow over the total volume of the packing material) and drain valve (excess biomass is removed from the biofilter).

The operation principle of the lab biofilter (Fig. 1) refers to the packing material in which microorganisms decompose gaseous emissions to CO_2 and H_2O. The polluted air is supplied to the biofilter through the air duct 100 mm in diameter (1). Air flow moves through the biofilter because of the ventilator arranged in the air duct for supplying the polluted air (3). The valve (2) is placed in the air duct of the polluted air and adjusts air flow and

supplied air rates. Next, the polluted air flow reaches the cassette of the biofilter (16) charged with the packing material made of porous plates. The perforated plate (15) is used for evenly distributing air flow throughout the total volume of the packing material. The polluted air moves between porous plates immersed in a liquid medium and placed in 4mm apart. The purified air stream passes through the cassette of the biofilter (16) charged with the packing material, reaches the purified air duct 100 mm in diameter and is removed to the environment. The cassette is attached to the device with the help of fasteners (7). The ducts of the polluted and purified air have installed sample slots (6). These places have the established air flow rate and temperature as well as the concentrations of pollutants supplied to and removed from the biofilter. The excess biomass is removed from the biofilter through the drain valve of the biomass (10). The required temperature of the supplied air flow is supported by an air duct heater (18) equipped with a control thermostat and sensor (4, 5). The temperature of the bio-medium is maintained using the heating element (14). To supply a nutrient-rich solution to the biofilter, a tank with adjustable valves (10, 12) and a hose (11) have been installed.

The main element of the biofilter is a cassette made of wavy lamellar plates onto which the biopacking material, i.e. not-woven caulking material and wood fiber, is attached. The dimensions of the cartridge make 900×200×200 mm.

The packing material consists of wavy porous plates vertically arranged next to each other and making a capillary humidifying effect of the packing material. Such an arrangement of the plates points to 6 mm spacing. The structure of the plates is shown in (Fig. 2).

Both sides of the wavy lamellar polymer plate have attached steam exploded birch fiber pellets. The steam explosion of birch fiber is necessary for maintaining its durability. Birch fiber is received through the steam explosion of birch sawdust in the reactor under the pressure of 32 bars and a temperature of 235 °C. Thus, changes in the chemical structure of wood prevent birch fiber from decay in a humid environment, and therefore the durability of the packing material of the biofilter increases. To enlarge the capillarity of the plate along with the uplift height of

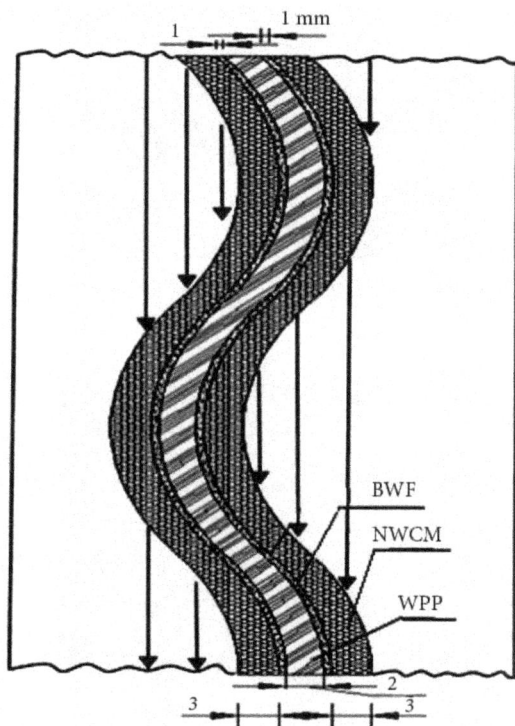

Fig. 2. The structure of the porous plate (WPP – wavy lamellar polymer plate, BWF – birch wood fiber, NWCM – not-woven caulking material); air flow direction

the bio-medium, not-woven caulking material is attached to birch fiber. The dimensions of the porous plate in the bio-filter make 900×200×10 mm.

The relative and absolute humidity of materials has been established with reference to the weight based on a decrease in the amount of mass. Porosity has been found out with the help of saturation, whereas density – employing the weighing method.

Choosing materials has been determined by their inner structure. The composition of the material has been defined applying the method of electron microscopy. Scanning electron microscopy has been done using field emission scanning electron microscope JEOL ISM – 7600 F magnifying from 25 to 1 000 000 times. The accelerating voltage of electrons varied from 0,1kV to 30 kV. Image resolution was up to 5120×3840 pixels. The structure of the investigated materials is shown in (Fig. 3) and (Fig. 4).

1.2. Maintaining humidity in the biofilter with a wavy lamellar structure

The biofilter is immersed in a solution saturated with biogenic elements (Fig. 1). The solution used for research purposes has been made of K_2HPO_4 – 1 g, KCl – 0.5 g,

$MgSO_4 \times 7H_2O$ – 0.5 g, $FeSO_4 \times 7H_2O$ – 0,1 g, $NaNO_3$ – 0.90 g and 1000 g distilled water (Baltrėnas, Zagorskis 2009; Trejo-Aguilar et al. 2005; Liao et al. 2008; Chang, Lu 2003; Wright 2005; Dorado et al. 2008; Mansour et al. 2011).

The depth of the soaked porous plates reached 55 mm. Due to the porous structure and spacing of wavy lamellar plates placed to each other within a distance of 6 mm, the capillary effect of humidifying the packing material takes place. The humidity of the packing material is found using measuring device M0290, the operation principle of which is based on the method for measuring electrical resistance. The interval of calculating humidity varies from 0 to 99.9%, and measurement error is ±0.1%.

1.3. Establishing pH and temperature of the bio-medium

The required pH and temperature of the solution (bio-medium) saturated with biogenic elements were maintained using buffer solutions (Baltrėnas, Zagorskis 2010). pH was found out with reference to LST ISO 10523 standard. To set the necessary temperature and pH, the Mettler Toledo

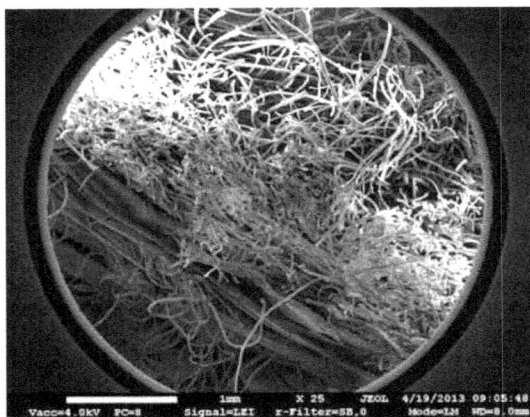

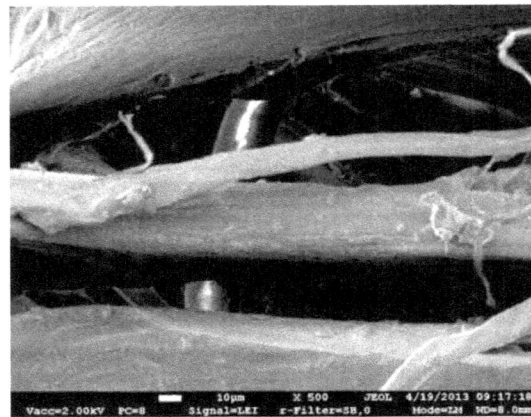

Fig. 3. The structure of not-woven caulking material, magnified 25 times (left) and 500 times (right)

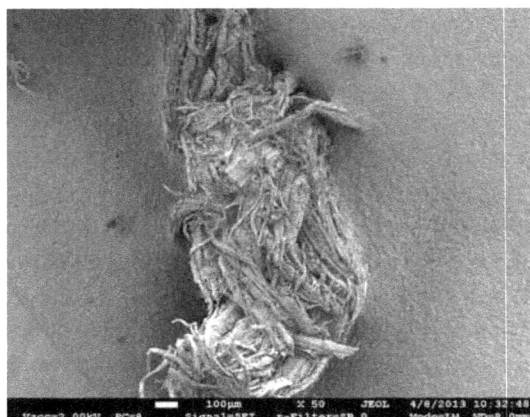

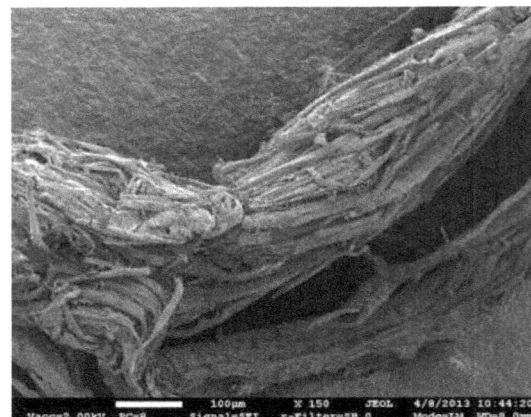

Fig. 4. The structure of birch wood fiber, magnified 50 times (left) and 150 times (right)

measuring device was used. The measurement interval of the device is from 0 to 14, and measurement error is ±0.01. The records of pH index and temperature were displayed on a daily basis.

To preserve a constant temperature of 30 °C in the bio-medium, a heating element of the bio-medium in the lower part of the biofilter was arranged. In order the air supplied to the biofilter should not cool down the bio-medium on the packing material, the duct supplying pollutants has an installed channel heater ensuring a constant 30 °C temperature of the air flow supplied to the biofilter.

1.4. Microbiological research

To identify microorganisms and to calculate their quantity, the flush (suspension dilution) method (presented in the developed methodology) was used. From each sample, 1g-sized piece is weighed and placed in a flask with 90 ml of 0.8% NaCl where suspension procedures take place. To compare different samples with each other, calculations are done drying the specimen material to a constant weight, and then, the number of the microorganisms of 1 g of a dry weight of the biofilter material is defined. In addition, the area of the weighed piece is measured, and the number of microorganisms per 1 cm² is calculated.

The composition of the nutrient agar includes water – 1000 g, agar – 15.0 g, peptone – 5.0 g, NaCl – 5.0 g, yeast extract – 5.0 g and meat extract – 1 g. Seeding takes place three times. Under a stagnant medium, the dish moves having a mix of the medium and suspension to make them uniformly spread at the bottom of the dish. At a later stage, Petri dishes, including microorganisms, are incubated in the thermostat.

Petri dishes with bacteria are incubated in the thermostat for 2–3 days and with micromycetes – 5–7 days at a temperature of 26–28 °C. The colonies of micromycetes and bacteria are calculated figuring out the amount of rudiments in 1g of the tested substance. The number of live cells is defined having multiplied the average number of colonies in the dishes and the coefficient of dilution.

Micromycetes were distinguished on the agarized beer wort. Seeds were incubated in Petri dishes under a temperature of +28 °C for 5–7 days.

Pure cultures of micromycetes are identified employing classic methods with reference to the definitions of micromycetes (Chaverri, Samuels 2003; Samson, Frisvad 2004; Domsh et al. 2007; Pečiulytė, Bridžiuvienė 2008, etc.).

Yeast were distinguished on the nutrient media of Sabouraud agar with chloramphenicol (Liofilfem, Italy) and Rose Bengal CAF agar (Liofilchem, Italy). Seeds took place in Petri dishes at a temperature of +28 °C for 3–4 days.

Yeast were discovered applying Api 20 C AUX (bio-Mérieux, France) identification systems.

Bacillus cereus media were agarized. Crop prepared bacterial suspension included 1:10, 1:100, 1:1000, 1:10.000, 1:100.000, 1:1.000.000, 1:10.000.000, 1:100.000.000, 1:1.000.000.000, 1:10.000.000.000.

The distinguished bacteria were identified according to their morphological and physiological properties and compared with data provided in literature. Bacteria were described according to Bergey's Manual of Systematic Bacteriology (Palleroni 1984; Garrity et al. 2005).

1.5. Activating the packing material and establishing the efficiency of the biodestruction process

The conducted research involved the air flow that passed through the packing material of the biofilter and was contaminated with acetone, xylene and ammonia vapours. The air flow rate between wavy lamellar plates reached, on average, 0,08 m/s and was established and observed on a daily basis applying precision humidity measuring instrument Testo 400 with a thermocouple, the precision of the measurements of which, under the air flow rate from 0 m/s to 2 m/s, makes ±0.01 m/s.

The initial concentration of pollutants reaches, on average, 5.7 mg/m³. The device was supplied with acetone vapour for 15 minutes 4 times a day. The next day, the concentration of the organic compound was increased 20±5 mg/m³ thus extending the delivery time of acetone vapour up to 1 hour. The research on supplying the air polluted with acetone vapour took 15 days. Afterwards, the pollutant was replaced with xylene, and investigation was carried on gradually increasing the concentration of acetone vapour delivered to the biofilter. The study on emitting the air polluted with xylene vapour took 5 days. Upon the completion of research on xylene, the air contaminated with ammonia vapours passed through the packing material. Investigation into ammonia lasted for 6 days.

The efficiency of pollutant biodegradation is calculated having found the concentration of the pollutant before and following the treatment device. Pollutant concentration is established with the help of portable VOC monitor MiniRae 2000, the measurement limits of which are in the range from 0 to 7000 mg/m³. Measurement accuracy, when pollutant concentration is in the range from 0 to 100 mg/m³, makes 0.1 mg/m³, and when pollutant concentration exceeds 100 mg/m³ – 1 mg/m³.

2. Results and discussion

The biopacking material is one of the basic elements of the biofilter. A crucial point before conducting investigation is the formation of the packing material upon which the physical qualities of the packing material depend and determine the efficiency of the biodegradation process. In this case, the porosity of the packing material and capillary play a leading role. It has been established that a higher porosity of the material causes better absorption of pollutants from the contaminated air (Beyaz *et al.* 2010). Herewith, a higher porosity of the material makes the capillary humidification of the packing material more effective. Due to a higher porosity of the packing material and effective capillary, the biodegradation process becomes more efficient.

Investigation into the structure of materials has revealed that the major part of not-woven caulking material is made from small filaments of 15 to 25 μm thick. The spaces between the filaments are 5 to 10 times larger than the thickness of the filaments themselves (Fig. 3, left). Such filament distribution allows making a biofilm thus avoiding anaerobic zones harmful to microorganisms. Chaotic filament distribution allows increasing the specific surface area of the material and the volume of the bio-medium in the material. The picture of the material magnified up to 500 times displays the filaments of 120–180 μm thick making the capillaries of 10–30 μm thick (Fig. 3, right). Steam exploded birch fiber also has an irregular surface, porous structure, and thus a higher specific surface area (Fig. 4). As indicated, birch wood fiber is made of like the many tiny "straws" arranged in parallel to each other and reaching 15–30 μm thick. Wood fiber in the packing material is required so that the microorganisms in the biomedia should take up organic carbon that is in the fiber.

Before starting the exploitation of biological air treatment equipment, the installed biopacking material is biologically activated emitting it through the air contaminated with organic pollutants (Baltrėnas *et al.* 2004). The biologically activated packing material is fully accepted as such when covered with a thin layer of the biofilm (5–30 μm thick) with a population of microorganisms. In this case, the packing material was activated up to the 10th day of the experiment. The packing material was activated supplying acetone to the polluted air thus increasing the concentration of acetone and monitoring the treatment efficiency of the biofilter.

Fig. 5 shows that the air treatment efficiency of the biofilter increased up to the 10th day of the experiment. This is exactly the time when the highest established treatment efficiency reached 94.7%, which was caused by a growing amount of bacteria up to $(1.0\pm0.2)\times10^{10}$ cfu/cm². The initial acetone concentration before treatment made 25 mg/m³, whereas at the end of the experiment it reached 700 mg/m³.

On the 10th day, 220 mg/m³ acetone vapour concentration was supplied to the biofilter. Under the pollutant concentration of 300 mg/m³, the air treatment efficiency of the biofilter made 93.4%.

Later, a growth in the amount of pollutant concentration in air up to 500 mg/m³ reduced the air treatment efficiency of the biofilter to 86%. On the 16th day of the

conducted investigations, the concentration of acetone was increased to 700 mg/m³, which resulted in the treatment efficiency of 84%.

The analysis of the below chart showing the treatment efficiency of acetone demonstrates that, under steady analysis (starting from the 10th day of the experiment), treatment efficiency was higher than 80%.

The average amount of bacteria after the 10th day of research was $(6.8\pm0.2)\times10^8$ cfu/cm², yeast – $(3.8\pm0.4)\times10^6$ cfu/cm² and micromycetes – $(3.2\pm0.6)\times10^7$ cfu cm².

Study was carried out under the treated air flow and made 1.08 l/s. Time for contact between the packing material and the polluted air reached 11.39 s.

A comparison of the research findings of other scientists such as Chang and Lu (2003) who investigated the air contaminated with acetone vapour and the results of our examination demonstrate that treating efficiency reaches from 80% to 85%, and the concentration of acetone vapour in the supplied air makes from 175 ± 10 mg/m³ to 700 ± 35 mg/m³. In our case, the treatment efficiency of 80% is available when the concentration of acetone in air gets to ~ 600 mg/m³.

Fig. 6 shows that treatment efficiency within the biodestruction process of xylene in all days of the performed investigation reached more than 80%. The initial pollutant

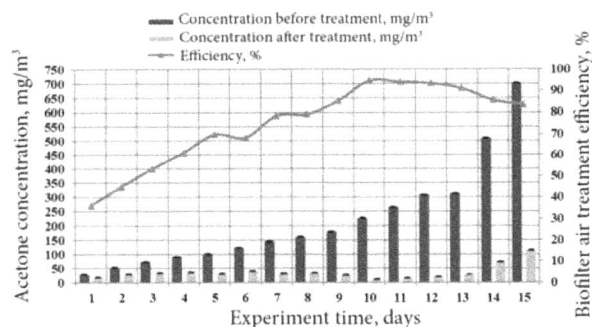

Fig. 5. Acetone concentration before and after treatment (mg/m³) and the efficiency of biofilter treatment under the application of wavy lamellar plates and the packing material made of wood fiber and not-woven caulking material

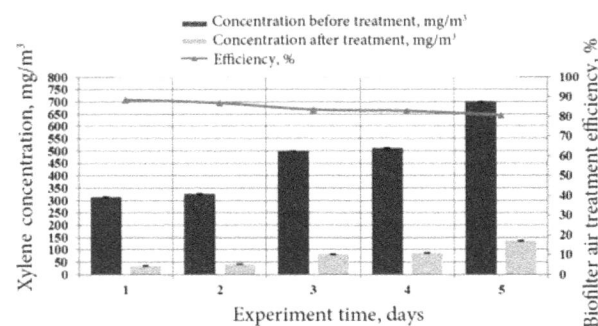

Fig. 6. Xylene concentration before and after treatment (mg/m³) and the efficiency of biofilter treatment under the application of wavy lamellar plates and the packing material made of wood fiber and not-woven caulking material

concentration before treatment was 313 mg/m³, whereas at the end of the experiment – 699 mg/m³. Fig. 6 also demonstrates that the air treatment efficiency of the biofilter, under the pollutant concentration of 300 mg/m³ (1st day of research), achieved 87.7%, under a concentration of 500 mg/m³ (3rd day of research) – 83.1% and under a concentration of 700 mg/m³ (5th day of research) – 80.6%.

An increase in pollutant concentration in the supplied air resulted in a gradual decrease of 3–4% in air treatment. A reduction of microorganisms in the packing material can also be attached to this phenomenon. At the beginning of the research, the average amount of bacteria was $(1.8\pm0.3)\times10^9$ cfu/cm², yeast – $(4.1\pm0.1)\times10^6$ cfu cm² and micromycetes – $(1.2\pm0,6)\times10^5$ cfu/cm², whereas at the end (5th day of research) – $(4.6\pm0.0)\times10^8$ cfu/cm², yeast – $(2.2\pm0.1)\times10^5$ cfu/cm² and micromycetes – $(1.4\pm0.3)\times10^6$ cfu/cm². Japanese scientists Jeong et. al. (2008) also investigated the biofilter supplying it with the concentrations of xylene vapour. The obtained results indicate that, under the biofilter productivity of 50 g/m³/h, treatment efficiency makes 80–85%. Wu et al. (2006) studied biofilter treatment efficiency under the device productivity

of the same 50 g/m³/h and found out that the biodestruction process of xylene amounted to 70%. As regards our research, under similar productivity of the biofilter, the efficiency of the biodestruction process makes 83%.

While treating the air contaminated with acetone vapour, the efficiency of the biodestruction process was higher than that in the case of xylene vapour. This could be determined by a lower coefficient of gas solubility in water. The dependence of gas solubility in the bio-medium is determined by Henry's Law (Miller, Allen 2005) which provides that, "at a constant temperature, the amount of a given gas that dissolves in a given type and volume of liquid is directly proportional to the partial pressure of that gas in equilibrium with that liquid". Thus, an increase in temperature decreases gas solubility in the medium while a decrease increases it (Miller, Allen 2004).

Figure 7 displays ammonia concentration in the air supplied to and removed from the device and the dependence of air treatment efficiency on time considering the air flow rate in the packing material (0.079 m/s). The research was carried out under the treated air flow making 1.1 l/s. Time for contact between the biopacking material and polluted air was 11.42 s.

Figure 7 also shows that the treatment efficiency of the ammonia biodestruction process, in all days of investigation, similarly to xylene, reached more than 80%. The air treatment efficiency of the biofilter, under 300 mg/m³ (1st day of research), in the concentration of ammonia vapour got to 88.1%, the one under 500 mg/m³ (3rd day of research) – 83.3% and that of 700 mg/m³ (6th day of research) – 80.9%.

The tendency towards the efficiency of removing ammonia from air, like xylene, remained similar – an increase in the concentration of ammonia in the supplied air resulted in a gradual decrease of 4–5% in the efficiency of taking away the pollutant. This influenced a reduction in microorganisms in the packing material. At the beginning of research, the average amount of bacteria was $(7.8\pm1.0)\times10^8$ cfu/cm², yeast – $(2.6\pm0.1)\times10^7$ cfu/cm² and micromycetes – $(5.6\pm0.9)\times10^6$ cfu/cm², whereas at the end of research (6th day) – $(1.6\pm0.2)\times10^8$ cfu/cm², yeast – $(6.1\pm0.2)\times10^6$ cfu/cm² and micromycetes – $(2.4\pm0.2)\times10^5$ cfu/cm². The biopacking material contained the largest established amount of bacteria and the smallest amount of micromycetes.

Figure 8 shows the use of the natural microorganism associations multiplied naturally on the biopacking material from ambient air, which is the reason for achieving a lower air treatment efficiency of the biofilter.

While conducting experimental investigations into wavy lamellar plates and using the packing material made of not-woven caulking material and wood fiber, both the self-propagated and selected microorganisms survived under the same maintained conditions: the average humidity

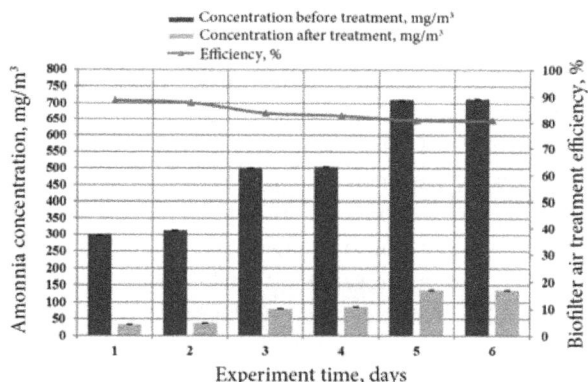

Fig. 7. Ammonia concentration before and after treatment (mg/m³) and the efficiency of biofilter treatment under the application of wavy lamellar plates and the packing material made of wood fiber and not-woven caulking material

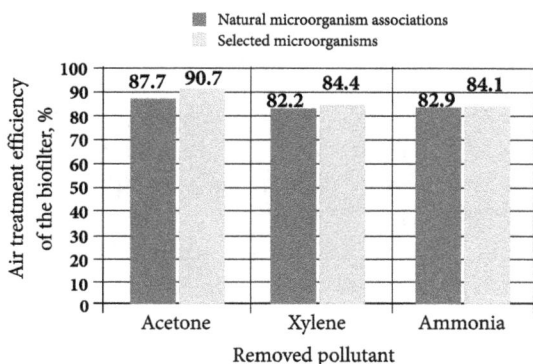

Fig. 8. A comparison of the air treatment efficiency of the biofilter using natural microorganisms and selected microorganisms

was from 70 to 80%, the humidity of the packing material made 60–65 %, air temperature reached 26–28 °C and air flow rate – 0.08 m/s. Contact time for the bio-medium remained the same thus making 11.25 s, pH index varied between 7.2 and 7.3 and the temperature of the medium was 30±1 °C. Thus, it can be accepted that the conditions for the development of the self-propagated and selected microorganisms were favourable enough, and therefore the findings of the performed research can be compared with each other.

The presented (Fig. 8) demonstrates that, along the entire investigation, the air treatment efficiency of the biofilter was more than 80%. The conducted research also found that the highest air treatment efficiency was observed using the selected microorganisms, and while treating the air contaminated with acetone vapour, it reached 90.7%. Study on air treatment efficiency under conditions when natural microorganism associations grew and multiplied on the biopacking material suggest the efficiency of 87.7%. Next to acetone, the air contaminated with xylene vapour was supplied to the biofilter, and air treatment efficiency was higher employing the selected microorganisms (84.4%) rather than in the case of the self-propagated ones (82.2%). The research on xylene and supplying the air contaminated with ammonia to the biofilter highlight the same tendency displaying higher air treatment efficiency under the use of the selected microorganisms, which made 84.1%, i.e. 1.2% more than the application of natural microorganism associations.

The species composition of microorganisms also has an impact on the air treatment efficiency of the biofilter. The efficiency of using the microorganisms selected and self-propagated on the packing material is higher, as the microorganisms themselves have been specially selected so that to adopt them to the biopacking material and to the three types of pollutants supplied to the biofilter. Thus, the conducted experimental investigation into air treatment has shown that air treatment efficiency, yet at the beginning of the experiment, reaches approximately 50%, whereas as regards natural microorganism associations, it makes only 50%. In the latter case, various species of the natural microorganism associations multiply on the biopacking material; however, only those able to adapt to the contaminant supplied to the biofilter and to its concentration survive.

The advantage of using the selected microorganisms for removing pollutants from the supplied air is that large concentrations of polluted vapours can be supplied to the biofilter, because the selected microorganisms unlike the self-propagated ones do not require adaptation, and therefore air treatment efficiency is higher.

The conducted research has demonstrated that acetone and ammonia rather than xylene are best removed from the contaminated air. This is due to the fact that

acetone and ammonia are highly soluble in water, whereas xylene is not. For their growth, microorganisms on the biopacking material absorb airborne pollutants through water. Therefore, microorganisms easier absorb ammonia and acetone rather than xylene, which allows making a conclusion that the higher is the solubility of the pollutant in water, the greater is treatment efficiency.

Conclusions

The conducted research has shown that the use of the capillary system for the humidification of the packing material made of porous polymer wavy lamellar plates with wood fiber and not-woven caulking material results in the air treatment efficiency of the biofilter, which is more than 80%.

The performed investigation demonstrates that the highest air treatment efficiency has been achieved employing the selected microorganisms. While treating the air contaminated with acetone vapour, efficiency reached 90.7%, xylene made 84.4% and ammonia – 84.1%. The average use of the selected microorganisms points to air treatment efficiency which is 2.4% higher than that employing natural microorganism associations.

Experimental investigation into the selected microorganisms applied for air treatment has revealed that air treatment efficiency, yet at the beginning of research, achieves about 50%, whereas in the case of the self-propagated ones, only 50% can be observed. In the latter case, various species of natural microorganism associations multiply on the biopacking material; however, only those able to adapt to the contaminant supplied to the biofilter and to its concentration survive.

The carried out research has established that acetone and ammonia rather than xylene are best removed from the contaminated air. This is due to the fact that acetone and ammonia are highly soluble in water, whereas xylene is not.

Bacteria – $(1.0±0.2)×10^{10}$ cfu/cm^2, yeast – $(6.1±0.2)×10^6$ cfu/cm^2 and micromycetes – $(3.2±0.6)×10^7$ cfu/cm^2 are among the cultures most frequently found between the selected microorganisms of the biopacking material.

Acknowledgements

The research has been carried out within the framework of the project *Applied research and technological development of plate type air treatment biofilter with a capillary humidification system for packing material "BIOFILTER"* (Project No. VP1-3.1-ŠMM-10-V-02-015) under the Operational Programme for the Development of Human Resources 2007–2013, priority axis 3 "Strengthening researchers abilities", measure VP1-3.1-ŠMM-10-V "Promotion of high level international research". The project has been funded by the European Social Fund, supported and co-funded by the European Union and the Republic of Lithuania. The beginning of project implementation was on 1 February 2013. Total duration of the project is 24 months. The aim of the project is high-level international research implementation in order to create a more efficient model of the plate type air treatment biofilter with a capillary humidification system for the packing material. The research was performed employing the facilities of the Open Access Centre of the Joint Nature Research Centre.

References

Baltrėnas, P.; Zagorskis, A. 2010. Investigation into the air treatment efficiency of biofilters of different structures, *Journal of Environmental Engineering and Landscape Management* 18: 23–31. http://dx.doi.org/10.3846/jeelm.2010.03

Baltrėnas, P.; Zagorskis, A. 2009. Investigation of cleaning efficiency of a biofilter with an aeration chamber, *Journal of Environmental Engineering and Landscape Management* 17: 12–19. http://dx.doi.org/10.3846/1648-6897.2009.17.12-19

Baltrėnas, P.; Zigmontienė, A.; Vaiškūnaitė, R. 2004. *Oro valymo biotechnologijos* [Biotechnology of air purification], Vilnius: Technika. 205 p.

Beyaz, S.; Darkrim Lamari, F.; Weinberger, B.; Langlois, P. 2010. Nanoscale carbon material porosity effect on gas adsorption, *International Journal of Hydrogen Energy* 35(1): 217–224. http://dx.doi.org/10.1016/j.ijhydene.2009.10.007

Boswell, J. 2010. Understanding biofilters, *Pollution Engineering* 42(2): 14–18.

Chang, K; Lu, C. 2003. Biofiltration of isopropyl alcohol and acetone mixtures by a trickle-bed air biofilter, *Process Biochemistry* 39: 415–423. http://dx.doi.org/10.1016/S0032-9592(03)00096-7

Chaverri, P.; Samuels, G. J. 2003. Hypocrea Trichoderma (Ascomycota, Hypocreales, Hypocreaceae): species with green ascospores, Studies in Mycology 48: 1–119.

Delhomenie, M.; Heitz, M. 2005. Biofiltration of air: a review, *Critical Reviews in Biotechnology* 25: 53–72. http://dx.doi.org/10.1080/07388550590935814

Domsh, K. H.; Gams, W.; Anderson, T. H. 2007. *Compendium of soil micromicetes*, 2 nd ed., taxonomically revised by W. Gams. Eching: IHW-Verlag. 672 p.

Dorado, A. D.; Baquerizo, G.; Maestre, J. P.; Gamisans, X.; Gabriel, D.; Lafuente, J. 2008. Modeling of a bacterial and fungal biofilter applied to toluene abatement: kinetic parameters estimation and model validation, *Chemical Engineering Journal* 140(1–3): 52–61. http://dx.doi.org/10.1016/j.cej.2007.09.004

Eldon, R.; Murthy, D. V. S.; Swaminathan, T. 2010. Effect of flow rate, concentration and transient-state operations on the performance of a biofilter treating xylene vapors, *Journal of Water, Air & Soil Pollution* 211(1–4): 79–93.

Garrity, G. M.; Bell, J. A.; Lilburn, T. 2005. Order VI. Legionellales ord. nov., in D. J. Brenner, N. R. Krieg, J. T. Staley (Eds.). *Bergey's manual of systematic bacteriology*, 2nd ed., vol. 2: The Proteobacteria, part B: The Gammaproteobacteria. Springer. 210 p.

Jeong, E.; Hirai, M.; Shoda, M. 2006. Removal of *p*-xylene with pseudomonas sp. NBM21 in biofilter, *Journal of Bioscience and Bioengineering* 102(4): 281–287. http://dx.doi.org/10.1263/jbb.102.281

Jeong, E.; Hirai, M.; Shoda, M. 2008. Removal of *o*-xylene using biofilter inoculated with Rhodococcus sp. BTO62, *Journal of Hazardous Materials* 152(1): 140–147. http://dx.doi.org/10.1016/j.jhazmat.2007.06.078

Jun, Y.; Wenfeng, X. 2009. Ammonia biofiltration and community analysis of ammonia-oxidizing bacteria in biofilters, *Bioresource Technology* 100(17): 3869–3876. http://dx.doi.org/10.1016/j.biortech.2009.03.021

Liao, Q.; Tian, X.; Chen, R.; Zhu, X. 2008. Mathematical model for gas–liquid two-phase flow and biodegradation of a low concentration volatile organic compound (VOC) in a trickling biofilter, *International Journal of Heat and Mass Transfer* 51(7–8): 1780–1792. http://dx.doi.org/10.1016/j.ijheatmasstransfer.2007.07.007

LST EN ISO 10523: 2012. Water quality – Determination of pH. Lietuvos standartizacijos departamentas.

Mansour, H. B.; Ghedira, K.; Barillier, D.; Ghedira, L. C.; Mosrati, R. 2011. Degradation and detoxification of acid orange 52 by Pseudomonas putida mt-2: a laboratory study, *Environmental Science and Pollution Research* 18(9): 1527–1535. http://dx.doi.org/10.1007/s11356-011-0511-7

Miller, M. J.; Allen, D. G. 2004. Transport of hydrophobic pollutants through biofilms in biofilters, *Chemical Engineering Science* 59: 3515–3525. http://dx.doi.org/10.1016/j.ces.2004.05.011

Miller, M. J.; Allen, D. G. 2005. Biodegradation of α-pinene in model biofilms in biofilters, *Environmental Science & Technology* 39(15): 5856–5863. http://dx.doi.org/10.1021/es048254y

Mohseni, M.; Allen, D. G. 2000. Biofiltration of mixtures of hydrophilic and hydrophobic volatile organic compounds, *Chemical Engineering Science* 55: 1545–1558. http://dx.doi.org/10.1016/S0009-2509(99)00420-0

Palleroni, N. J. 1984. Genus I. Pseudomonas Migula 1984, 237AL, in N. R. Krieg, J. G. Holt (Eds.). *Bergey's manual of systematic bacteriology*, vol. 1, 141–199. Baltimore: Williams & Wilkins.

Paulauskienė, T.; Zabukas, P.; Vaitiekūnas, P.; Žukauskaitė, A.; Kvedaras, V. 2011. Investigation of volatile organic compound (VOC) emission beyond the territory of oil terminals during different seasons, *Journal of Environmental Engineering and Landscape Management* 19(1): 44–52. http://dx.doi.org/10.3846/16486897.2011.558994

Pečiulytė, D.; Bridžiuvienė, D. 2008. *Lietuvos grybai. II tomas. Skurdeniečiai ir pelėsiečiai*. Botanikos institutas, Vilnius. 264 p.

Pielech-Przybylska, K.; Ziemiński, K.; Szopa, J. S. 2006. Acetone biodegradation in a trickle-bed biofilter, *International Biodeterioration & Biodegradation* 57(4): 200–206. http://dx.doi.org/10.1016/j.ibiod.2005.12.005

Samson, R. A.; Frisvad, J. 2004. Polyphasic taxonomy of Penicillium subgenus Penicillium. A guide to identification of food and air-borne terverticillate Penicillia and their mycotoxins, Studies in Mycology 49: 1–174.

Shareefdeen, Z.; Herner, B.; Webb, D.; Wilson, S. 2003. Biofiltration eliminates nuisance chemical odors from industrial air streams, *Journal of Industrial Microbiology and Biotechnology* 30: 168–174.

Trejo-Aguilar, G.; Revah, S.; Lobo-Oehmichen, R. 2005. Hydrodynamic characterization of a trickle bed air biofilter, *Chemical Engineering Journal* 113(2–3): 145–152. http://dx.doi.org/10.1016/j.cej.2005.04.001

Wright, W. F. 2005. Transient response of vapor-phase biofilters, *Chemical Engineering Journal* 113(2–3): 161–173. http://dx.doi.org/10.1016/j.cej.2005.04.009

Wu, D.; Quan, X.; Zhao, Y.; Chen, S. 2006. Removal of *p*-xylene from an air stream in a hybrid biofilter, *Journal of Hazardous Materials* 136(2): 288–295. http://dx.doi.org/10.1016/j.jhazmat.2005.12.017

Yang, C.; Chen, H.; Zeng, G.; Yu, G.; Luo, S. 2010. Biomass accumulation and control strategies in gas biofiltration, *Biotechnology Advances* 28(4): 531–540.

http://dx.doi.org/10.1016/j.biotechadv.2010.04.002

Zigmontienė, A.; Baltrėnas, P. 2004. Biological purification of air polluted with volatile organic compounds by using active sludge recirculation, *Journal of environmental engineering and landscape management* 12(2): 45–52.

Zigmontienė, A.; Žarnauskas, L. 2011. Investigation and analysis of air-cleaning biofilter hybrid biocharge quantitative and qualitative parameters, *Journal of Environmental Engineering and Landscape Management* 19(1): 81–88. http://dx.doi.org/10.3846/16486897.2011.557472

Kęstutis MAČAITIS. Bachelor of Science (Roads and Railways) from VGTU (2011). Master of Environmental and Climate Engineering from the Department of Environmental Engineering, Vilnius Gediminas Technical University (VGTU), 2013. Publications: the author of 3 research papers. Research interests: air pollution, electromagnetic fields, waste management technologies, physical pollution.

Antonas MISEVIČIUS. Bachelor of Science (Environmental Engineering) from VGTU (2009). Research interests: Master of Environmental Management and Cleaner Production from the Department of Environmental Engineering, Vilnius Gediminas Technical University (VGTU), 2011. Publications: the author of 5 research papers. Research interests: waste management technologies, waste management, recycling, air pollution.

Algimantas PAŠKEVICIUS. Head of the Laboratory of Biodeterioration Research at the Institute of Botany of Nature Research Centre, Doctor of Science (Biomedicine Sciences). 110 scientific publications, 6 invention patents. Research interests: fungal physiology and biochemistry, epidemiology and biology of pathogenic micromycetes, biotechnologies.

Vita RAUDONIENĖ. Dr, Researcher at the Laboratory of Biodeterioration Research at the Institute of Botany of Nature Research Centre. 36 scientific publications. Main fields of scientific investigation: microscopic micromycetes under various ecological conditions and their ecological and physiological peculiarities, the effect of ecological factors on the activity of microorganisms, significance of micromycetes for the degradation of the plant remnant in the lignin-cellulose complex.

Jūratė REPEČKIENĖ. Dr, Researcher at the Laboratory of Biodeterioration Research at the Institute of Botany of Nature Research Centre. 71 scientific publications. Main fields of scientific investigation: relationship of micromycetes and bacteria population in soil and their physiological peculiarities. Scientific research interests include the ability of microorganisms to degrade different waste and pollutant and to produce secondary metabolites.

19

KINETIC AND ISOTHERM STUDY OF CUPPER ADSORPTION FROM AQUEOUS SOLUTION USING WASTE EGGSHELL

Ayben Polat, Sukru Aslan

Department of Environmental Engineering, Cumhuriyet University, 58140 Sivas, Turkey

Abstract. The sorption of Cu^{2+} ions from aqueous solutions by eggshell was investigated in a batch experimental system with respect to the temperature, initial Cu^{2+} concentrations, pH, and biosorbent doses. The adsorption equilibrium was well described by the Langmuir isotherm model with the maximum adsorption capacity of 5.05 mg Cu^{2+}/g eggshell at 25 °C. The value of q_e increased with increasing the temperature while also increases the release of Ca^{2+} and HCO^-_3 ions from the eggshell. The highest sorption of Cu^{2+} onto the waste eggshell was determined at the initial pH value of 4.0. The results confirming that the adsorption reaction of Cu^{2+} on the eggshell was thought to be endothermic. A comparison of the kinetic models such as pseudo first and second-order kinetics, intraparticle diffusion, and Elovich on the sorption rate demonstrated that the system was best described by the pseudo second-order kinetic model.

Keywords: eggshells, adsorption, copper, water cleaning technologies.

Reference to this paper should be made as follows: Polat, A.; Aslan, S. 2014. Kinetic and isotherm study of cupper adsorption from aqueous solution using waste eggshell, *Journal of Environmental Engineering and Landscape Management* 22(02): 132–140.

Introduction

Industrial wastewaters contained various kinds of pollutants including heavy metals are commonly produced from many industrial processes. High concentrations of heavy metals affect negatively human, animal and vegetation in the water body. Because of unique characteristics of heavy metals, which are non-biodegradable and accumulated by living organisms, are the main environmental concerns (Ghazy *et al.* 2011).

Treatment of the wastewater including heavy metal ions became particularly difficult due to implementation of more restrict law regulations that control the concentration of pollutants in effluents discharged into waters and soil on the level lower than 1 mg/kg (Chojnacka 2005). The traditional treatment processes such as chemical precipitation and coagulation-flocculation for the removal of metal ions became inefficient to achieve below this concentration.

Several methods such as ion exchange, solvent extraction, phytoextraction, ultrafiltration, reverse osmosis, and adsorption have been widely used in order to remove heavy metals from industrial wastewaters. However, adsorption method is widely applied to eliminate heavy metals due to the some limitations like requirements of pretreatments, low removal efficiency and high capital cost of the other methods (Jai *et al.* 2007).

The application of adsorption is one of the effective, simple and low cost methods to remove low concentration heavy metal from industrial wastewater. Low cost sorbents are investigated for heavy metal elimination. The most frequently studied biosorbents are bacteria, fungi, and algae (Yeddou, Bensmaili 2007), grape stalks, crop milling waste, olive stone, sawdust (Zheng *et al.* 2007; Ozacar, Sengil 2005), peanut hull pellets, dried sunflower leaves, sugar beet pulp, *capsicum annuum* seeds (Slijvic *et al.* 2009), fish bones (Kizilkaya *et al.* 2010), and bone char (Cheung *et al.* 2000).

One cheap and easily available material having possibilities as suitable sorbent for heavy metal is eggshell. Due to their low cost and high calcium content, after these materials have been expended, they can be disposed without

---

Corresponding author: Sukru Aslan
E-mail: saslan@cumhuriyet.edu.tr

expensive regeneration. Disposal of eggshells is also a serious problem for egg processing industries due to stricter environmental regulations and high landfill costs (Rao *et al.* 2010).

Copper is considered as one of the most toxic metal and poses a potential threat to the human health and environment, even at low concentrations (Ahmad *et al.* 2010). However, low concentrations of cupper is essential for living organisms and additionally deficiency of it may cause effects on human health like, anemia osteoporosis, decreased glucose tolerance, arthritis, cardiac arrhythmias and neurological problems. On the contrary, high concentrations of cupper causes toxicity, as the redox properties, essential for its function in excess copper in cells causes cuproenzymes, can also result in marked reactive oxygen species formation that can damage lipids, nucleic acids and proteins (Sljivic *et al.* 2009). Because of the toxic effects on the living organisms, copper containing wastewaters from industries require treatment before their discharge into the environment.

The main objective of the experimental study was to investigate the adsorption of Cu^{2+} ions in the synthetic wastewater by using waste eggshells. The removal efficiency of Cu^{2+} on the eggshell was investigated as a function of temperature, pH, contact time, initial Cu^{2+} concentrations, and adsorbent doses. Various models were applied to determine the adsorption isotherms with the best fit to the experimental data. The kinetic models, pseudo first and second order kinetics, intraparticle diffusion, and Elovich were applied in order to investigate the mechanisms of eggshell sorption.

1. Materials and methods

1.1. Preparation of eggshell

The chicken eggshells were collected from bakeries in Sivas, Turkey. After the eggshells were rinsed several times with tap and distilled water to remove impurities like organics and salts, it was dried at 60 °C for 24 hours in an oven. The eggshells were crushed and screened through a set of sieves to get the size of 106–250 μm.

1.2. Sorption studies

All solutions were prepared from analytical reagent chemicals. The synthetic solutions were prepared by diluting Cu^{2+} standard stock solutions (250 mg/L) obtained by dissolving $CuCl_2$ in the distilled water. Fresh dilutions of the synthetic wastewater were used in the experiments.

The sorption studies of Cu^{2+} from aqueous solution onto the eggshells were carried out using batch equilibrium techniques. Experiments were performed in 250 mL Erlenmeyer flasks containing Cu^{2+} and eggshells. Cu^{2+} analyses were performed in the initial solutions and clear samples at the end of batch tests. A sample of 0.25 g

eggshell was added to 100 mL solution which was contained desired concentrations of Cu^{2+}.

The effects of experimental parameters such as, initial Cu^{2+} ion concentration (10–50 mg/L), pH (2.0–5.0), adsorbent dosage (0.05–0.5 g/L) and temperature (25–50 °C) on the removal of Cu^{2+} ions were studied. Initial pHs of the solutions were adjusted using H_2SO_4 or NaOH solutions. Various concentrations of Cu^{2+} solutions (15, 25, and 35 mg/L) at a constant initial pH value ($\cong$ 5.0), and adsorbent dosage (2.5 g/L) were used for the kinetic experiments. The batch units were agitated in an orbital incubator shaker (Gerhardt) for a contact time varied in the range 0–2880 min at a speed of 150 rpm at 25 °C. The samples were then centrifuged in NUVE Centrifuge NF800 at 4000 rpm for 10 min to separate the solution from the adsorbent.

Calcination was carried out by increasing the temperature of furnace (REF 150 model, REFSAN) at a rate of 4 °C/min to 1000 °C after crushing the sample.

The initial and final concentrations of Cu^{2+} in the aqueous solutions were determined by using a Merck Spectraquant analytical Cu^{2+} kit (14767) with a Merck photometer PHARO100. The other measurements such as alkalinity, Ca^{2+}, HCO_3^-, etc. were carried out using the APHA (1998).

1.3. Calculations

The amounts of Cu^{2+} sorbed by eggshells were calculated from the differences between Cu^{2+} quantity added to the sorbent and Cu^{2+} concentration of the supernatant by using following equation:

$$q_e \left(mg|g \right) = \left(C_o - C_e \right) \left(mg|L \right) \times V / M \left(mL|g \right), \quad (1)$$

The efficiency of Cu^{2+} removal (E) (%) is calculated by using Eqn (2):

$$E \left(\% \right) = \frac{C_0 - C_e}{C_0} \times 100, \quad (2)$$

where: q_e (mg/g) is the maximum amount of Cu^{2+} adsorbed at equilibrium; C_o and C_e (mg/L) are the initial and equilibrium concentrations of Cu^{2+} in the solution, M is the mass of eggshell (g); and V is the volume of the solution, respectively.

Sorption experiments were performed in triplicate and the average values of samples were presented. Also, blank samples (without Cu^{2+}) were used to compare the results through all batch procedures. Data presented are the mean values from the experiments, standard deviation ($\leq 6\%$) and error bars are indicated in figures.

2. Results and discussion

The surface of adsorbent characterized by scanning electron microscopy (SEM) was determined in the

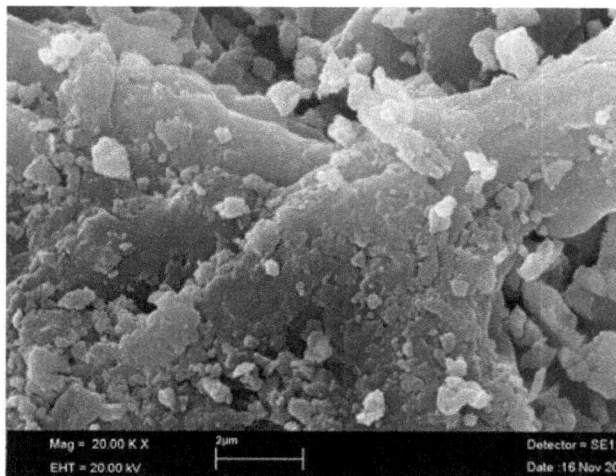

Fig. 1. SEM micrograph of waste eggshell

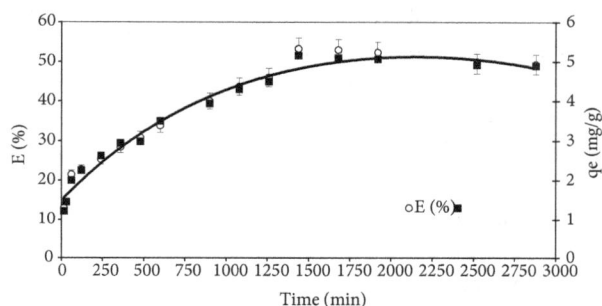

Fig. 2. Effect of contact time on Cu^{2+} sorption onto waste eggshell

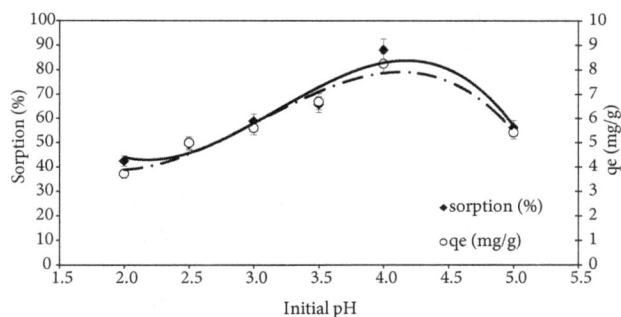

Fig. 3. Effect of initial pH on Cu^{2+} sorption onto eggshell

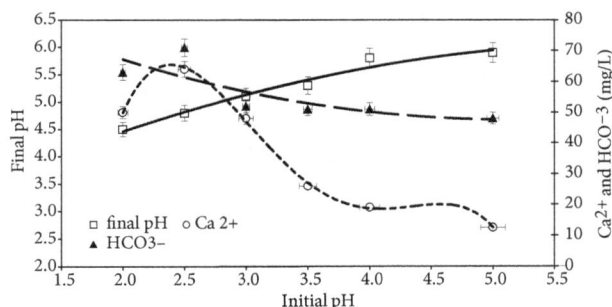

Fig. 4. Effect of initial pH on the release of Ca^{2+} and HCO^{3-} ions from the eggshell

laboratory of Kayseri Teknokent. SEM micrograph shows that eggshell has considerable numbers of pores where the Cu^{2+} ions can be adsorbed (Fig. 1).

2.1. Effect of contact time

The effects of contact time for the initial concentration of 25.0 mg Cu^{2+}/L were studied. The data obtained from the experiments showed that the contact time of 1440 min was sufficient to achieve equilibrium, because the adsorption reached a plateau at this time (Fig. 2). At this point the highest Cu^{2+} sorption efficiency (about 55%) and adsorption value (q_e = 5.2 mg/g) were achieved. Further increase the contact time, the sorption did not take place. Due to the active sites of the eggshell availability and the highest driving force for the mass transfer, rapid Cu^{2+} uptake onto the sorbent was observed at the beginning (zero to 60 min) of the sorption experiments. After this period, Cu^{2+} sorption was slower because of the occupancy of eggshell active sites and the lower concentrations of Cu^{2+} in the solution. Long mixing time was necessary in order to achieve the equilibrium time, due to the removal rate of Cu^{2+} was quite slow.

2.2. Copper sorption at different initial solution pH

The acidity of solution is an important parameter for the sorption of heavy metals from aqueous solutions since the value of pH is responsible for protonation of metal binding sites, calcium carbonate solubility and Cu^{2+} speciation in the solution (Chojnacka 2005). The uptake of Cu^{2+} was investigated as the function of pH in the range of 2.0 to 5.0 with an increment of 0.5 pH units.

Sorption of Cu^{2+} after interaction of Cu^{2+} and eggshell sorbents are presented in Figure 3. It was found that Cu^{2+} uptake by eggshells was a function of initial solution pH. The lowest adsorption efficiency of 42.5% was observed at the pH value of 2.0. Increasing the pH value from 2.0 to 4.0, sorption capacities (q_e) and the removal efficiencies of Cu^{2+} increased significantly from 3.7 mg/g to 8.2 mg/g and 42.5% to 88% respectively. Further increase the pH value to 5.0, the q_e value and removal efficiency decreases to about 5.4 mg/g and 56%, respectively. The ionization degree of heavy metal and the surface property of the eggshell may be affected by the pH. The same experimental results were also observed by Li and Wua (2010) and Ahmad et al. (2010). The optimum initial pH value for Cu^{2+} sorption by eggshell was determined to be 4.0. This results are expected as it is established that eggshell operate more efficiently under acidic conditions (Chojnacka 2005; Rao et al. 2010).

Experimental results showed that all the studied initial pH values were gradually increased and the highest final pH of about 6.0 was determined at the end of experiments (Fig. 4). Due to the release of HCO_3^- ions from the eggshells, the pH in the solutions increased. As can be seen in the Figure 4 that the initial pHs significantly affects the release of Ca^{2+} and HCO_3^- ions from the eggshell.

The chemical composition of eggshell was mainly calcium carbonate (Jai et al. 2007; Tsai 2006, 2008; Arunlertaree et al. 2007). In order to determine the CaO contents of the eggshell, calcinations experiments were carried out. The following reaction was happened in the calcinations (Zhang et al. 2011):

$$CaCO_3 \rightarrow CaO + CO_2 \uparrow. \qquad (3)$$

After calcinations of 550 g pre-treated eggshell, the sample weight was 300 g and corresponding mass loss was as 45.5%. As most of the impurities such as organics and humidity were removed by the pretreatment process, it was assumed that a major composition (55.5%) of the eggshell was identified as CaO when the temperature was ascending to 1000 °C. A similar result was determined by Jai et al. (2007).

The chemical composition (by weight) of by-products eggshell has been reported as follows $CaCO_3$ (94%), magnesium carbonate (1%), calcium phosphate (1%), and organic matter (4%) (Tsai 2008). The principal components of eggshell are $CaCO_3$ and HCO_3^-, Ca^{2+}, $CaHCO_3^+$, and $CaHO^+$. They are in the solution and their proportion is dependent on the pH value (Ghazy et al. 2011).

It was expected that any water equilibrated with the eggshell became basic that confirmed with following mechanisms (Arunlertaree et al. 2007):

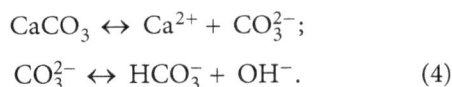

$$CaCO_3 \leftrightarrow Ca^{2+} + CO_3^{2-};$$
$$CO_3^{2-} \leftrightarrow HCO_3^- + OH^-. \qquad (4)$$

As can be seen in the Eqn (4), the solution has become more basic due to the hydrolysis reaction of $CaCO_3$ which gives OH^- and Ca^{2+} content of the solution is also increased.

In general, sorption of divalent metal cations on metal oxides, hydroxides and oxyhydroxides is known to be promoted by increasing pH. When divalent metal cations adsorb on these materials, the cations undergo a reaction with a surface hydroxyl group (Kuh, Kim 2000). Adsorption or precipitation mechanisms involve characteristic reactions of some metals with CaO and MgO surfaces, with adsorption occurring at low concentration of metals solution, and precipitation dominating at high concentrations (Pehlivan et al. 2009). Carbonates formed by $CaCO_3$ dissolution increase the pH values in the solution and therefore may be formation of precipitate form of cupper precipitate near the surface of eggshell and then these forms adsorb on the eggshell (Kuh, Kim 2000).

Because of the eggshell composition, the final pHs of the solutions were higher than the initial value. The precipitation forms of Cu^{2+} are formed when the pH value is higher than 6.0. However, the final pHs of the solutions were lower than 6.0 in this study.

2.3. Effect of temperature

It was found that the value of q_e increases with increasing the temperature while also increase the release of Ca^{2+} from the eggshell. When the temperature increases from 25 to 50 °C, the adsorption capacity increased from 5.16 to 9.94 mg/g indicating that the adsorption was endothermic in nature. At the temperature of 25 °C and 50 °C, the removal efficiency of Cu^{2+} ion at equilibrium was 54.5% and 97.5%, respectively (Fig. 5). It was probably related with the increase of Ca^{2+} release from the eggshells at higher temperature. Elevating the temperature from 25 to 50 °C, the release of Ca^{2+} ions into the aqueous solution was increased (about two times). Results might be attributed to the creation of some new active sites on the eggshell and increase in collision frequency between adsorbent and Cu^{2+} ions at high temperatures. In addition to that, the rise of adsorption with temperature may enlarge the pore size to some extent which may also affect the adsorption capacity (Demirbas et al. 2009).

As mentioned at above, eggshells are composed mainly of calcium carbonate. Calcium ions are bound via ion-exchange and can be thus exchanged by other cations – in this case Cu^{2+}. The experimental results and previous studies confirming that the release of Ca^{2+} from the various adsorbents were a part of the sorption mechanisms (Sljivic et al. 2009; Arunlertaree et al. 2007; Cheung et al. 2000; Kuh, Kim 2000). This situation can be explained by the fact that at higher temperature, the kinetic energy of Cu^{2+} is high; therefore, contact between Cu^{2+} and the eggshell is sufficient, leading to an increase in adsorption efficiencies. The results are consistent with the results of Ghazy et al. (2011). Results indicating that the adsorption of Cu^{2+} ions was favored at higher temperatures.

2.4. Effect of sorbent amount

The dosage of adsorbent is an important parameter in the sorption studies because it provides the capacity of an adsorbent for a given initial concentration of the adsorbate. Figure 6 shows that the absorbability diminished as the adsorbent dosages increased, resulting in that the

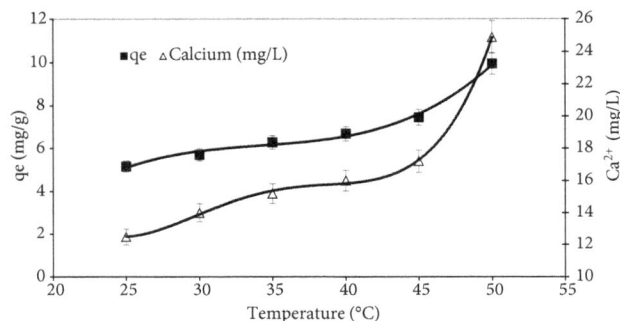

Fig. 5. Temperature effects on Cu^{2+} adsorption onto eggshell

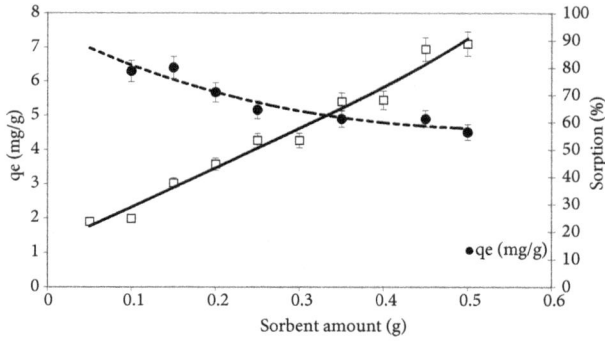

Fig. 6. The effect of sorbent amount on Cu^{2+} adsorption

amount of Cu^{2+} adsorbed per unit mass of eggshell decreased. While the amount of Cu^{2+} sorbed increase from about 24% to 89% with an increase in sorbent dosage from 0.05 to 0.5 g, the absorbability of Cu^{2+} was decreased from about 6.5 to 4.5 mg/g. The decrease in unit adsorption with increase in the dosage of adsorbent was due to adsorption sites remaining unsaturated during the adsorption process (Demirbas et al. 2009). This leads to make a suggestion that, higher Cu^{2+} concentrations should be tested in conjunction with an appropriate adsorbent dosage in order to determine the optimal eggshell dosage.

2.5. Modeling of sorption equilibrium depending on Cu^{2+} concentrations

Equilibrium relationships between adsorbent and adsorbate are described by adsorption isotherms. After determining the data with respect to the initial concentrations of Cu^{2+}, the results were verified with the Langmuir, Freundlich, Temkin, Dubinin-Raduskevich (D-R) adsorption isotherm models.

Langmuir, (Eqn (5)) Freundlich, (Eqn (7)), Temkin (Eqn (8)), D-R (Eqn (10)) isotherms were plotted by using standard straight-line equations and corresponding two parameters for Cu^{2+} were calculated from their respective graphs.

A basic assumption of the Langmuir theory is that sorption takes place at specific homogenous sites within the sorbent (Baig et al. 2010). The Langmuir isotherm equation is represented by the following equation (Tsai et al. 2008).

$$q_e\,(mg\,/\,g) = q_m \frac{K_L C_e}{1 + K_L C_e},\qquad(5)$$

where q_m indicates the monolayer sorption capacity of adsorbate (mg/g).

In order to predict the affinity between the waste eggshells and Cu^{2+} ions, the Langmuir parameters of the dimensionless separation factor R_L can be used. The value of R_L can be calculated by the following equation:

$$R_L = \frac{1}{1 + K_L C_0}.\qquad(6)$$

The value of R_L indicates that the shape of the sorption process is; unfavorable ($R_L > 1$), linear ($R_L = 1$), favorable ($0 < R_L < 1$) or irreversible ($R_L = 0$) (Kilic et al. 2011; Sljivic et al. 2009).

The Freundlich isotherm model is considered to be appropriate for describing both multilayer sorption and sorption on heterogeneous surfaces (Coles, Yong 2006). The Freundlich isotherm equation is represented by the following equation (Tsai et al. 2008).

Freundlich

$$q_e\,(mg\,/\,g) = K_{Fi} C_e^{\frac{1}{n}}.\qquad(7)$$

Temkin and Pyzhev considered the effects of indirect adsorbent/adsorbate interactions on adsorption isotherms (Kilic et al. 2011).

Temkin

$$q_e(mg/g) = B\ln A_T + B\ln C_e.\qquad(8)$$

The linear form of Temkin isotherm equation is as follows:

$$q_e(mg/g) = B\ln K_T + \ln C_e,\qquad(9)$$

where: $B = RT/b_T$ (Temkin constant related to heat of sorption, J/mol); $1/b_T$ indicates the adsorption potential of the adsorbent.

The experimental data were also analyzed using the Dubinin-Radushkevich (D–R) isotherm model to determine the nature of biosorption processes as physical or chemical (Baig et al. 2010) by applying the following equation:

$$\ln q_e = \ln q_{max} - \beta\varepsilon^2,\qquad(10)$$

and ε can be correlated as:

$$\varepsilon = RT\ln\left(1 + \frac{1}{C_e}\right).\qquad(11)$$

The constants β and E are the mean free energy and sorption per molecule of the sorbate, respectively. They can be computed using the following relationship (Kose, Kivanc 2011).

$$E = \frac{1}{\sqrt{-2\beta}}.\qquad(12)$$

Sorption parameters for the isotherms are as follows: K_L (L/mg) Langmuir constant related to the energy of sorption; K_{Fi} (L/mg) Freundlich constant related to sorption capacity of adsorbent; q_{max} (mg/g) is the maximum biosorption capacity of D–R. b_T and A_T (L/mg) Temkin isotherm parameters; R is the gas constant (8.314 joule.mol/K); T is the absolute temperature (K).

The constants of all isotherms equation are presented in Table 1. As a result of the experiments, the highest correlation coefficient of 0.999 was determined using the Langmuir model than the others; it is suggesting that the Cu^{2+} ions were adsorbed onto the eggshell in a monolayer. Additionally, the calculated q_{cal} value of the Langmuir model equation corresponded well with the experimentally obtained. The amounts of sorbed Cu^{2+} increased with the increase of initial Cu^{2+} concentration in the solution until the equilibrium was achieved. Further increase, the removal of Cu^{2+} became independent of the initial cation concentration, due to occupancy of all active sites on the adsorbent surface (Sljivic et al. 2009). The experimentally obtained maximum capacity for monolayer saturation was 5.05 mg Cu^{2+}/g eggshell at 25±1 °C. The calculated R_L values were between 2.27×10^{-3} and 11×10^{-2} which indicated that the Cu^{2+} sorption by waste eggshell sample was favorable.

Table 1. Correlation coefficient and sorption parameters for various models

Model	Equation	Sorption Parameters	
Freundlich	$q_e(mg/g) = K_{Fi}C_e^{\frac{1}{n}}$	R^2	0.6958
		n	15.9
		K_F	4.08
Langmuir	$q_e(mg/g) = q_m \dfrac{K_L C_e}{1 + K_L C_e}$	R^2	0.9995
		R_L	11×10^{-2} 2.27×10^{-3}
		q_m	5.05
		K_L	70,7
Temkin	$q_e (mg/g) = B_T \ln A_T + B_T \ln C_e$	R^2	0.720
		b_T	7485
		A_T (L/g)	248
R–D	$\ln q_e = \ln q_{max} - \beta\varepsilon^2$	R^2	0.994
		q_0 (mg/g)	5.083
		β (mol²/j²)	−0.063
		E (kj/mol)	2.82

2.6. Kinetics of sorption

In order to determine the uptake rate of adsorbate at the solid-phase interface, adsorption kinetics study is important. Various kinetic models including, pseudo first and second order kinetics, intraparticle diffusion, and Elovich were applied to the experimental data in order to investigate the mechanisms of eggshell sorption.

The adsorption kinetic models were investigated at three adsorbate dosages of 15, 25, and 35 mg Cu^{2+}/L. Lagergren and Annadurai and Krishan presented the first (Eqn (13)) and second order (Eqn (14)) rates expression for the first and second pseudo order kinetics (Chiou, Li 2002; Rao et al. 2010; Chairat et al. 2005):

$$\log(q_e - q_t) = \log q_e - \frac{k_1}{2.303}t; \qquad (13)$$

$$\frac{1}{q_t} = \frac{1}{k_2 q_e^2} + \frac{t}{q_e}, \qquad (14)$$

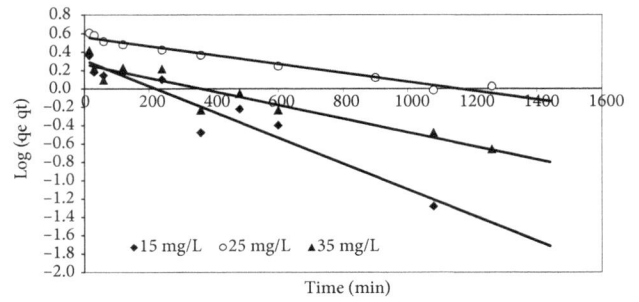

Fig. 7. Pseudo-first order kinetics of Cu^{2+} adsorption onto the eggshell at various adsorbate amounts

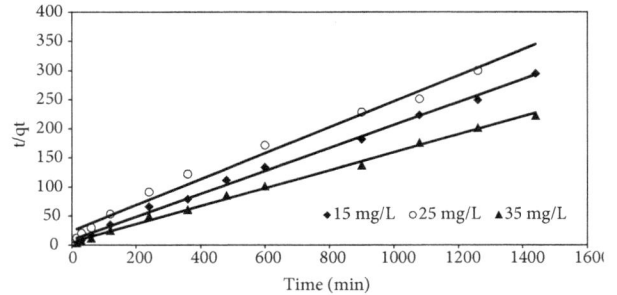

Fig. 8. Pseudo-second order kinetics of Cu^{2+} adsorption onto the eggshell at various adsorbate amounts

where: q_e and q_t (mg/g) are the amount of Cu^{2+} adsorbed onto the eggshell at equilibrium and at time t (min), respectively, the first and second order rate constants $(min)^{-1}$ are k_1 and k_2 (mg/g·min), respectively. In order to determine the value of k_1 and q_e, the plot of log $(q_e - q_t)$ against t is employed for the first pseudo order kinetic constants. The second pseudo order kinetics constants k_2 and q_e are calculated by the slope and intercept of (t/qt) versus t (Figs 7 and 8).

The initial sorption rate h (mg/g.min) is determined by using the Eqn (15):

$$h = k_2 \times q_e^2. \qquad (15)$$

The uptake of adsorbate by the sorbent from solutions involves bulk, film, and intraparticle diffusion in the solid phase and within the pores, and finally adsorption on the sites. In order to determine the rate-controlling step, intraparticle diffusion model was applied to adsorption kinetic data by applying the Eqn (16) (Ghasemi et al. 2012).

$$q_t = k_{id}t^{\frac{1}{2}} + C, \qquad (16)$$

where: q_t is the amount of Cu^{2+} ions adsorbed onto the eggshell at time t and k_{id} (g/mg.min) is the intraparticle diffusion rate constant and C presents an idea on the thickness of the boundary layer (Ahmad et al. 2010).

The value of rate constant of Morris–Weber transport, K_{id}, calculated from the slope of the linear plot are shown in Figure 9.

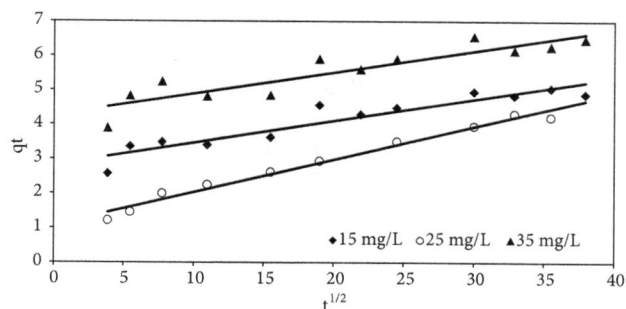

Fig. 9. Intraparticle diffusion for the adsorption of Cu^{2+} onto the eggshell at various adsorbate amounts

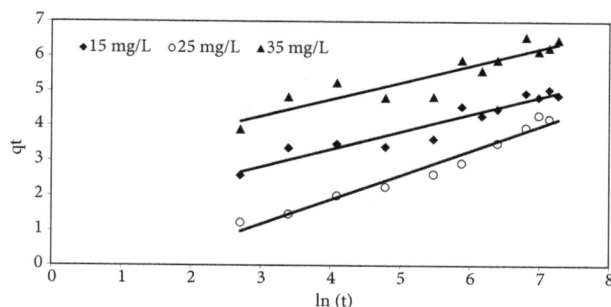

Fig. 10. Elovich Model for Cu^{2+} adsorption onto the eggshell at various adsorbate amounts

The integrated Elovich equation is given as:

$$q_t = \frac{1}{\beta}\ln\alpha\beta + \frac{1}{\beta}\ln t, \qquad (17)$$

where: α (mg/g.min) is the initial sorption rate and β (g/mg) is related to the extent of surface coverage and activation energy for chemisorption (Ozacar, Sengil 2005). A plot of q_t versus $\ln(t)$ should yield a linear relationship with a slope of $(1/b)$ and an intercept of $(1/b)\ln(\alpha b)$ (Demirbas et al. 2009). The values of rate constants of Elovich model are shown in Figure 10.

The values of q_e, k_1, k_2, α, and β with the correlation coefficient (R^2) for various eggshell amounts of 15, 25, and 35 mg/L were calculated by their respected plots and the results are presented in Table 2.

Table 2. Kinetic parameters for the sorption of Cu^{2+} onto the eggshell

Conc	$q_{e,exp}$	Pseudo-first-order			Pseudo-second-order			
mg/L	mg/g	$q_{e,cal}$	$k_1 \cdot 10^{-3}$	R^2	$q_{e,cal}$	$k_2 \cdot 10^{-3}$	R^2	h
15	4.88	2.035	3.22	0.92	5.08	4.15	0.99	0.107
25	5.26	3.62	1.15	0.97	4.49	2.04	0.98	0.041
35	6.48	1.836	1.61	0.88	6.51	4.13	0.99	0.175

Conc	$q_{e,exp}$	Intraparticle diffusion		Elovich		
mg/L	mg/g	k_p	R^2	α	β	R^2
15	4.88	0.0632	0.911	6.83	1.99	0.913
25	5.26	0.0949	0.981	2.634	1.423	0.966
35	6.48	0.0624	0.819	36.2	2.03	0.838

On changing the initial concentration of Cu^{2+} in the solutions from 15 to 35 mg/L, the amount of Cu^{2+} adsorbed onto the eggshell increased experimentally. The correlation coefficient for the pseudo-second orders were relatively higher than the other kinetic models and the experimental q_e (4.88 mg/g and 6.48 mg/g) values are also very close to the calculated q_e values (5.08–6.51 mg/g). The rate constant slightly decreases with an increasing of initial Cu^{2+} concentration while the initial sorption rate increases with an increasing of initial Cu^{2+} concentration for the pseudo second-order model. The similar

phenomena have also reported in sorption of Cu^{2+} onto the biowaste materials (Sljivic et al. 2009; Kizilkaya et al. 2010; Ahmad et al. 2010; Demirbas et al. 2009). Increase of q_e value is a result of the increase in the driving force of the concentration gradient with the increase in the initial Cu^{2+} concentration. Therefore a higher initial concentration of Cu^{2+} ions may increase the adsorption capacity of eggshell. It means that the biosorption is highly dependent on initial concentration of metal ion.

The results are indicated that the sorption perfectly complies with pseudo-second order reaction and the sorption of Cu^{2+} onto the eggshell appeared to be controlled by the chemisorption process.

Several biosorbents have been used to remove Cu^{2+} from aqueous solutions. A comparison of the adsorbent capacity is presented in Table 3. As can be seen from the table that eggshell shows the comparable sorption capacity for Cu^{2+} with respect to the other biosorbents.

Table 3. Comparison of various biosorbent for Cu^{2+} removal

Biosorbent	mg/g	References
Fish bones	150.7	Kızılkaya et al. 2010
Green alga Spirogyra	133.3	Gupta et al. 2006
Garden grass	58.34	Hossain et al. 2012
Iron oxide coated eggshell powder	44.843	Ahmad et al. 2010
Sunflower shell	30.30	Onal et al. 2008
Sawdust	8.452	Larous et al. 2005
Rice straw	8.14	Rocha et al. 2009
Modified mangrove barks	6.950	Rozaini et al. 2010
Eggshell	6.48	This study
Soybean straw	5.40	Šciban et al. 2008
Eggshell	5.03	Vijayarghavan et al. 2005
Orange skin	4.96	Onal et al. 2008
Barley straws	4.64	Pehlivan et al. 2009
Wheat straw	4.448	Šciban et al. 2008
Corn stalk	3.749	Šciban et al. 2008
Corn cob	2.16	Šciban et al. 2008

Conclusions

The present experimental study results showed that the waste eggshells might be applicable successfully as a sorbent of cupper ions from aqueous solution. As a conclusion:

1) The adsorption of cupper onto the waste eggshell was found to be initial Cu^{2+} concentration, pH, temperature, mixing time, and adsorbate dosage depended.

2) The optimum pH value for the experimental study was determined as 4.0.

3) Increasing the temperature, q_e value increased with the increase the release of Ca^{2+} from the eggshell.

4) Sorption of eggshell onto the eggshell was well described by Langmuir model.

5) Results indicated that pseudo second-order kinetic model, which is an agreement with a chemisorption mechanism, provided the best correlation of the experimental data.

Acknowledgment

This study was supported by The Research Fund of Cumhuriyet University (CUBAP) under Grant No. M-459, Sivas, Turkey.

References

Ahmad, R.; Kumar, R.; Haseeb, S. 2010. Adsorption of Cu^{2+} from aqueous solution onto iron oxide coated eggshell powder: evaluation of equilibrium, isotherms, kinetics, and regeneration capacity, *Arabian Journal of Chemistry* 5(3): 353–359. http://dx.doi.org/10.1016/j.arabjc.2010.09.003

APHA, 1998. *Standard methods for the examination of water and wastewater*, 20th ed. American Public Health Association/ American Water Works Association/Water Environment Federation, Washington, DC, USA.

Arunlertaree, C.; Kaewsomboon, W.; Kumsopa; Pokethitiyook, P.; Panyawathanakit, P. 2007. Removal of lead from battery manufacturing wastewater by egg shell, *Songklanakarin Journal of Science and Technology* 29(3): May – Jun: 857868.

Baig, J. A.; Kazi, T. G.; Shah, A. Q.; Kandhro, G. A.; Afridi, H. I.; Khan, S.; Kolachi, N. F. 2010. Biosorption studies on powder of stem of *Acacia nilotica*: removal of arsenic from surface water, *Journal of Hazardous Materials* 178(1–3): 941–948. http://dx.doi.org/10.1016/j.jhazmat.2010.02.028

Chairat, M.; Rattanaphani, S.; Bremner, J. B.; Rattanaphani, V. 2005. An adsorption and kinetic study of lac dyeing on silk, *Dyes and Pigments* 64(3): 231–241. http://dx.doi.org/10.1016/j.dyepig.2004.06.009

Cheung, C. W.; Porter, J. F.; McKay, G. 2000. Sorption kinetics for the removal of copper and zinc from effluents using bone char, *Separation and Purification Technology* 19(1–2): 55–64. http://dx.doi.org/10.1016/S1383-5866(99)00073-8

Chiou, M. S.; Li, H. S. 2002. Equilibrium and kinetic modeling of adsorption of reactive dye on cross-linked chitosan beads, *Journal of Hazardous Materials* 93(2): 223–248. http://dx.doi.org/10.1016/S0304-3894(02)00030-4

Chojnacka, K. 2005. Biosorption of Cr (III) ions by eggshells, *Journal of Hazardous Materials* 121(1–3): 167–173. http://dx.doi.org/10.1016/j.jhazmat.2005.02.004

Coles, C. A.; Yong, R. N. 2006. Use of equilibrium and initial metal concentrations in determining Freundlich isotherms for soils and sediments, *Engineering Geology* 85(1–2): 19–25. http://dx.doi.org/10.1016/j.enggeo.2005.09.023

Demirbas, E.; Dizge, E.; Sulak, M. T.; Kobya, M. 2009. Adsorption kinetics and equilibrium of copper from aqueous solutions using hazelnut shell activated carbon, *Chemical Engineering Journal* 148(2–3): 480–487. http://dx.doi.org/10.1016/j.cej.2008.09.027

Ghasemi, Z.; Seif, A.; Ahmadi, T. S.; Zargar, B.; Rashidi, F.; Rouzbahani, G. M. 2012. Thermodynamic and kinetic studies for the adsorption of Hg(II) by nano-TiO_2 from aqueous solution, *Advanced Powder Technology* 23(2): 148–156. http://dx.doi.org/10.1016/j.apt.2011.01.004

Ghazy, S. El-S.; El-Asmy, A. A.-H.; El-Nokrashy, A. M. 2011. Batch removal of nickel by eggshell as a low cost sorbent, *International Journal of Industrial Chemistry* 2(4): 242–252.

Gupta, V. K.; Rastogi, A.; Saini, V. K.; Jain, N. 2006 Biosorption of copper(II) from aqueous solutions by *Spirogyra* species, *Journal of Colloid and Interface Science* 296(1): 59–63. http://dx.doi.org/10.1016/j.jcis.2005.08.033

Hossain, M. A.; Ngo, H. H.; Guo, W. S.; Setiati, T. 2012. Adsorption and desorption of copper(II) ions onto garden grass, *Bioresource Technology* 121: 386–395. http://dx.doi.org/10.1016/j.biortech.2012.06.119

Jai, P. H.; Wook, J. S.; Kyu, Y. J.; Gil, K. B.; Mok, L. S. 2007. Removal of heavy metals using waste eggshell, *Journal of Environmental Science* 19(12): 1436–1441. http://dx.doi.org/10.1016/S1001-0742(07)60234-4

Kilic, M.; Varol, E. A.; Putun, A. E. 2011. Adsorptive removal of phenol from aqueous solutions on activated carbon prepared from tobacco residues: equilibrium, kinetics and thermodynamics, *Journal of Hazardous Materials* 189(1–2): 397–403. http://dx.doi.org/10.1016/j.jhazmat.2011.02.051

Kizilkaya, B.; Tekinay, A. A.; Dilgin, Y. 2010. Adsorption and removal of Cu (II) ions from aqueous solution using pretreated fis bones, *Desalination* 264(1–2): 37–47. http://dx.doi.org/10.1016/j.desal.2010.06.076

Kose, T. E.; Kivanc, B. 2011. Adsorption of phosphate from aqueous solutions using calcined waste eggshell, *Chemical Engineering Journal* 178: 34–39. http://dx.doi.org/10.1016/j.cej.2011.09.129

Kuh, S. E.; Kim, D. S. 2000. Removal characteristics of cadmium ion by waste egg shell, *Environmental Technology* 21(8): 883–890. http://dx.doi.org/10.1080/09593330.2000.9618973

Larous, S.; Meniai, A.-H.; Lehocine, M. B. 2005. Experimental study of the removal of copper from aqueous solutions by adsorption using sawdust, *Desalination* 185(1–3): 483–490. http://dx.doi.org/10.1016/j.desal.2005.03.090

Li, S. Z.; Wua, P. X. 2010. Characterization of sodium dodecyl sulfate modified iron pillared montmorillonite and its application for the removal of aqueous Cu(II) and Co(II), *Journal of Hazardous Materials* 173(1–3): 62–70. http://dx.doi.org/10.1016/j.jhazmat.2009.08.047

Onal, O.; Ozcelik, E.; Benli, S., *et al.* 2008. Adsorption of Fe^{3+} and Cu^{2+} on orange skin and sunflower shell, in *4th European BioRemediation Conference*, 3–6 September, 2008, Chania, Crete, Greece.

Ozacar, M.; Sengil, I. A. 2005. A kinetic study of metal complex dye sorption onto pine sawdust, *Process Biochemistry* 40(2): 565–572. http://dx.doi.org/10.1016/j.procbio.2004.01.032

Pehlivan, E.; Ozkan, A. M.; Dinc, S.; Parlayici, S. 2009. Adsorption of Cu^{2+} and Pb^{2+} ion on dolomite powder, *Journal of Hazardous Materials* 167(1–3): 1044–1049. http://dx.doi.org/10.1016/j.jhazmat.2009.01.096

Rao, H. J.; Kalyani, G.; Rao, K. V.; Kumar, T. A.; Mariadas, K.; Kumar, Y. P.; Vijetha, P.; Pallavi, P.; Sumalatha, B.; Kumaraswamy, K. 2010. Kinetic studies on biosorption of lead from aqueous solutions using egg shell powder, *International Journal of Biotechnology and Biochemistry* 6: 957–968.

Rocha, G. C.; Zai, D. A. M.; Alfaya, R. V. S. 2009. Use of rice straw as biosorbent for removal of Cu(II), Zn(II), Cd(II) and Hg(II) ions in industrial effluents, *Journal of Hazardous Materials* 166(1): 383–388. http://dx.doi.org/10.1016/j.jhazmat.2008.11.074

Rozaini, C. A.; Jain, K.; Oo, C. W.; Tan, K. W.; Tan, L. S.; Azraa, A.; Tong, K. S. 2010. Optimization of nickel and copper ions removal by modified mangrove barks, *International Journal of Chemical Engineering and Applications* 1(1): 84–89. http://dx.doi.org/10.7763/IJCEA.2010.V1.14

Šciban, M.; Klašnja, M.; Škrbic, M. 2008. Adsorption of copper ions from water by modified agricultural by-products, *Desalination* 229(1–3): 170–180. http://dx.doi.org/10.1016/j.desal.2007.08.017

Sljivic, M.; Smiciklas, I.; Plecas, I., *et al.* 2009. The influence of equilibration conditions and hydroxyapatite physico-chemical properties onto retention of Cu^{2+} ion, *Chemical Engineering Journal* 148(1): 80–88.

Tsai, W. T.; Yang, J. M.; Lai, C. W.; Cheng, Y. H.; Lin, C. C.; Yeh, C. W. 2006. Characterization and adsorption properties of eggshells and eggshell membrane, *Bioresource Technology* 97(3): 488–493. http://dx.doi.org/10.1016/j.biortech.2005.02.050

Tsai, W.-T.; Hsien, K.-J.; Hsu, H.-C. ; Lin, C.-M.; Lin, K.-Y.; Chiu, C.-H. 2008. Utilization of ground eggshell waste as an adsorbent for the removal of dyes from aqueous solution, *Bioresource Technology* 99(6): 1623–1629. http://dx.doi.org/10.1016/j.biortech.2007.04.010

Vijayaraghavan, K.; Jegan, J.; Palanivelu, K., *et al.* 2005. Removal and recovery of copper from aqueous solution by eggshell in a packed column, *Mineral Engineering* 18(5): 545–547. http://dx.doi.org/10.1016/j.mineng.2004.09.004

Yeddou, N.; Bensmaili, A. 2007. Equilibrium and kinetic modeling of iron adsorption by eggshells in a batch system: effect of temperature, *Desalination* 206(1–3): 127–134. http://dx.doi.org/10.1016/j.desal.2006.04.052

Zhang, S.; Guo, Z.; Xu, J., *et al.* 2011. Effect of environmental conditions on the sorption of radiocobalt from aqueous solution to treated eggshell as biosorbent, *Journal of Radioanalytical and Nuclear Chemistry* 288(1): 121–130. http://dx.doi.org/10.1007/s10967-010-0895-8

Zheng, W.; Li, X.-M.; Yang, Q., *et al.* 2007. Adsorption of Cd(II) and Cu(II) from aqueous solution by carbonate hydroxylapatite derived from eggshell waste, *Journal of Hazardous Materials* 147(1–2): 534–539. http://dx.doi.org/10.1016/j.jhazmat.2007.01.048

Ayben POLAT. She is an Environmental Engineer at the Department of Environmental Engineering, Cumhuriyet University, Sivas. Her research interest includes adsorption.

Sukru ASLAN. Dr Lecturer at the Department of Environmental Engineering, Cumhuriyet University, Sivas. His research interests include biological nutrient removal, sorption.

PERMISSIONS

LIST OF CONTRIBUTORS

Ričardas Butkus, Alvidas Šarlauskas and Gediminas Vasiliauskas
Institute of Agricultural Engineering and Safety, Aleksandras Stulginskis University, Studentų g. 15b, 53361 Akademija, Kaunas distr., Lithuania

Mihaela Budianu
Environmental Protection Agency Vaslui, Str. Calugareni, no.63, 730149 Vaslui, Romania

Valeriu Nagacevschi and Matei Macoveanu
Department of Environmental Engineering and Management, Faculty of Chemical Engineering, Technical University of Iasi, 71A D. Mangeron Bd., 700050 Iasi, Romania

Petras Venckus and Jolanta Kostkevičienė
Department of Botany and Genetics, Faculty of Natural Sciences, Vilnius University, M. K. Ciurlionio 21/27, LT-03101 Vilnius, Lithuania

Vida Bendikienė
Department of Biochemistry and Molecular Biology, Faculty of Natural Sciences, Vilnius University, M. K. Ciurlionio 21/27, LT-03101 Vilnius, Lithuania

Nuri İlgürel and Neşe Yüğrük Akdağ
Department of Architecture, Yıldız Technical University, D-107, Beşiktaş, İstanbul, 34349 Turkey

Ali Akdağ
Hidrotek Architecture and Engineering Ltd., İstanbul, Turkey

Ahmed M. Azzam
aEnvironmental Researches Department, Theodor Bilharz Research Institute, 30-12411, Imbaba, Giza, Egypt

Ahmed Tawfik
Department of Environmental Engineering, Egypt-Japan University of Science and Technology (E-JUST), New Borg El Arab City, 21934, Alexandria, Egypt

Laura Masilionytė, Stanislava Maikštėnienė, Aleksandras Velykis and Antanas Satkus
Joniškėlis Experimental Station, Lithuanian Research Centre for Agriculture and Forestry, 39301 Joniškėlis, Lithuania

Gülay Zorer Gedik and Neşe Yüğrük Akdağ
Faculty of Architecture, Yildiz Technical University, Istanbul, 34349, Turkey

Fatih KİRAZ
Faculty of Fine Arts and Design, Nuh Naci Yazgan University, Kayseri, 38040, Turkey

Bekir ŞENER
Naval Architecture and Maritime Faculty, Yildiz Technical University, Istanbul, 34349, Turkey

Raşide ÇAÇAN
Yildiz Technical University, Istanbul, 34349, Turkey

Petras Vaitiekūnas, Egidijus Petraitis, Albertas Venslovas and Aleksandras Chlebnikovas
Institute of Environmental Protection, Vilnius Gediminas Technical University, Saulėtekio al. 11, 10223 Vilnius, Lithuania

Huasai Simujide, Chen Aorigele and Bai Manda
The College of Animal Science, Inner Mongolia Agricultural University, 010018 Hohhot, P. R. China

Chun-Jie Wang
College of Veterinary Medicine, Inner Mongolia Agricultural University, 010018 Hohhot, P. R. China

Jun-E Yu
Ulanqab Vocational College, 012000 Ulanqab, P. R. China

Ma Lina
College of Life Sciences, Inner Mongolia Agricultural University, 010018 Hohhot, P. R. China

Sukru Aslan, Ayben Polat and Ugur Savas Topcu
Department of Environmental Engineering, Cumhuriyet University, 58140, Sivas, Turkey

Žilvinas Venckus, Albertas Venslovas and Mantas Pranskevičius
Institute of Environmental Protection, Vilnius Gediminas Technical University, Saulėtekio al. 11, LT-10223 Vilnius, Lithuania

Muhammad Irfan
Department of Civil Engineering, University of Engineering and Technology, Lahore, Pakistan

Muhammad Imran Khan
Department of Civil Engineering, University of Engineering and Technology, Lahore, Pakistan
Department of Civil Engineering and Applied Mechanics, McGill University, Montreal, Canada

Mubashir Aziz
Department of Civil Engineering, Al Imam Mohammad Ibn Saud Islamic University, Riyadh, Saudi Arabia

Ammad Hassan Khan
Department of Transportation Engineering and Planning, University of Engineering and Technology, Lahore, Pakistan

Elshad Gurbanov
Department of Botany, Faculty of Biology, Baku State University, Baku, Azerbaijan

Naglaa Youssef
Department of Botany, Faculty of Biology, Baku State University, Baku, Azerbaijan
Department of Botany, Faculty of Sciences, Sohag University, Sohag, Egypt

Bernd Markert and Simone Wünschmann
Environmental Institute of Scientific Neworks (EISN-Institute), Fliederweg 17, 49733 Haren, Germany

Haciyeva Sevnic
Department of Ecological Chemistry, Faculty of Chemistry, Baku State University, Baku, Azerbaijan

Vera Suzdalenko and Martins Gedrovics
Institute of Energy System and Environment, Riga Technical University, Riga, Latvia

Jurgita Malaiškienė, Romualdas Mačiulaitis and Raminta Mikalauskaitė
Department of Building Materials, Vilnius Gediminas Technical University, Saulėtekio al. 11, 10223 Vilnius, Lithuania

Gintaras Svecevičius
Nature Research Centre, Akademijos g. 2, 08412 Vilnius, Lithuania

Raimondas Leopoldas Idzelis and Eglė Mockutė
Vilnius Gediminas Technical University, Saulėtekio al. 11, 10223 Vilnius, Lithuania

Eglė Abrutytė, Audronė Žukauskaitė, Rima Mickevičienė, Vytenis Zabukas and Tatjana Paulauskienė
Faculty of Marine Engineering, Klaipėda University, Bijūnų g. 17, 91225 Klaipėda, Lithuania

Emine Elmaslar Özbaş and Nilgün Balkaya
Istanbul University, Faculty of Engineering, Environmental Engineering Department, 34320 Avcilar, Istanbul, Turkey

Kęstutis Mačaitis and Antonas Misevičius
Environmental Institute, Vilnius Gediminas Technical University, Saulėtekio al. 11, LT-10223 Vilnius, Lithuania

Algimantas Paškevičius, Vita Raudonienė and Jūratė Repečkienė
Nature Research Centre, Vilnius University, Akademijos g. 2, LT-08412 Vilnius, Lithuania

Ayben Polat and Sukru Aslan
Department of Environmental Engineering, Cumhuriyet University, 58140 Sivas, Turkey

Index